A History of Natural Philosophy

Natural philosophy encompassed all natural phenomena of the physical world. It sought to discover the physical causes of all natural effects and was little concerned with mathematics. By contrast, the exact mathematical sciences – such as astronomy, optics, and mechanics – were narrowly confined to various computations that did not involve physical causes. Natural philosophy and the exact sciences functioned independently of each other. Although this began slowly to change in the late Middle Ages, a much more thoroughgoing union of natural philosophy and mathematics occurred in the seventeenth century and thereby made the Scientific Revolution possible. The title of Isaac Newton's great work, *The Mathematical Principles of Natural Philosophy*, perfectly reflects the new relationship. Natural philosophy became the "Great Mother of the Sciences," which by the nineteenth century had nourished the manifold chemical, physical, and biological sciences to maturity, thus enabling them to leave the "Great Mother" and emerge as the multiplicity of independent sciences we know today.

Edward Grant is Distinguished Professor Emeritus of the History and Philosophy of Science at Indiana University, Bloomington. He is the author or editor of twelve books, one of which has been translated into eleven languages and one into three languages. He is also the author of approximately ninety articles on the history of science and natural philosophy. He was Vice President and President of the History of Science Society and was awarded the prestigious George Sarton Medal of that society.

A History of Natural Philosophy

From the Ancient World to the Nineteenth Century

EDWARD GRANT

Indiana University, Bloomington

CAMBRIDGE
UNIVERSITY PRESS

CAMBRIDGE
UNIVERSITY PRESS

32 Avenue of the Americas, New York NY 10013-2473, USA

Cambridge University Press is part of the University of Cambridge.

It furthers the University's mission by disseminating knowledge in the pursuit of
education, learning and research at the highest international levels of excellence.

www.cambridge.org
Information on this title: www.cambridge.org/9780521869317

First published 2007
Reprinted 2008

A catalogue record for this publication is available from the British Library

Library of Congress Cataloguing in Publication data
Grant, Edward, 1926–
A history of natural philosophy : from the ancient world to the nineteenth century /
Edward Grant.
p.   cm.
Includes bibliographical references and index.
ISBN-13: 978-0-521-86931-7
ISBN-10: 0-521-86931-5
ISBN-13: 978-0-521-68957-1 (pbk.)
ISBN-10: 0-521-68957-0 (pbk.)
I. Physics – History.   2. Science – History.   I. Title.
QC7.G73   2007
509 – dc22   2006012920

ISBN 978-0-521-86931-7 Hardback
ISBN 978-0-521-68957-1 Paperback

In Remembrance of
Marshall Clagett (1916–2005)
Teacher and Dear Friend

Contents

Contents

Preface

Most of my publications over many years have in one way or another been about, or concerned with, natural philosophy. In all those years, however, neither I nor anyone else has seen fit to write a history of that discipline. Although numerous histories of science have been published, and will undoubtedly continue to appear, I am unaware of any history of natural philosophy. It occurred to me that an account of the historical evolution of natural philosophy should prove helpful to a better understanding of the development of the history of science itself. Indeed, as readers will discover, the historical relationship between natural philosophy and science is by no means straightforward. Opinions about their association and interconnections have often been controversial and sometimes quite elusive.

Once I determined to write a history of natural philosophy, I had to decide whether that history should be all encompassing – from its origins to its general replacement by modern science – or whether it should be confined to one or two historical periods. Because my area of specialization has been the late Middle Ages, it seemed plausible to begin with the origins of natural philosophy in the ancient world and conclude at about 1500, when medieval natural philosophy reached the height of its development. Around 1998 I became aware of an opinion that claimed that natural philosophy was always about God, even when God is not discussed or mentioned; and, consequently, that natural philosophy could not be science, because the latter was never about God. Although Dr. Andrew Cunningham, the scholar who proposed this interpretation (see Chapters 9 and 10), focused on Isaac Newton and the seventeenth century, his claims applied to all of natural philosophy, including the Middle Ages. This view of natural philosophy was so utterly contrary to my own understanding of that ancient discipline that I decided to extend my historical range, not only to the seventeenth century but also to the nineteenth century by the end of which natural philosophy had largely passed from the scene, although, in many older universities and colleges, some academic scientists continue to the present to hold the title "Professor of Natural Philosophy."

My study is not intended as a highly detailed description of all aspects of the history of natural philosophy. That would be a very formidable task.

For example, I have said almost nothing about natural philosophy in the fifteenth and eighteenth centuries, largely because I do not believe any dramatic changes occurred in those periods. Nor have I included discussions about nineteenth-century *Naturphilosophie*, associated with the names of Schelling, Fichte, Hegel, and others. My objective, rather, has been to describe the general characteristics of natural philosophy in the different historical periods and to trace the major transformations that occurred over the centuries. As readers will observe, the most profound change in natural philosophy occurred in the seventeenth century. It involved a union of the exact sciences and natural philosophy, a phenomenon that has received relatively little attention in the vast literature about the meaning and causes of the Scientific Revolution. Without that fusion, however, it is doubtful that the Scientific Revolution could have occurred in the seventeenth century. One major result of this coming-together was that natural philosophy, once regarded as largely independent and isolated from mathematics and the exact sciences, became significantly mathematized. In this mathematized form, natural philosophy became synonymous with the term science, which came into use in the nineteenth century. As the reader will see, it was because of natural philosophy's capacity for absorbing sciences and expanding their horizons that, in the seventeenth century, Sir Francis Bacon, with great insight and vision, designated natural philosophy as the "Great Mother of the Sciences."

By virtue of the enormous role Aristotle's works have played in the history of natural philosophy, there are many quotations from, and references to, his treatises. In my numerous references to his works, I have followed the usual conventions. Citations of passages in Aristotle's works almost always follow the page numbering of his Greek texts, which were edited in the nineteenth century by Immanuel Bekker and published by the Berlin Academy (1831–1870). Every reference to a passage in Aristotle's works consists minimally of a page number, a column letter, and a line number. There is no confusion about page numbers, because the pagination of the Bekker edition is completely sequential from volume to volume. For example, *Physics* 184a.10–15 is a reference to page 184, column a, lines 10–15 of Aristotle's *Physics* in the Bekker edition; or *Metaphysics* 1067b.25–30, which refers to page 1067, column b, lines 25–30 of the Bekker edition. Vernacular translations of Aristotle's works have adopted Bekker's page, column, and line numbering, so that when one cites a passage from the Oxford English translation (which is used throughout this volume), the reference is equally valid for both the Greek edition and the English translation. This is evident from the fact that the beginning of every page in the Greek (Bekker) edition is printed in the margin alongside the corresponding line in the English translation. The lines are then numbered in multiples of five until the end of that column (*a* or *b*) is reached, at which point the next page and column number are printed, and so on through every page. The first treatise in the revised Oxford English translation is the *Categories*, which has 1a1 (actually 1ᵃ1, but I shall not use superscripts for columns *a* and *b*) alongside the first line of the treatise, signifying page 1, column a, line 1; some 30 lines below, we find 1b1 in the margin, and 30 lines later, we have 2a1 in the margin, and so on until we reach 1462b1, or page 1462, column b, line 1, which coincides with the last page of Aristotle's *Poetics* in the Greek edition.

One also can broaden a reference by adding book and chapter numbers. For example, *On the Heavens* 2.14.296b.12–23, indicates a reference to book 2, chapter 14 in the English translation of *On the Heavens*, and, more specifically to that part of the English translation that corresponds to page 296, column b, lines 12–23 of the Greek text.

I wish to express a considerable debt of gratitude to the librarians and staff of the Herman B. Wells Library at Indiana University, who, as always, facilitated my work in countless ways. I am also grateful to Indiana University for research funds that enabled me to obtain essential equipment and supplies. Once again, I wish to express my deep gratitude to my weekly luncheon companions, colleagues, and longtime friends – Frederick Churchill, H. Scott Gordon, Noretta Koertge, Jack Moore, and John Walbridge – who responded to my numerous queries with their usual insight and concern. I was fortunate to have had two diligent and intelligent anonymous readers of my manuscript who offered many helpful suggestions to improve its quality. To each of them I wish to express sincere and grateful thanks. Finally, as so often in the past, I thank my wife, Sydelle, for her patience in listening to numerous problems relevant to my books and for always responding with helpful, intelligent suggestions.

My book is dedicated to the memory of a great scholar and dear friend, Professor Marshall Clagett, who passed away on October 21, 2005, at the age of eighty-nine. Since 1964, Marshall Clagett was a Professor of the History of Science in the School of Historical Studies of the Institute for Advanced Study, Princeton, New Jersey. I had the great good fortune to have him as my major professor in the Department of the History of Science at the University of Wisconsin–Madison, in the years 1951–1957. His textual scholarship was extraordinary and the range of his publications awesome. He published extensively on ancient Greek science; medieval Latin physics, which included *Archimedes in the Middle Ages*, a five volume work in ten parts published between 1964 and 1984; and, finally, a few years before his retirement from the Institute in 1986, he came to focus his research talents exclusively on ancient Egyptian science, for which purpose he traveled to Egypt many times and learned to read Egyptian hieroglyphics. The result of these extraordinary activities occurred between 1989 and 1999 when Marshall Clagett published three volumes (in four tomes) on Egyptian science. These volumes ranged over cosmology, astronomy, and mathematics and were largely comprised of source documents, many of which Clagett translated. A fourth volume on medicine was in process at the time of his death.

There can be little doubt that Marshall Clagett was one of the greatest and most unusual scholars of the twentieth century. His like will not soon come again.

1

Ancient Egypt to Plato

THE PRELITERATE BEGINNINGS

Natural philosophy began with no name to designate it, and in its embryonic phase it included just about anything relevant to nature. Until the time of Aristotle, who shaped the discipline of natural philosophy for the following two thousand years, the study of nature may be said to have embraced all inquiries and questions about the physical world. On what is such a claim based? Surely it is not based on anything said or recorded. But we may reasonably interpret the earliest form of natural philosophy as embracing "all inquiries about the physical world" because we have no reason not to do so. Natural philosophy may be said to have begun with the first efforts to understand the world by the earliest human beings in their fight for survival. Thus, it extends to preliterate societies, which, for thousands of years, amassed knowledge about the world, which they passed on to subsequent generations.

Members of preliterate societies learned by empirical methods about the habits of this or that animal, or this or that plant, or devised explanations, either magical or natural, about this or that individual natural phenomenon. They must have gleaned knowledge about nature from hunting and from the earliest kinds of agriculture in which they engaged. "But to have the idea of the nature of some particular object is not to have the general conception of a *domain of nature* encompassing all natural phenomena."[1] The idea of a "*domain of nature* encompassing all natural phenomena" was probably not arrived at, or invented, by the Greeks.[2] Nature was not invented. It was a given. The first humans must have been aware of nature, which was all around them and which was involved in everything they did. What the Greeks seem to have invented were instructive ways of talking about nature. They consciously pursued ways of studying and explaining the nature that surrounded them and in which they were immersed. But long before the Greeks, the ancient civilizations of Egypt and Mesopotamia learned much about nature and its actions.

[1] From G. E. R. Lloyd, "The Invention of Nature," in G. E. R. Lloyd, *Methods and Problems in Greek Science* (Cambridge: Cambridge University Press, 1991), 418.
[2] As Lloyd would have it in his article cited in the preceding note.

ANCIENT EGYPT AND MESOPOTAMIA

The first written evidence of anything that we might appropriately characterize as natural philosophy appears in the two great contemporaneous river-valley civilizations of Egypt and Mesopotamia, commencing sometime around 3500 to 3000 BC. Because each of the two civilizations developed their own form of writing – hieroglyphics for the Egyptians, who wrote on papyrus and on tomb walls and monuments, and cuneiform or wedge-shaped characters for the Mesopotamians, who wrote on clay – they left written records that modern scholars have deciphered. The surviving literature reveals a great emphasis on mythology and religion as the means of explaining the creation of the world and its operations. There is also a rather practical interest in the physical world that manifested itself primarily in the areas of astronomy, mathematics, and medicine.[3]

In his splendid multivolume work on Egyptian science, Marshall Clagett explains that what passed for natural philosophy among the ancient Egyptians was never distinct from religion and magic.[4] It is not surprising that in ancient Egypt and Mesopotamia, the initial interest in anything resembling a physical question was focused on how the world came to be. It is here where religion, myth, magic, and gross observation fused together to provide a variety of answers to perplexing questions. The idea of creation from nothing (ex nihilo) did not occur in the ancient world until the rise of Christianity. Before that, it was always assumed that the world came out of something. The Egyptians, for example, assumed that the world was created out of Nun, who was regarded as the primitive waters, or abyss, out of which things emerged. Out of Nun came a variety of creator gods, for example, the sun, or Ptah, or a cluster of gods called the primitive Eight (the Ogdoad). Before they could create anything, however, they had first to create themselves.[5]

A version of Babylonian creation myths appears in the Enuma Elish, which has striking similarities to the first two chapters of the Book of Genesis.[6] It

---

[3] Accounts of Egyptian and Mesopotamian science can be found in George Sarton, A History of Science: Ancient Science through the Golden Age of Greece (Cambridge, MA: Harvard University Press, 1952), chs. 2–3, 19–99; Marshall Clagett, Greek Science in Antiquity (London: Abelard-Schuman, 1957), ch. 1, 3–20. For a briefer presentation, see David C. Lindberg, The Beginnings of Western Science: The European Scientific Tradition in Philosophical, Religious, and Institutional Context, 600 B.C. to A.D. 1450 (Chicago: University of Chicago Press 1997), ch. 1, 13–20.

[4] Marshall Clagett, Ancient Egyptian Science: A Source Book, 3 vols. (Philadelphia: American Philosophical Society, 1989–1999), vol. 1, tome 1, ch. 2 ("The World and Its Creation: Cosmogony and Cosmology"), 263.

[5] See Clagett, Ancient Egyptian Science, vol. 1, tome 1, 264–265.

[6] On the Enuma Elish, I follow the translation and summary account in Alexander Heidel, The Babylonian Genesis: The Story of Creation, second edition (Chicago: University of Chicago Press, 1951).

tells of the relations between the gods before the creation of the world. It describes the time when only Apsû, the primeval sweet-water ocean, and Ti'âmat, the saltwater ocean, existed along with their son, Mummu, who seems to represent "the mist rising from the two bodies of water and hovering over them."[7] After an unknown time, the god Ea and the goddess Damki[na] produced a child, Marduk, who would eventually become "the wisest of the gods."[8] Indeed, so beloved was he that many of his fellow gods made him the supreme god in the pantheon. His rival was Ti'âmat, who was thought to be invincible. In an epic showdown battle, Marduk killed Ti'âmat and then split her immense body in two, using one-half of her corpse to fashion the sky, and used the other half to make the earth.[9] Thus was our world created.

Egyptians and Mesopotamians viewed the world as a place where magic was essential for survival. It was used to explain virtually all phenomena that we would regard as natural. This is nowhere better exemplified than in medicine, where the diagnosis and treatment of internal ailments relied heavily on magic. For obvious reasons, medicine is probably the first discipline to be developed by any people. But Egyptian medicine is the first to have left a written legacy in the form of seven or more papyri documents that convey a good idea of the level of their medical knowledge. Among these, the most important are the documents known as the Ebers and Smith papyri, which probably date from the seventeenth and sixteenth centuries BC but reflect knowledge and practices that were in use centuries earlier.[10] The Ebers papyrus, approximately five times larger than the Smith papyrus, was composed as a guide for physicians. The Egyptians believed that internal ailments were caused by the presence of demons in the body. To restore the body to health, it was essential to drive the demon from the body or to drive out the poisons it may have injected into the body. To do this, Egyptian physicians usually recited threatening spells and incantations against the demons and used amulets and other efficacious objects to protect the patient. They also used drugs and medicines, some of which proved helpful over time. It was, indeed, in the domain of drugs and medicines that Egyptian physicians acquired a reputation in the ancient world.

In the treatment of internal ailments, Mesopotamian medicine was similar to that of the Egyptians, relying on spells and incantations to cure the patient. But the Egyptians produced one medical text that far exceeds all the medical texts of Mesopotamia. The Edwin Smith Surgical Papyrus is an extraordinary medical treatise. It includes forty-eight cases, which are all about wounds to the body. Although incomplete, the forty-eight cases were organized from head to toe, but the last case extends only to the spinal column, where the

---

[7] Heidel, ibid., 3.   [8] Ibid., 5.   [9] Ibid., 8–9.
[10] For brief, but lucid, descriptions of the Ebers and Smith papyri, see Sarton, *A History of Science*, 44–48, and Clagett, *Greek Science in Antiquity*, 6–9.

treatise terminates. Each of the forty-eight cases is subdivided systematically into five parts as follows:

1. Title of case.
2. Examination.
3. Diagnosis and opinion.
4. Treatment.
5. Glosses to explain possible obscure terminology.

Because the wounds were easily observable, the Smith papyrus had little recourse to magic, although it is not wholly devoid of it. The cases bore titles of the following kind:

Case 4:   "Instructions concerning the gaping wound in his head, penetrating to the bone, and splitting his skull."

Case 6:   "Instructions concerning a gaping wound in his head, penetrating to the bone, smashing his skull, and rending open the brain of his skull."[11]

In thirteen of the forty-eight cases – including Case 6 – the author warns that they are untreatable. Although there is no evidence that the Egyptian practice of mummification added anything to their knowledge of anatomy, the Smith papyrus reveals a high level of knowledge and understanding. The forty-eight cases range from the head to the spinal column. Most cases involve broken bones, each of which is systematically investigated. The Smith papyrus also includes mention of the pulse, as well as the first extant attempt to describe the brain. In thirteen of the forty-eight cases, physicians are advised not to treat the wounds because they are inevitably fatal.[12] The Smith papyrus, and to a lesser extent the Ebers papyrus, give us a very favorable idea of the medicine, anatomy, and physiology of the Egyptians, and of the scientific outlook they had obtained at least two thousand years before Hippocrates.[13]

If the ancient Egyptians showed a greater aptitude for medicine than did their Mesopotamian contemporaries, there is little doubt that the Mesopotamians were superior to their Egyptian contemporaries in astronomy and mathematics.

Great strides were made in astronomy. Although the Egyptians devised a solar calendar in which the year was divided into three parts of four months each, their most significant achievement was a civil calendar of exactly 365 days formulated sometime around 2900 BC. The civil calendar was not based on any astronomical phenomena; nor, indeed, did it have any

[11] Sarton, *A History of Science*, 46. For the English translation of Case 6, see Clagett, *Greek Science in Antiquity*, 8–9.

[12] Clagett, *Greek Science in Antiquity*, 7–8.   [13] Sarton, *A History of Science*, 47–48.

astronomical function. It consisted of twelve months of thirty days each, plus five festival days, to yield 365 days. The civil calendar was a great achievement because it played a significant economic, social, and scientific role. Because the civil calendar always began on the same day of the year, which was not true of the Egyptian lunar calendar, the precise day when debts fell due could be calculated easily for many years into the future. Similarly, the exact number of days intervening between different festivals was readily determined. But the civil calendar found its most enduring role in astronomy. From Claudius Ptolemy (fl. 150 AD) to Nicholas Copernicus (1473–1543), some astronomers (if not many) found it convenient to record astronomical observations in the Egyptian civil calendar, as it enabled them to determine the exact interval between any two observations of a celestial body.

It was, however, the Babylonians and Assyrians in Mesopotamia who brought astronomy to its greatest heights in the period to approximately 500 BC. Our great debt to them becomes evident when we realize that they applied their sexagesimal numerical system – that is, a number system based on sixty and its subdivisions – to the sky. Around 500 BC, the Babylonians introduced the concept of the ecliptic, which was the circle traced out by the Sun's apparent path around the earth. They assigned 360 degrees to the ecliptic and divided it into twelve divisions of thirty degrees. The twelve divisions formed the signs of the zodiac. The Babylonians were fine observers of the heavens and by 300 BC knew how to predict the length of a month – whether it was twenty-nine or thirty days. They could do this because they recorded their observations in tables that proved useful for their needs and also for the future of Greek astronomy.[14]

Babylonian, or Mesopotamian, astronomy reached a sophisticated level because it could utilize an exceptionally well-developed mathematics. Their flexible and powerful sexagesimal number system enabled them to express all numbers with only two symbols and to carry out all arithmetic operations with ease, a status the Egyptians never attained. They could use only two symbols because they arrived at the concept of place notation, whereby the value of a symbol depended on its place in the number. Thus they had the same kind of mathematical flexibility as we have with our decimal system. Sometime around 300 BC, the Babylonians introduced the idea of zero and used it in their astronomical calculations. Finally, the Babylonians carried out arithmetic operations on fractions in exactly the same manner as they did on their integers.

With a powerful and sophisticated number system, the Babylonians made and utilized all manner of numerical tables. As a consequence, they attained

[14] For a detailed study of Mesopotamian mathematics and astronomy, see O. Neugebauer, *The Exact Sciences in Antiquity* (Princeton, NJ: Princeton University Press, 1952).

a high level of achievement in algebra, solving many kinds of quadratic equations.[15]

Because they were the first to leave written records of their achievements, there can be little doubt that scholars in ancient Egypt and Mesopotamia began the human process of understanding the operations of the natural world. They did so largely in the areas of medicine, astronomy, and mathematics, but also inevitably began to gather information about natural history. What they began and developed would become a legacy for the Greeks, who arrived on the scene long after their Egyptian and Mesopotamian predecessors.

EARLY GREEK NATURAL PHILOSOPHY AND MEDICINE

Although they made significant contributions toward a better understanding of nature, the Egyptians and Mesopotamians were heavily reliant on explanations rooted in magic, mythology, and the supernatural. Their role was nonetheless significant because they began the lengthy quest for understanding the workings of our world. The interplay between natural and supernatural explanations of observed effects in the physical world took a dramatic turn toward the natural around 600 BC, when the ancient Greeks appeared on the scene and left traces of their earliest speculations during the period between 600 and 400 BC, a period that laid the foundations of Greek science and natural philosophy for the next six hundred years.

The Pre-Socratic Natural Philosophers

Greek thought blossomed in the city-states that Greeks had founded along the coast of Asia Minor in the seventh and sixth centuries BC. Of these, the most important was the city of Miletus, which produced some of the most famous early thinkers, such as Thales, Anaximenes, and Anaximander, collectively known as Milesians.

The years from 600 to 400 BC usually are called the pre-Socratic period – the period of philosophical activity before the time that Socrates (469–399 BC), the teacher of Plato, lived – and the philosophers of whom we have any record are identified collectively as "Pre-Socratics." None of their works is known to have survived; only bits and pieces, mere fragments that were preserved by subsequent authors who quoted from their works. For example, Theophrastus, who succeeded Aristotle as head of the Lyceum, had before him the works of various Pre-Socratics and wrote a treatise titled *Opinions of the Physicists* in sixteen or eighteen books, of which only the last has survived, bearing the title *On Sensation*. In order to evaluate the

[15] For a brief, clear account of the Babylonian number system and Babylonian contributions in algebra, see Clagett, *Greek Science in Antiquity*, 16–19.

thought of pre-Socratic philosophers, Hermann Diels searched Greek literature for actual quotations from Pre-Socratic authors. In 1903, he published *The Fragments of the Pre-socratics (Die Fragmente der Vorsokratiker)*. To aid in the interpretation of these fragmentary thoughts, scholars use the doxographic tradition that derives from Theophrastus' work and consists of opinions of later authors on pre-Socratic thinkers. To this, we must add Aristotle's important discussion of the Pre-Socratics in the first book of his *Metaphysics*. Aristotle regarded these early philosophers as his predecessors and thought it important to describe their views about the nature of the physical world.

To convey a flavor of the fragments that Diels published, it will be useful to cite a few from the translation by Kathleen Freeman. Although Anaximenes of Miletus is regarded as one of the most important pre-Socratic philosophers, and is known to have written one book, only one authentic sentence survives, in which he declares: "As our soul, being air, holds us together, so do breath and air surround the whole universe."[16] Among the more than three hundred fragments attributed to Democritus of Abdera (fl. 420 BC) is this important and somewhat lengthy one:

9. Sweet exists by convention, bitter by convention, colour by convention; atoms and Void (*alone*) exist in reality.... We know nothing accurately in reality, but (*only*) as it changes according to the bodily condition, and the constitution of those things that flow upon (*the body*) and impinge upon it.[17]

These two fragments are reasonably intelligible, but many others are little more than snippets, as, for example, when Democritus asserts that:

145. Speech is the shadow of action.
147. Pigs revel in mud.
151. In a shared fish, there are no bones.[18]

Despite the enormous difficulties of interpreting fragments that have no proper context, modern scholars have recognized their great, and even overwhelming, significance. These early Greek thinkers mark a dramatic break with all that went before in the Greek and non-Greek worlds. G. E. R. Lloyd sees two basic innovations in their thought: "First, there is what may be described as the discovery of nature, and second the practice of rational criticism and debate."[19] By "discovery of nature," Lloyd means "the appreciation of the distinction between the 'natural and the 'supernatural', that is the recognition that natural phenomena are not the products of random or arbitrary influences, but regular and governed by determinable

---

[16] Kathleen Freeman, *Ancilla to the Pre-Socratic Philosophers: A Complete Translation of the Fragments in Diels, "Fragmente der Vorsokratiker"* (Oxford: Basil Blackwell, 1948), 19.
[17] Freeman, ibid., 93. [18] Ibid., 105.
[19] G. E. R. Lloyd, *Early Greek Science: Thales to Aristotle* (New York: W. W. Norton, 1970), 8.

sequences of cause and effect."[20] Consequently, although "the idea of the
divine often figures in their cosmologies, the supernatural plays no part in
their explanations."[21]

There can be no doubt that this was a monumental change of outlook.[22]
It was a new approach that was added to the mythological explanations of
the world that had characterized earlier Greek descriptions of physical phenomena by the likes of Hesiod and Homer. Pre-Socratics no longer explained
natural phenomena, such as earthquakes, lightning, storms, and eclipses, as
the actions of happy or angry gods, but as the actions of natural forces that
regularly produced such effects. Thus, Thales of Miletus, who is regarded
as the first of the Greek investigators into nature, is said to have declared
that "the world is held up by water and rides like a ship, and when it is
said to 'quake' it is actually rocking because of the water's movement."[23]
Rather than attribute earthquakes to Poseidon, god of the sea, as Greeks had
done for centuries, Thales chose to give a natural explanation, as did all the
Pre-Socratics who followed him.

Not only did the Pre-Socratics eliminate the gods as the causes of natural
phenomena and replace them with natural causes, but they also adopted a
number of different approaches to explain the apparent diversity and change
they observed in the world around them. In the process, they enunciated
some of the most basic problems that would shape the discipline that was
eventually known as physics, or natural philosophy. The first wave of Pre-
Socratics is often called monists because they sought to explain changes in the
world in terms of a single substance, or stuff. They coped with what has been
called the one-many problem, in which they sought to explain how the many
things that we see and experience could come from one basic substance or
stuff. Thus, Thales is said to have taken water as the basic substance, whereas
Anaximander (ca. 610–ca. 547 BC) assumed the existence of an indeterminate
substance called the *apeiron*, or boundless, out of which things came and
to which they returned. Anaximander introduced an idea that became an
integral part of Greek explanations of change. He regarded change as the
product of an interchange of opposite qualities, namely, hot and cold, which
came out of the basic substance – which he called "the boundless" – and

---

[20] Ibid.   [21] Ibid., 9.

[22] For three excellent accounts of the substance and significance of pre-Socratic contributions to
the physical inquiry about the world, see Clagett, *Greek Science in Antiquity*, chs. 2 ("Greek
Science: Origins and Methods"), 21–33, and 3 ("Science and Early Natural Philosophy"),
34–38; G. E. R. Lloyd, *Early Greek Science: Thales to Aristotle*, chs. 2 ("The Theories of
the Milesians"), 16–23; 3 ("The Pythagoreans"), 24–35; and 4 ("The Problem of Change"),
36–49; and David C. Lindberg, *The Beginnings of Western Science*, 25–35.

[23] This attribution to Thales is by the Roman philosopher, Seneca, in his *Natural Questions*,
III, 14, and translated by G. S. Kirk and J. E. Raven, *The Presocratic Philosophers: A Critical
History with a Selection of Texts* (Cambridge: Cambridge University Press, 1957), 92. See
also Lloyd's discussion in *Early Greek Science*, 9.

returned to it. An eternal motion of the boundless produces hot and cold that together form many worlds. Anaximander also used the principle of sufficient reason, or insufficient reason, when he argued that the earth lies unsupported, but motionless, because it is equidistant from everything and therefore has no reason or desire to move toward anything. Perhaps, as Lloyd explains, Anaximander "appreciated that Thales' view, and views like it, run into an obvious difficulty: if water holds the earth up, what holds the water up?"[24]

Anaximenes of Miletus (fl. 546 BC), like Thales, chose a sensible element, air, as the basic substance out of which all things emerged. Simplicius, an important commentator on Aristotle in the sixth century AD, reports that for Anaximenes the physical mechanism that causes the air to change is rarity and density. "Being made finer it becomes fire, being made thicker it becomes wind, then cloud, then (when thickened still more) water, then earth, then stones: And the rest come into being from these. He, too, makes motion eternal, and says that change, also comes about through it."[25]

From the Greeks who colonized the west and had come to Italy and Sicily, great contributions were forthcoming during the pre-Socratic period. In the course of the fifth century BC, Pythagoras and his followers, known as the Pythagorean School, formed a school in Italy that was largely religious in character. We know little about the contributions of Pythagoras himself, who was born on the island of Samos, off the coast of Asia Minor, and later migrated to Italy, and only a little more about the members of his school, which seems to have had a continuous existence for some centuries after the death of Pythagoras. A major source for our knowledge of the earliest Pythagoreans is Aristotle, who rarely refers to Pythagoras, the man, but usually speaks of the Pythagoreans as a group. From Aristotle, we learn that the Pythagoreans did not opt for a material cause as the basic substance of the world but assigned that role to number. The Pythagoreans focused their interests on mathematics, although precisely how the Pythagoreans understood a world in which number is the basis of all material things is a mystery.[26]

As the substratum underlying our world, however, they chose to emphasize a formal, rather than material, aspect. The idea of mathematics as the basis of nature would have a long history and represents another contribution by these early Greek thinkers.

---

[24] Lloyd, *Early Greek Science*, 20–21. For a detailed description of Anaximander's views, see Kirk and Raven, *The Presocratic Philosophers*, 99–142. Anaximander's ideas about the earth's centrality and immobility are reported by Aristotle in *On the Heavens* 2.13.295b.10–15.

[25] The translation is from Simplicius's *Commentary on Aristotle's Physics*, 149.32, and appears in James B. Wilbur and Harold J. Allen, *The Worlds of the Early Greek Philosophers* (Buffalo, NY: Prometheus Books, 1979), 44.

[26] Later in this chapter, the reader will find a passage from Aristotle on the Pythagorean attitude toward mathematics.

The underlying idea in the monist approach to nature was that the material – or, in the case of the Pythagoreans, formal – substratum that underlay all change was itself permanent and indestructible, an idea that would play a vital role in later natural philosophy. But the change that the monists assumed as self-evident was attacked by those philosophers who began to question the reliability of the senses and came to regard the whole notion of change as an illusion. Parmenides of Elea (ca. 515–ca. 450 BC), one of the giants of Western thought, was the foremost critic of monist thought based on a continually changing world. Writing in hexameter verse, Parmenides left a poem that divides into an introduction followed by two distinct parts. In the first major part, called the Way of Truth, Parmenides argues that change is impossible. He insists that the Way of Truth is a logical way of talking about things, because it claims only that what is is. That which exists could not have had a beginning and is therefore ungenerable and indestructible. What exists cannot have had a beginning, because it would have had to come from something that is not-being, which implies that a change took place from not-being to being, which is impossible. By a similar argument, what exists cannot be destroyed and come to an end, because that could only occur if what exists passed from being to not-being, which is impossible. Parmenides explains that

One way only is left to be spoken of, that it *is*; and on this way are full many signs that what *is* is uncreated and imperishable, for it is entire, immovable and without end. It was not in the past, nor *shall* it be, since it *is* now, all at once, one, continuous; for what creation wilt thou seek for it? How and whence did it grow? Nor shall I allow thee to say or to think, "from that which is not"; for it is not to be said or thought that it is not.[27]

To reinforce his argument that being could not have come into existence, Parmenides invokes the principle of sufficient reason, asking, "What need would have driven it on to grow, starting from nothing, at a later time rather than an earlier?"[28]

Thus did Parmenides argue that change is an illusion and is logically impossible; therefore, motion is impossible. Nothing can come into existence or pass out of existence. The only thing that exists is what is. Parmenides distinguished three ways of thinking:

1.   That what you can think must exist.[29]
2.   What you cannot think cannot possibly exist.[30]

[27] Translation by Kirk and Raven, *The Presocratic Philosophers*, frag. 347, 273.
[28] Kirk and Raven, ibid.
[29] "For it is the same thing to think and to be," Parmenides argues. See Freeman, *Ancilla to the Pre-Socratic Philosophers*, 42.
[30] Parmenides encapsulates the first two ways of speaking in the following passage: "Come, I will tell you – and you must accept my word when you have heard it – the ways of inquiry

3. The third way of speaking combines the first two ways of thought and is unavoidably used in common speech. It is the way of "the two-faced mortals," as Simplicius, to whom we owe the preservation of Parmenides's poem, put it, because it combines contraries.[31] What Parmenides means is that we might say that the tomato is green and later, when the tomato has been seen to ripen, we may say that it is red, or not-green. Despite a temporal interval between these statements, they are nevertheless contradictory. People who speak this way believe that change exists in the universe.

Parmenides, however, insists that change is impossible and an illusion. You cannot provide a logical explanation for change. Parmenides is prepared to follow the logic of his argument, even if it means violating his senses. Thus did he subordinate the senses to reason. This was a major development in the history of thought. Parmenides of Elea showed that the one-many problem was impossible. The many cannot be derived from the one, because there is only the one thing in existence, namely, being. From this standpoint, Parmenides is the ultimate monist, one who denied change and motion.

Zeno of Elea, Parmenides's fellow townsman and most important follower, defended the master's ideas by formulating a series of paradoxes that sought to prove that change, plurality, and motion were logically impossible. Among his most famous arguments are four against the possibility of motion that Aristotle reports in his *Physics* (6.9.239b.10–240a.16), in which he informs us at the outset that Zeno's four arguments about motion "cause so much trouble to those who try to answer them." The first argument, known as the "dichotomy," or "bisection," declares that, in order to traverse any distance, you must first arrive at the halfway point before you reach the end. By extrapolation, we see that you also will have to traverse the first quarter of the distance, before you reach the halfway point, and so on. Given this infinite regressive divisibility of the distance into half distances, you will never begin your journey and therefore motion is impossible.[32]

In an analogous thrust, Zeno proposes, as his second paradox, the famous Achilles argument, which, Aristotle tells us, "amounts to this, that in a race the quickest runner can never overtake the slowest, since the pursuer must

---

which alone are to be thought: the one that IT IS, and that it is not possible for IT NOT TO BE, is the way of credibility, for it follows Truth; the other, that IT IS NOT, and that IT is bound NOT TO BE: this I tell you is a path that cannot be explored; for you could neither recognise that which is NOT, nor express it." Freeman, ibid.

[31] See Kirk and Raven, *The Presocratic Philosophers*, 271.

[32] Although this is the more popular interpretation of the first argument, Gregory Vlastos argues that Zeno did not mean to show that you do not get started, but, rather, that you never reach the goal even if you are moving toward it progressively; that is, even if you arrive at the first half of the distance, then reach half of the remaining distance and so on, you will never traverse the entire distance. See Gregory Vlastos, "Zeno's Race Course," *Journal of the History of Philosophy*, vol. 4, Nr. 2 (April 1966), 95–96.

A A A A

B B B B → →

← ← C C C C

Figure 1.1.

first reach the point whence the pursued started, so that the slower must always hold a lead."[33] A suppressed premise in this argument is that the slowest runner begins the race with a head start. Aristotle regards this argument as similar to the first argument, except "that the spaces with which we have successively to deal are not divided into halves." In this argument, Achilles, the quickest runner cannot overtake the slowest runner, usually assumed to be a tortoise. Whenever Achilles reaches a place in which the tortoise has been, the tortoise will have ambled on a bit. Although Achilles will continually narrow his distance from the tortoise, he cannot catch it, because the conditions of the race preclude such an outcome. Zeno's major point in this paradox is simply that if Achilles, the swiftest runner in the world cannot catch the slowest runner, the concept of motion must be absurd.[34]

In his very brief description of Zeno's third paradox, called the Flying Arrow, Aristotle announces the paradox immediately when he declares that "the flying arrow is at rest," because "time is composed of moments."[35] Kirk and Raven reconstruct this argument as follows: "An object is at rest when it occupies a space equal to its own dimensions. An arrow in flight occupies, at any given moment, a space equal to its own dimensions. Therefore an arrow in flight is at rest."[36]

In the fourth, or Stadium paradox, Zeno assumes that equal bodies are moving past other equal bodies in a stadium. In Figure 1.1.[37] the first B is in motion. During that motion, it lies opposite two As and four Cs in the same time interval, as is obvious from Figure 1.2. Because B has been moving past both As and Cs, and all As and Cs are equal, the first B should have passed just as many As as Bs. Therefore, the first B should have passed four As but only passed two As; consequently, 4As = 2As, which is preposterous. The

---

[33] *Physics* 5.9.239b.14–16.

[34] For an excellent analysis of this argument, see G. E. L. Owen, "Zeno and the Mathematicians," *Proceedings of the Aristotelian Society*, 1957–1958, 199–222; reprinted in Wesley C. Salmon, *Zeno's Paradoxes* (Indianapolis: The Library of Liberal Arts, Bobbs-Merrill, 1970), 139–163.

[35] *Physics* 5.9.239b.30–32.

[36] Kirk and Raven, *The Presocratic Philosophers*, 294–295. Aristotle argues that if we reject Zeno's assumption that time is composed of moments, the conclusion does not follow, namely, that the arrow is at rest.

[37] The two diagrams were made by Simplicius, a Greek commentator on the works of Aristotle who lived in the sixth century AD.

A A A A

B B B B →

← C C C C

Figure 1.2.

paradox lies in the fact that motion must be something absurd if one and the same object, namely B, passes two As and four As in the same time interval.

For all of these paradoxes, and others that Zeno directed against the possible existence of a plurality of things, alternative interpretations have been proposed. Despite these disagreements, scholars are unanimous in believing that Zeno had propounded powerful paradoxes to subvert the idea that motion and change are intelligible. Bertrand Russell, the great twentieth-century philosopher, paid tribute to Zeno when he declared that "Zeno's arguments, in some form, have afforded grounds for almost all the theories of space and time and infinity which have been constructed from his day to our own."[38]

Parmenides and Zeno would readily have acknowledged that we continually observe and experience change and motion. But they were equally convinced that reason could not demonstrate the existence of change; on the contrary, they were convinced that they had demonstrated the illogicality, and therefore the impossibility, of change and motion. The arguments of Parmenides had a profound effect on his contemporaries and successors. They abandoned monism and agreed with Parmenides that something could not come into being from nothing, or not-being. To avoid the dilemmas and paradoxes enunciated by Parmenides and Zeno, they became pluralists. That is, they assumed the existence of a plurality of basic substances, or elemental bodies, from which the world and the things in it were composed. By entering into combinations, the elements form the bodies we observe, and by separating from one another, they cause those same bodies to pass away, thus freeing their former constituent elements to combine and form different bodies. Thus did they account for our changing and diverse world.

Among the most important figures to explain the world and its phenomena in this pluralistic mode were Empedocles of Acragas (fl. 450 BC) and the two atomists, Leucippus of Elea (or Miletus) (fl. 450 BC) and Democritus of Abdera (fl. 420 BC), who are usually cited together as Leucippus and Democritus, as their individual contributions cannot be distinguished.[39]

---

[38] Bertrand Russell, *Our Knowledge of the External World* (New York: Mentor Books, New American Library, 1960),140. These words and more are quoted by James B. Wilbur and Harold J. Allen, *The Worlds of the Early Greek Philosophers* (Buffalo, NY: Prometheus Books, 1979), 130–131.

[39] On their dates, I follow Sarton, *A History of Science: Ancient Science through the Golden Age of Greece*, 251.

Empedocles was an extraordinary individual. He wrote on medicine, religion, and nature. Simplicius, the sixth century AD commentator on Aristotle's *Physics*, who found many occasions to preserve the opinions of Pre-Socratics, succinctly described Empedocles' interpretation of physical change. Empedocles, Simplicius declares, "makes the material elements four in number, fire, air, water, and earth, all eternal, but changing in bulk and scarcity through mixture and separation; but his real first principles, which impart motion to these, are Love and Strife. The elements are continually subject to an alternate change, at one time mixed together by Love, at another separated by Strife; so that the first principles are, by his account, six in number."[40] Thus did Empedocles assume the existence of four basic elements, earth, water, air, and fire, out of which everything was formed in varying proportions. He further postulated two causal mechanisms for producing and dissolving bodies, which he called Love and Strife. With the formulation of the four element theory, Empedocles launched one of the most basic and long-lived physical theories ever proposed, a theory that was still in use in the eighteenth century, largely because Aristotle adopted the same four elements as his basic building blocks for the physical world.

Perhaps the most significant theory of the fifth century BC was the atomic theory formulated first by Leucippus and developed further by Democritus. As presented by Democritus, the atomic theory was a reply to the theory of Parmenides with its denial of change and motion. Assuming the existence of change and motion, Democritus explained them by asserting the existence of two fundamental, invariant entities: atoms and the void. Atoms were infinite in number but shared certain properties. They were all uncreated and eternal, as well as homogeneous, indivisible, and impenetrable. But each atom was nevertheless unique. Atoms possessed an infinity of sizes and shapes, but each had its own unique size and shape and each had its own unique position in space. Indeed, even if two atoms had the same size and shape, they would nevertheless differ because their impenetrability guaranteed that each would occupy a different portion of space.

The macro bodies of the world are comprised of these atoms, which, infinite in number, are spread through an infinite void. "They are, moreover, in continuous motion, and their movements give rise to constant collisions between them. The effects of such collisions are two-fold. Either the atoms rebound from one another, or if the colliding atoms are hooked or barbed, or their shapes otherwise correspond to one another, they cohere and thus form compound bodies. Change of all sorts is accordingly interpreted in terms of the combination and separation of atoms. The compounds thus formed possess various sensible qualities, such as colour, taste, temperature and so on, but the atoms themselves remain unaltered in substance."[41] In Democritus's

---

[40] Translated in Kirk and Raven, *The Presocratic Philosophers*, frag. 426, 329–330.
[41] Lloyd, *Early Greek Science*, 46. For Aristotle's brief account of the Democritean atomic theory, see his *Metaphysics* 1.4.985b.4–20.

system, each atom was like a Parmenidean being in the sense that it was eternal and unchanging; but, unlike the being described by Parmenides, it had a capacity for motion. Indeed, each atom moved constantly with a jostling motion. If it collided with another atom, it acquired a derived motion as it moved off in another direction. But in what did it move?

It moved in an infinite void. Democritus assumed the existence of two real entities: atoms and the void. The latter is a spatial incorporeality, for which reason Democritus regarded it as akin to something unreal, that is, something that is "not-being." This may have been the first time that something characterized as "not-being" was considered as a real existent entity.

Indeed, perhaps also for the first time, Democritus assumes the existence of an infinite universe with infinite worlds. He proclaimed the existence of

innumerable worlds, which differ in size. In some worlds there is no sun and moon, in others they are larger than in our world, and in others more numerous. The intervals between the worlds are unequal; in some parts there are more worlds, in others fewer; some are increasing, some at their height, some decreasing; in some parts they are arising, in others failing. They are destroyed by collision one with another. There are some worlds devoid of living creatures or plants or any moisture.[42]

Judging from the titles of some fifty-two works on numerous subjects (including works on mathematics, physics, cosmology, music, zoology, botany, and medicine) that were attributed to Democritus but have not survived, it is obvious that he was one of the most prolific authors in the ancient world.[43]

The Pythagoreans also sought to explain the world of nature. They were followers of Pythagoras, who was born around the middle of the sixth century on the island of Samos and later moved to southern Italy, where he founded a religious brotherhood, whose members believed in the immortality and transmigration of souls.[44] Over the centuries, despite the attribution of numerous strange beliefs and practices to the Pythagoreans,[45] they enjoyed a reputation for mathematical achievements, as the Pythagorean theorem attests. Indeed, Aristotle informs us that

the Pythagoreans, as they are called, devoted themselves to mathematics; they were the first to advance this study, and having been brought up in it they thought its principles were the principles of all things. Since of these principles numbers are by nature the first, and in numbers they seemed to see many resemblances to the things that exist and come into being – more than in fire and earth and water (such and such a modification of numbers being justice, another being soul and reason, another being opportunity – and similarly almost all other things being numerically expressible); since, again, they saw that the attributes and the ratios of the musical scales were

[42] Translated in Kirk and Raven, *The Presocratic Philosophers*, frag. 564, 411.
[43] For lists of the subjects on which Democritus wrote, see Lloyd, *Early Greek Science*, 48, and Kirk and Raven, *The Presocratic Philosophers*, 404.
[44] Lloyd, *Early Greek Science*, 24.
[45] For the evidence about the Pythagoreans, see Kirk and Raven, *The Presocratic Philosophers*, 218–227.

expressible in numbers; since, then, all other things seemed in their whole nature to be modelled after numbers, and numbers seemed to be the first things in the whole of nature, they supposed the elements of numbers to be the elements of all things, and the whole heaven to be a musical scale and a number.[46]

Aristotle was apparently convinced that the Pythagoreans believed things are somehow actually composed of numbers. He could make little sense of these beliefs. Throughout their history, Pythagoreans were not only associated with mystical religious beliefs but also were regarded as strong proponents of number mysticism. It is nevertheless significant that they regarded the universe as in some meaningful sense mathematical, seeing mathematical harmonies in music and in the heavens. They regarded the number ten as a perfect number, and, therefore, as Aristotle explains

they say that the bodies which move through the heavens are ten, but as the visible bodies are only nine, to meet this they invent a tenth – the "counter-earth."[47]

In *On the Heavens*, Aristotle provides the context for the Pythagorean counter-earth. He says that most people place the earth at the center of the world, but not the Pythagoreans.

At the centre, they say, is fire, and the earth is one of the stars, creating night and day by its circular motion about the centre. They further construct another earth in opposition to ours to which they give the name counter-earth.[48]

Thus, the Pythagoreans seem to have assumed the existence of a mass of fire at the center of the world. Next in the order of the things, as we move away from the center, is the counter-earth, which moves around the central fire. Keeping pace with the counter-earth is the earth itself, the next, or second, celestial body, which, like the counter-earth, moves around the central fire. Beyond the earth, lie the moon, sun, and the other planets, all moving around the central fire.[49] Aristotle vigorously disagrees with the Pythagoreans but says that numerous individuals have denied to the earth the central place in the universe.

Their view is that the most precious place befits the most precious thing; but fire, they say, is more precious than earth, and the limit [more precious] than the intermediate, and the circumference and the centre are limits. Reasoning on this basis they take the view that it is not the earth that lies at the centre of the sphere, but rather fire. The Pythagoreans have a further reason. They hold that the most important part of the world, which is the centre, should be most strictly guarded, and name the fire, which occupies that place the "Guard-house of Zeus," as if the word "centre" were

---

[46] Aristotle, *Metaphysics* 1.5.985b.23–986a.1. Oxford English translation by W. D. Ross.
[47] Aristotle, ibid., 986a.7–11.
[48] Aristotle, *On the Heavens* 2.13.293a.20–24. Oxford English translation by J. L. Stocks.
[49] For a brief account of Pythagorean views on the cosmos, see Lloyd, *Early Greek Science*, 27–30.

quite unequivocal, and the centre of the mathematical figure were always the same with that of the thing or the natural centre.[50]

Because of the attribution of such opinions and beliefs to the Pythagoreans, the term "Pythagorean" became synonymous with the heliocentric system of Copernicus during the sixteenth and seventeenth centuries, especially in Galileo's struggle with the Church.

Along with their reputation for contributions to geometry and their deep concern for number theory,[51] the Pythagoreans played a role in the emergence of mathematics and science, although that role will always be obscure and vaguely documented.

The meager and fragmentary writings the Pre-Socratics left to posterity make them a shadowy group. We might first ask who they were? Biographical information is largely absent, but later Greek writers, often many centuries later, provide bits and pieces of information, which may be unreliable. But scholars have speculated on their roles in society. Those whom we call Pre-Socratics were known to the Greeks as *physikoi*, or those concerned with the physical world. Although they shared an interest in the physical world, they pursued different careers, among which were physician, sophist, or mathematician.[52] The majority of them were probably well off economically and many apparently also were teachers.

What was the legacy of the pre-Socratic natural philosophers? Even a cursory glance at the fragments makes it evident that something momentous had occurred. Perhaps the most striking feature of pre-Socratic thought is the emphasis on rational analysis of problems and the general avoidance of appeals to divine intervention for the explanation of natural phenomena. During the fifth century BC, the Pre-Socratics gradually substituted natural causation for divine causation.

The absence of divine intervention produced another important feature of early Greek thought. Many pre-Socratic thinkers believed that the world had no beginning and would have no end, or, if they believed in the existence of an infinity of worlds, they assumed, as did Democritus, that worlds were always coming into existence and passing away. But worlds always existed and would always exist in the future. And, with the exception of Parmenides and Zeno, whether they assumed only one world, or an infinity of worlds, they believed there was an unchanging substance or substances that under-lay observed changes and motions. The object was to determine the causal mechanisms producing the incessant changes in the world. In studying the operations of the physical world, their aim was not to control nature but

---

[50] Aristotle, *On the Heavens* 2.13.293a.29–293b.6. I have added the bracketed words.

[51] Lloyd presents a summary account of Pythagorean mathematics in *Early Greek Science*, 31–35.

[52] Lloyd, ibid., 125. For a valuable account of the overall contributions made by the early Greek natural philosophers, see Lloyd's "Conclusion," 125–146.

to understand and explain it. In coming to understand and explain it, they rarely used careful observational data, or experiments, in support of their claims. Nevertheless, the problems that the pre-Socratic philosophers identified, and with which they grappled, largely by abstract, rational arguments, formed the basis of natural philosophy as it would be shaped in the fourth century BC by Aristotle.

Hippocratic Medicine

Much less vague and shadowy in the early history of natural philosophy is the contribution to an understanding of the physical world and the human body made by a collection of medical treatises grouped under the name of Hippocrates, the almost legendary physician, who was born in the middle of the sixth century BC on the island of Cos. Although mentioned by Plato and Aristotle, little is known about Hippocrates. He was apparently a teacher in a medical school on Cos and had disciples. Approximately seventy works have been attributed to Hippocrates, although which, if any, were actually written by him is a contentious matter among scholars. One thing seems certain: the Hippocratic works were written by different individuals, perhaps including Hippocrates. Many of them, however, were composed sometime after the death of Hippocrates, in the period between 440 BC and 350 BC.[53]

The diversity of works with regard to purpose, viewpoint, and emphasis indicates that a multiplicity of authors produced the Hippocratic corpus. Within this disparate collection of treatises, some authors emphasize theory and derive medical knowledge from the nature of the universe. Others are empirical in character and describe symptoms and the course of ailments and diseases. In the treatise titled *Epidemics* (I.1–3, case 2), the physician presents numerous clinically detached, impersonal reports, among which is this:

Case II
Silenus lived on broadway near the place of Eualcidas. After over-exertion, drinking, and exercises at the wrong time he was attacked by fever. He began by having pains in the loins, with heaviness in the head and tightness of the neck. From the bowels on the first day there passed copious discharges of bilious matter, unmixed, frothy,

---

[53] Lloyd declares (*Early Greek Science*, 50) that "although the Corpus as a whole came to be named after the great fifth-century physician Hippocrates, it is nowadays thought unlikely that he wrote any of the treatises himself." By contrast, Robert Joly explains that "it was certainly not true for his contemporaries that Hippocrates was a name with no writings attached to it, and it is true for us to only a limited degree, since we possess many medical works from the time and from the school of Cos. It is very probable that some of the most outstanding of these are by Hippocrates." See Robert Joly, "Hippocrates of Cos," in Charles C. Gillispie, *Dictionary of Scientific Biography*, 18 vols. (New York: Charles Scribner's Sons, 1970–1990), vol. 6, 419.

and highly coloured. Urine black, with a black sediment; thirst; tongue dry; no sleep at night.

Second day. Acute fever, stools more copious, thinner, frothy; urine black; uncomfortable night; slightly out of his mind.

Third day. General exacerbation; oblong tightness of the hypochondrium, soft underneath, extending on both sides to the navel; stools thin, blackish; urine turbid, blackish; no sleep at night; much rambling, laughter, singing; no power of restraining himself.

Fourth day. Same symptoms.

Fifth day. Stools unmixed, bilious smooth, greasy; urine thin, transparent; lucid intervals.

Sixth day. Slight sweats about the head; extremities cold and livid; much tossing; nothing passed from the bowels; urine suppressed; acute fever.

Seventh day. Speechless; extremities would no longer get warm; no urine.

Eighth day. Cold sweat all over; red spots with sweat, round, small like acne, which persisted without subsiding. From the bowels with slight stimulus there came a copious discharge of solid stools, thin, as it were unconcocted, painful. Urine painful and irritating. Extremities grow a little warmer; fitful sleep; coma; speechlessness; thin, transparent urine.

Ninth day. Same symptoms.

Tenth Day. Took no drink; coma, fitful sleep. Discharges from the bowels similar; had a copious discharge of thickish urine, which on standing left a farinaceous, white deposit, extremities again cold.

Eleventh Day. Death.

From the beginning the breath in this case was throughout rare and large. Continuous throbbing of the hypochondrium; age about twenty years.[54]

There are numerous similarly striking descriptions of illnesses and diseases, most of which end in death. Thus, the Hippocratic School emphasized, among other things, careful attention to the symptoms and behavior of their patients.

Finally, some Hippocratic treatises emphasized a use of theory and experience in combination. The emphasis on theory, whether by itself or in combination with experience, reveals an influence of philosophy on medicine. Most physicians were akin to craftsman and therefore were unphilosophical. But the philosophically minded physicians were the ones who raised the level of medical understanding. For it was they who investigated the larger issues, such as the course of disease, distinctions between types of diseases, the causes of diseases, and prognosticating the course of a disease from observed symptoms. Galen (ca. 129–ca. 200), the greatest and most famous physician of the ancient world, followed the early tradition and placed great emphasis on philosophy in medicine. But there were those who sought to divorce medicine from philosophy, because they regarded medicine as an art

---

[54] Cited from Morris R. Cohen and I. E. Drabkin, *A Source Book in Greek Science* (New York: McGraw-Hill Book Company, Inc., 1948), 492–493. The translation is by W. H. S. Jones.

and were hostile to philosophy, as was the author of the Hippocratic treatise, *On Ancient Medicine*, who attacked those who introduce hypotheses as the basis of their medicine:

– heat, cold, moisture, dryness, or anything else they may fancy – who narrow down the causal principle of diseases and of death among men, and make it the same in all cases, postulating one thing or two, all these obviously blunder in many points even of their statements, but they are most open to censure because they blunder in what is an art, and one which all men use on the most important occasions, and give the greatest honors to the good craftsmen and practitioners in it.

Later in the treatise, our author takes on the philosophers directly when he declares:

Certain physicians and philosophers assert that nobody can know medicine who is ignorant what man is; he who would treat patients properly must, they say, learn this. But the question they raise is one for philosophy; it is the province of those who, like Empedocles, have written on natural science, what man is from the beginning, how he came into being at the first, and from what elements he was originally constructed. But my view is, first, that all that philosophers or physicians have said or written on natural science no more pertains to medicine than to painting. I also hold that clear knowledge about natural science can be acquired from medicine and from no other source, and that one can attain this knowledge when medicine itself has been properly comprehended, but till then it is quite impossible – I mean to possess this information, what man is, by what causes he is made, and similar points accurately.[55]

The most important contribution that emerges from the diverse Hippocratic works is the strong emphasis on rational, scientific medicine. It is this more than anything else that links Greek medicine to natural philosophy. In numerous places in the works of Hippocrates, authors attack the use of magic in medicine and incessantly emphasize rational observation and techniques. Hippocratic authors sought to treat disease as a natural phenomenon produced by natural causes,[56] rather than as something caused by gods or magic. It was this sound attitude that gave rise to a famous remark about epilepsy, which often was regarded as brought on by divine intervention. In the treatise called *On the Sacred Disease*, the author declares:

I am about to discuss the disease called "sacred" [i.e. epilepsy]. It is not, in my opinion, any more divine or more sacred than other diseases, but has a natural cause, and its supposed divine origin is due to men's inexperience, and to their wonder at its peculiar character.... But if it is to be considered divine just because it is wonderful, there will be not one sacred disease but many, for I will show that other diseases are no less wonderful and portentous, and yet nobody considers them sacred.[57]

[55] The two passages from *Ancient Medicine* are quoted from Sarton, *A History of Science: Ancient Science through the Golden Age of Greece*, 366.
[56] See Lloyd, *Early Greek Science*, 54.
[57] Translated in Sarton, *A History of Science: Ancient Science through the Golden Age of Greece*, 355.

The author goes on to explain that:

those who first attributed a sacred character to this malady were like the magicians, purifiers, charlatans, and quacks of our own day, men who claim great piety and superior knowledge. Being at a loss, and having no treatment that would help, they concealed and sheltered themselves behind superstition, and called this illness sacred, in order that their utter ignorance might not be manifest.[58]

Numerous statements in a variety of Hippocratic treatises bear witness to their rationalistic approach to medicine. In *Airs, Waters, Places*, a famous treatise concerned with the effects of climate and geography on health, the author declares, "Each disease has a nature of its own, and none arises without its natural cause."[59] To coordinate a large number of observations, Hippocratic physicians urged their colleagues "to make a synthesis of all the data concerning an illness in order to determine the similarities, then to establish between the latter new differences in order to arrive finally at a unique similarity."[60] In *Joints*, a treatise that deals with fractures, the author describes an unsuccessful treatment and then adds: "I relate this on purpose, for it is also valuable to know what attempts have failed and why they have failed."[61]

Early Greek medicine as exemplified in the treatises from the school of Hippocrates was itself often archaic and seemingly the practice of folk medicine. But that is hardly surprising.[62] These reservations and qualifications about the Hippocratic works cannot, however, overshadow or stain the significant level of achievement they attained. The explicit methodological standards were high and far more rigorous than anything that had preceded. The Hippocratic approach to nature added to the contributions of the pre-Socratic natural philosophers and prepared Greek society for the enormous advances and contributions that Aristotle would make. But we cannot move on to Aristotle before noting the attitude toward nature of his great teacher, Plato, who will appropriately conclude the pre-Aristotelian phase of this study.

PLATO

With the appearance of Socrates (d. 399 BC) in the second half of the fifth century BC, Greek philosophy made two important shifts. Instead of focusing on cosmology and physics, as did most of his predecessors, Socrates emphasized ethical and moral philosophy; and he did so in Athens, which became the intellectual center of Greece, superseding Ionia in Asia Minor and the Greek colonies in southern Italy and Sicily. As a teacher in Athens, Socrates

---

[58] Sarton, ibid., 356.
[59] From *Airs, Waters, Places* 22, as cited in Joly, "Hippocrates of Cos," *Dictionary of Scientific Biography*, vol. 6, 425.
[60] *Epidemics* VI, 3, 12 as translated in Joly, "Hippocrates of Cos," ibid., 424.
[61] Cited from Joly, ibid., 424–425.   [62] For examples, see Joly, ibid., 425–429.

attracted a number of disciples. Among these was Plato (427–348/47 BC), who was then approximately twenty years of age. Born into an aristocratic family and given a good education, Plato remained a pupil of Socrates for eight years, until the latter's execution in 399. Following this traumatic event, Plato traveled for the next twelve years and visited other parts of Greece, Egypt, Italy, and Sicily, and toward the end of his travels was even captured by pirates who released him on payment of a ransom. In 387, he settled back in Athens and opened the Academy, his famous school of philosophy, which endured until 529 AD, when it was closed by Emperor Justinian because he regarded it as a center of pagan learning. Although he himself made no specific contributions to science, and seems to have had no particular interest in it, Plato attracted eminent mathematicians and astronomers to the Academy, which gave it the appearance of a scientific center.

Plato represents a reaction to the kind of cosmic thinking characteristic of his pre-Socratic predecessors. This emerges most forcefully in his last known work, the *Laws*. In Book X of that treatise, Plato regards the natural philosophers, namely, the pre-Socratic philosophers we have already discussed, as materialists who are dangerous opponents of religion. Why did he regard them as materialists? Because he believed that in their philosophies of nature they derived all natural objects, both organic and inorganic, from dead, mindless matter. In fact, some Pre-Socratics regarded matter as alive but denied to it any conscious, intelligent action. For them, objects in nature come into being by chance.[63]

By contrast, Plato laid emphasis on art (technê [τεχνη]) as opposed to nature (phusis [φυσισ]). Art is a product of mind and intelligence and is superior to the blind, mindless operations of nature.[64] In Book X of the *Laws*, the Athenian Stranger, who speaks for Plato, declares:

The facts show – so they [i.e., the natural philosophers] claim – that the greatest and finest things in the world are the products of nature and chance, the creations of art being comparatively trivial. . . .

Becoming more specific, the Athenian Stranger goes on to declare:

They maintain that fire, water, earth and air owe their existence to nature and chance, and in no case to art, and that it is by means of these entirely inanimate substances that the secondary physical bodies – the earth, sun, moon and stars – have been produced. These substances moved at random, each impelled by virtue of its own inherent properties, which depended on various suitable amalgamations of hot and cold, dry and wet, soft and hard, and all other haphazard combinations that inevitably resulted when the opposites were mixed. This is the process to which all the heavens and everything that is in them owe their birth, and the consequent establishment of

---

[63] This assessment appears in Friedrich Solmsen, *Plato's Theology* (Ithaca, NY: Cornell University Press, 1942), 134–135.
[64] Solmsen, ibid., 135.

the four seasons led to the appearance of all plants and living creatures. The cause of all this, they say, was neither intelligent planning, nor a deity, nor art, but – as we've explained – nature and chance.[65]

To counter what he regarded as the false materialist interpretation of the world, Plato had some years before, written the *Timaeus*, his one major treatise on cosmology and natural philosophy. To appreciate its significance, it is necessary to describe briefly Plato's theory of ideas, which lies at the heart of his philosophy. In his theory of ideas, or forms, Plato exalts reason over sense perception. He did this because our sense perception was about things that were always suffering change. Indeed, it was the very nature of the physical world to be in perpetual flux. Not only is everything in flux but things that we call by the same name are never identical. We call many creatures by the name "cat," but we know that although they resemble each other, each cat has one or more differences from all other cats. The same applies to all other living creatures and to all inanimate objects. Thus, we cannot generalize about cats, because they are all different, however slightly some may differ. Because things that were called the same thing differed from one another, Plato was convinced that we could not derive real knowledge by means of observation and sense perception. The physical world was a domain of becoming and change.

Plato assumed that because we can apply the term "cat" to many different creatures, it follows that we can do this because cats have something in common. They have a common "cattyness." Bertrand Russell explains that:

An animal is a cat, it would seem, because it participates in a general nature common to all cats. Language cannot get on without general words such as "cat," and such words are evidently not meaningless. But if the word "cat" means anything, it means something which is not this or that cat, but some kind of universal cattyness. This is not born when a particular cat is born, and does not die when it dies. In fact it has no position in space or time; it is "eternal."[66]

Thus, a form, or idea, of cat exists, but not in space or time. A cat that we perceive is an imperfect copy of the ideal form of the cat. There is a unique, ideal, and perfect form of every kind of animal and every kind of object, such as beds and tables and chairs; there also are ideal forms of all mathematical figures, such as triangles, squares, and rectangles. Indeed, there are ideal forms of everything possible, including qualities and relations, and good and bad things.

Plato regarded these forms as uncreated, eternal, incorporeal, and changeless. We can know them only by thought and reason; and because they are

---

[65] *Plato: The Laws*, translated with an Introduction and Notes by Trevor J. Saunders, Preface by R. F. Stalley (London: Penguin Books, 1970; Preface 2004), Book Ten, 373–374.
[66] Bertrand Russell, *A History of Western Philosophy* (New York: Simon and Schuster, 1945), 121.

eternal and unchanging, we also can have knowledge about them, whereas we can have only opinion about the multiplicity of changeable physical objects and living creatures that we perceive by our senses. Because our souls once gazed on the ideal forms, perceiving their pale copies in the world of sense causes humans to vaguely recollect the ideas that they once viewed directly. Human knowledge results when objects of sense cause our souls to recollect the ideal forms of the perceived objects.[67]

Plato's sole extant attempt to produce a natural philosophy appears in the *Timaeus*, one of his last works. In this treatise, Plato speaks through Timaeus, who presents a probable account of the creation of the world in the form of a myth. "If we can furnish accounts no less likely than any other," Timaeus declares to Socrates,

we must be content, remembering that I who speak and you my judges are only human, and consequently it is fitting that we should, in these matters, accept the likely story and look for nothing further.[68]

Plato did not believe that we could arrive at a precise description of the physical world. The material objects of our world are too changeable and transient to produce true knowledge.[69]

In the account of the structure of the world that Plato gives in the *Timaeus*, he makes the theory of ideas or forms the basis of the world as we experience it. He does this by showing that our material world is but a copy or image of the real, eternal world. Plato assumes that our unique world was fashioned by a god, or Demiurge, as he is usually called. The Demiurge, Plato declares through the words of Timaeus,

was good; and in the good no jealousy in any matter can ever arise. So, being without jealousy, he desired that all things should come as near as possible to being like himself. That this is the supremely valid principle of becoming and of the order of the world, we shall most surely be right to accept from men of understanding. Desiring, then, that all things should be good and, so far as might be, nothing imperfect, the god took over all that is visible – not at rest, but in discordant and unordered motion – and brought it from disorder into order, since he judged that order was in every way better.[70]

The Demiurge is therefore not a creator god, making a world from nothing, but he is, rather, a divine craftsman who makes a world from chaotic materials already available. In order to make the best possible world from

[67] See Eduard Zeller, *Outlines of the History of Greek Philosophy.* Thirteenth edition, revised by Wilhelm Nestle and translated by L. R. Palmer, Trinity College, Cambridge (New York: Meridian Books, 1955; thirteenth edition revised and published 1931), 148.

[68] *Timaeus* 29C in *Plato's Cosmology: The "Timaeus" of Plato translated with a running commentary* by Francis MacDonald Cornford (New York: The Liberal Arts Press, 1957), 23.

[69] See Lloyd, *Early Greek Science*, 71.   [70] *Timaeus* 29E (Cornford translation), 33.

this preexisting, chaotic, disordered matter, the Demiurge turns to the Idea or Form of Living Creature[71] and copies every species of perfectly existent entities that it contains. Although he is a god, the Demiurge can only make imperfect copies of these ideal forms, because the matter at his disposal is naturally intractable. Despite his best efforts, there will always be an element of irrationality in the world.

Plato devotes most of the *Timaeus* to a description of the way in which the Demiurge made the world. He explains how he made the body of the world from four primary bodies, fire, air, water, and earth, and explains that these four elements are ultimately composed of two triangles, the right-angled scalene triangle and the equilateral triangle. From these two triangles, the Demiurge shapes four three-dimensional geometric figures, each of which forms one of the primary elements: the tetrahedron produces fire; the octahedron yields air; the icosahedron water; and the cube forms earth. Plato describes the production of the world soul and explains that each immortal soul is made of the same material. He also explains the making of the human body and its various powers and parts, as well as the formation of animals. After the completion of his lengthy exposition, Plato declares:

Here at last let us say that our discourse concerning the universe has come to its end. For having received in full its complement of living creatures, mortal and immortal, this world has thus become a visible living creature embracing all that are visible and an image of the intelligible, a perceptible god, supreme in greatness and excellence, in beauty and perfection, this Heaven single in its kind and one.[72]

Although Plato is famous for his emphasis on a mathematically structured world, he had little confidence that we could know the always-changing material world, a world of ultimate instability. What we learned about the physical world he would classify as opinion, in contrast to the true knowledge that we obtained when we applied our reason to the eternal, unchanging forms that had served as the models for their material counterparts. Thus did he stress reasoned, abstract analysis, rather than observation and reliance on the senses. Plato's analyses and explanations are largely teleological, as he obviously viewed the world as the work of a divine intelligence.

Plato did not present his opinions about the world by posing problems and resolving them after consideration of numerous alternatives and possibilities.

[71] "Let us rather say," declares Timaeus, "that the world is like, above all things, to that Living Creature of which all other living creatures, severally and in their families, are parts. For that contains and embraces within itself all the intelligible living creatures, just as this world contains ourselves and all other creatures that have been formed as things visible. For the god, wishing to make this world most nearly like that intelligible thing which is best and in every way complete, fashioned it as a single visible living creature, containing within itself all living things whose nature is of the same order." *Timaeus* 30C–31A (Cornford translation, 40).

[72] *Timaeus* 92C (Cornford translation, 359).

Nor does he offer much by way of empirical evidence. His style is rather imperious. He tells us how things are, because he thought that is the way they had to be. We might well ponder whether a tradition of serious natural philosophy could have derived from Plato's approach to the world. His exclusive use of the dialogue form was an advance over the hexaemeral verse method used by Parmenides and others, but it was nevertheless not well suited to the advance of natural philosophy, although others would occasionally use it.[73] But although Plato's way of doing natural philosophy and his attitudes toward the world had some impact in the seventeenth century, it was destined to be superseded by the contributions of his great pupil, Aristotle, to whom we must now turn.

[73] For example, Adelard of Bath in his *Natural Questions* in the twelfth century.

2

Aristotle (384–322 BC)

LIFE

Aristotle was born in 384 BC, in the town of Stagira, which lay in Macedonia in northern Greece. His father was Nicomachus, a physician, in the service of King Amyntas of Macedon; his mother was Phaestis, a woman of independent wealth.[1] In 367, as a lad of seventeen, Aristotle moved to Athens to study with Plato in the Academy,[2] where he remained for twenty years, until the death of Plato in 347. It is plausible to assume that during those twenty years, Aristotle heard, and participated in, important philosophical discussions involving some of the greatest minds of the time. The themes that were debated must surely have ranged across issues that were dear to Plato, such as metaphysics, ethics, logic, politics, and epistemology. And although physics and cosmology were not themes to which Plato devoted much time and effort, Aristotle must surely have had occasion to engage in discussions about those subjects.

With the death of Plato in 347 and the emergence in Athens of anti-Macedonian sentiment, Aristotle, who never became an Athenian citizen, departed Athens and traveled to the coast of Asia Minor. There, he lived first in Assos, where he married Pythias, the niece of Hermias, the tyrant of Assos.[3] He moved next to Mytilene, on the island of Lesbos, where he met Theophrastus, who became an important friend and future colleague. It is likely that during his approximately four years in this region, Aristotle studied marine biology and used what he learned in his biological treatises.

In 343, after a brief return from Mytilene to his home in Stagira, Aristotle received an invitation from Philip II, King of Macedon, and son of Amyntas,

---

[1] See Jonathan Barnes, "Life and Work," in Jonathan Barnes, ed., *The Cambridge Companion to Aristotle* (Cambridge: Cambridge University Press, 1995), 3. In the brief life of Aristotle that follows, I am indebted to Barnes's account on pages 3–6. For a more detailed version, see Sarton, *A History of Science: Ancient Science through the Golden Age of Greece*, 470–473.

[2] Why he did so and why Plato admitted him are unknown, as is the degree of philosophical knowledge to which he may have attained by the time he left for the Academy.

[3] Aristotle had a daughter by Pythias who also was named Pythias. After the death of Pythias, Aristotle took a second wife, Herpyllis, who bore him a son named Nicomachus. See Sarton, *A History of Science*, 473.

to tutor his son, the future Alexander the Great. "Thus," declares Jonathan Barnes, "began the association between the most powerful mind of the age and the most powerful man." Although one could all too easily imagine how these two great figures might have mutually interacted and shaped each others thoughts and actions, it is well to keep in mind Barnes's admonition that "what Aristotle said to Alexander the Great, and Alexander to him, we do not know."[4]

After a long absence from Athens, Aristotle returned in 335 and established the Lyceum, a rival school to Plato's Academy. It was here where most of Aristotle's extant works were written. On the occasion of the death of Alexander the Great in 323, Aristotle, once again, fell victim to anti-Macedonian rage. In 322, he fled to Chalcis on the island of Euboea, where he died the same year at the age of sixty-two.

So that we may breathe some life into this skeletal account, here is Jonathan Barnes's vivid description of Aristotle the man:

Of Aristotle's character and personality little is known. He came from a rich family. He was a bit of a dandy, wearing rings on his fingers and cutting his hair fashionably short. He suffered from poor digestion, and is said to have been spindle-shanked. He was a good speaker, lucid in his lectures, persuasive in conversation and he had a mordant wit. His enemies, who were numerous, made him out to be arrogant and overbearing. His will, which has survived, is a generous and thoughtful document. His philosophical writings are largely impersonal; but they suggest that he prized both friendship and self-sufficiency, and that, while conscious of his place in an honourable tradition, he was properly proud of his own attainments. As a man, he was, I suspect, admirable rather than amiable.[5]

WORKS: ARISTOTLE'S WRITINGS AND THEIR PRESERVATION

Aristotle left behind a substantial legacy of written works. In *The Complete Works of Aristotle, The Revised Oxford Translation*, forty-six works are included, of which sixteen are deemed spurious by modern scholars.[6] In the tightly packed printing format of the Oxford Translation, the forty-six works occupy 2,383 pages.[7] Drawing on different ancient sources in which both lost and extant works are mentioned, it appears that Aristotle may have written

---

[4] Barnes, "Life and Work," *Companion to Aristotle*, 5. In parentheses, Barnes adds: "(It is in vain that historians look for Aristotelian influence on the bloody career of Alexander; and philosophers will find nothing – or virtually nothing – in Aristotle's political writings which betrays interest in the fortunes of the Macedonian empire.)" Ibid.

[5] Jonathan Barnes, *Aristotle* (Oxford: Oxford University Press, 1982), 1.

[6] *The Complete Works of Aristotle: The Revised Oxford Translation*, edited by Jonathan Barnes, 2 vols. (Bollingen Series X 2; Princeton, NJ: Princeton University Press, 1984).

[7] A list of fragments attributed to Aristotle follows the forty-six printed works. See ibid., vol. 2, 2384–2385, 2389–2465. The fragments conclude with Aristotle's will.

as many as two hundred works.[8] For obvious reasons, however, scholars have focused their attention on the thirty works regarded as genuinely by Aristotle. Let us see how these works are regarded.

At the outset, we confront a puzzle about the true nature of the Aristotelian texts. On Aristotle's death, his library, which included a large number of treatises ostensibly written by him, passed on to his friend, Theophrastus, who became the new head of Aristotle's school, the Lyceum, and who subsequently passed the library to his nephew, Neleus of Scepsis, who took it to Scepsis, a city in Asia Minor. There, Neleus buried the collection in a cave. For two centuries, the manuscripts lay rotting in that cave and were then rediscovered and taken to Athens, then to Rome, where they came into possession of Andronicus of Rhodes (first century BC). Andronicus, who was an Aristotelian philosopher, edited the manuscripts around 70 BC, and in so doing prepared the basic edition that is still in use today. Jonathan Barnes has asked: "What did Andronicus do? How did his edition – how does our edition – differ from what Aristotle actually wrote? The answer, roughly put, is probably this: Andronicus himself composed the works which we now read."[9]

How are we to understand this startling statement? Andronicus did not actually write the works, but by editing them, he gave them the form they have today. The works in their present form were thus not written by Aristotle, or even by his pupils and colleagues in the Lyceum. Who, then, did compose the original works that Andronicus edited? What did Aristotle actually write? In response to this question, Scott Montgomery has a ready reply that is worthy of full citation:

It is something of a shock, perhaps, to discover that "Aristotle" as an author, probably never existed. The evidence is very strong in this regard. It indicates that the works placed under his name were, at the earliest stage of their history, compilations of notes, recordings, collections of facts, and other fragments, mainly from his lectures at the Lyceum, which were assembled, amended, and very often written by his students. They were, in short, communal creations. Moreover, their contents were apparently never looked upon as final in any sense, but were instead continually updated and replaced during Aristotle's lifetime and thereafter, in accordance with the changing levels of discussion, insight, student participation, and so forth. This gave them an evolving, organic type of reality, one rather at odds with the modern concept of an authorial end product, intended for a receiving but nonparticipatory audience. At some point, no doubt, these communal works, never set down for an outside readership, were effectively frozen by the death of Aristotle or a succeeding teacher,

---

[8] For the list, see ibid., vol. 2, 2386–2388.

[9] I follow the account by Jonathan Barnes, "Life and Work," *Companion to Aristotle*, 10–11. Much the same account appears in Scott Montgomery, *Science in Translation: Movements of Knowledge through Culture and Time* (Chicago: University of Chicago Press, 2000), 7. In what follows in this section on the preservation of Aristotle's works, I am heavily indebted to the article by Jonathan Barnes cited here.

whose own direct participation came to an end. With such a death, the book came to life. Indeed it is only through this loss of the "father" that such works entered their fate into the greater journal of outside history.[10]

Montgomery goes on to assert that "'Aristotle' is therefore, in concrete terms, a fiction, or rather a construct. The Aristotle we have today, the one that has existed since the beginning, is a classroom assembly rather than a text-book."[11]

On this analysis, the body of works attributed to the philosopher Aristotle seems to evaporate. His works are the end product of a vague collective effort over many years and even when eventually "frozen by the death of Aristotle" we can detect, in the works attributed to him, no final, definitive character. To the opinion argued so ably by Scott Montgomery, let us oppose another that preserves Aristotle and his works in a traditional sense, and that seems more plausible. If we assume that Aristotle is the author of the works attributed to him, we immediately become aware of the fact that those works are unpolished and difficult to read. This is puzzling, because Aristotle had a reputation in the ancient world as an eloquent writer, in the mold of Plato. Unfortunately, these writings have perished. Nevertheless, we are left with the puzzle that an author, who had written dialogues for an educated public in the manner of Plato and who was admired for his style, has left to posterity only treatises that have been characterized in the following manner: "the syntax is spare, ornamentation is rare, transitions are abrupt, and connections opaque: the language rarely seems to have been chosen with any aesthetic aim, and often enough the intellectual aim is hard to discern – reading Aristotle, as the poet Thomas Gray put it, is like eating dried hay."[12]

The answer to the style of Aristotle's extant works may lie not in a collective authorship but in objectives and purposes. Jonathan Barnes proposes two plausible interpretations. Since the beginning of the twentieth century, numerous scholars have argued that the style of Aristotle's extant treatises derive from the fact that they were his lecture notes. Changes inserted over the years may explain the final versions that have survived. Barnes gives a vivid description of how such changes could have yielded the end product. To begin with, we should assume a basic first lecture course for a given subject:

but this basic layer will have been overwritten in numerous places and to different effects – some passages will have been deleted and replaced by paragraphs maintaining an entirely different thesis; other passages will have been modified in more subtle ways, the thesis or argument being qualified to meet objections; other passages again will have received additions which reinforce rather than change or destroy the

[10] Montgomery, *Science in Translation*, 7–8.   [11] Ibid., 9.
[12] Barnes, "Life and Work," *Companion to Aristotle*, 12.

original text; and so on. And there will have been several layers of this sort of thing, the text of the first revision being itself replaced by a second revision.... Moreover, Aristotle will not always have deleted the earlier material as it became dated; and his manuscripts – and therefore ultimately the texts which we read – will have contained "doublets": both X and Y will be printed, even though Aristotle intended Y to supplant rather than to supplement X.[13]

Barnes finds this argument seductive but judges it wanting because it "rests on the perilous supposition that Aristotle taught and worked in much the same way as a twentieth century professor of philosophy might teach and work."[14] As an alternative, he suggests that we view Aristotle's texts as working drafts rather than lecture notes.[15] We are reminded that one cannot read Aristotle as one might read Plato or Descartes or Kant.

When you pick up the *Metaphysics* or the *Nicomachean Ethics*, you are not picking up a finished philosophical text, comparable to the *Theaetetus* or the *Meditations* or the *Critique of Pure Reason*. It is proper to assume that you are picking up a set of papers united by a later editor; and it is proper to assume that you are reading a compilation of Aristotle's working drafts. In any case, you should surely read Aristotle's drafts in the manner in which you would read the notes that a philosopher had written for his own use. The sentences are crabbed – sometimes telegrammatic: you must expand them and illustrate them. The arguments are enthymematic – or mere hints: you must supply the missing premises. The transitions are sudden – and often implicit: you must articulate and smooth and explain.[16]

Either of these two interpretations – that Aristotle's works were lectures or working-drafts – is suitable as an explanation for the status of Aristotle's texts. But we are not done with problems about Aristotle's textual legacy. Assuming that scholars have proper criteria for determining the authentic from the inauthentic texts, there remains the problem of the chronological order in which the works attributed to Aristotle were written. We can confidently assume that Aristotle's works do have a chronology. But can this be determined with any reasonable degree of confidence? In a recent discussion, Jonathan Barnes concludes: "It makes no sense to provide a chronology of Aristotle's writings."[17] To begin with, Aristotle did not date his works. Although attempts have been made to sequence Aristotle's works on the basis of a developing philosophical maturity, this has proven implausible. Werner Jaeger assumed that Aristotle's development moved from an initial Platonism to an emphasis on empiricism and then organized the works

---

[13] Ibid., 13.   [14] Ibid.
[15] This is indeed also the judgment of Thomas Case, author of an excellent article on "Aristotle" in *The Encyclopaedia Britannica, A Dictionary of Arts, Sciences, Literature and General Information*, eleventh edition (Cambridge: Cambridge University Press, 1910), vol. 2: *Andros to Austria*, 501–522. For Case's discussion of Aristotle's works as lectures and as working drafts, see 506–509.
[16] Barnes, *Companion to Aristotle*, 14–15.   [17] Ibid., 21.

accordingly. But this method has failed to gain a consensus and one might find counterinstances to show that Aristotle moved in the reverse direction – from empiricism to Platonism. Another stumbling block is the problem of revision.

Suppose that work A was begun in 350, heavily revised a couple of years later, lightly retouched in about 340, and finally rethought a decade later. Suppose that work B was begun in 345, revised carefully in 335, looked at again a year or so later, and then abandoned. Well, which was written first, A or B? If you are going to produce a chronology of Aristotle's writings, will you put A before B (on the grounds that the first version of A preceded the first version of B) or will you put B before A (on the grounds that the final – let us not say definitive – version of B was later than the final version of A)? Pretty clearly, you will say neither of these things; for pretty clearly it is absurd to talk about chronology in these terms at all. If Aristotle's texts were subject to revisions of the sort I have sketched, then it makes no sense to ask whether A came before B – and hence it makes no sense to attempt to provide a chronology of Aristotle's writings.[18]

Questions about Aristotle's texts – authorship, chronology, and authenticity – emerged in the late nineteenth century and have attracted the attention of scholars ever since. But in a study about the history of natural philosophy, in which Aristotle is the paramount figure, those problems are of little consequence. During the lifetime of Aristotelianism as a viable philosophy, and even beyond into the nineteenth century, such considerations were never raised. It did not occur to medieval scholars, for example, to question the authorship of such texts. Even if they had known that Andronicus of Rhodes edited the works of Aristotle, they would not have found that any reason to question the veracity of Aristotle's authorship. Moreover, they were never concerned about the consistency of this or that particular treatise. Nor did they inquire whether someone – Andronicus, for example – might have rearranged and interpolated passages. Nor did anyone suggest that Aristotle's treatises were really the collective product of many minds at the Lyceum. Such problems and questions never arose. Until the late nineteenth century, the great philosopher Aristotle, regarded as the dominant intellect of antiquity and the Middle Ages, was assumed, without question, to have written the texts attributed to him.

In fact, no one to my knowledge raised questions about the chronology of Aristotle's works. In the modern age, it has been rightly assumed that Aristotle's ideas must have evolved and that this would somehow be reflected in his works. If statement A in one work conflicts with statement B in another treatise, modern scholars assume that Aristotle changed his mind. Thus, if one can determine that statement B was made subsequent to statement A, it would follow that the section containing B was written after the section

[18] Ibid.

containing *A*, and, therefore, perhaps the work including section *B* was written after the work that includes section *A*.

Questions about the chronology of Aristotle's works did not really arise until the late nineteenth century. His works were approached as if they had been composed in a timeless manner. Inconsistencies of the kind described in the preceding paragraph would have been resolved by somehow reconciling the two conflicting statements or by ignoring them. And, above all, despite the legitimate concerns of modern scholars about the nature of Aristotle's authorship, or whether he should even be regarded as the author of the many treatises attributed to him, philosophers and scientists who used them in late Greek antiquity, in Islam, and in Western Europe from the late Middle Ages to the end of the nineteenth century, assumed, without qualms, that the philosopher Aristotle was the undoubted author of them all.

ARISTOTLE'S ACHIEVEMENTS

Aristotle is probably the most significant intellectual figure in the history of Western thought up to the end of the sixteenth century. The range of topics he treated in his extant writings is extraordinary and the wisdom and insight he reveals is rather amazing for someone who lived in the fourth century BC. We can best appreciate Aristotle's contributions when we view them against the background of his role in Western thought. As G. E. R. Lloyd puts it: "To attempt to cover the history of Aristotle's influence on subsequent thought in full would be not far short of undertaking to write the history of European philosophy and science, at least down to the sixteenth century."[19] As if that were not enough, Lloyd also rightly declares: "The idea of carrying out systematic research is one that we in the West owe as much to Aristotle and to the Lyceum as to any other single man or institution."[20] Aristotle's research programs are exhibited in his biological and political works. Indeed, he is the founder of biology as a discipline. Not only did he do pioneering work in biological classification, but his description of the habits and behavior of certain species still elicit the admiration, and even awe, of modern zoologists.

In the *History of Animals*, we see Aristotle at his best as an observer and recorder of animal behavior. Among the numerous splendid descriptions especially noteworthy are those of the torpedo fish,[21] or stingray, the breeding habits of catfish, and bees,[22] as well as embryological data about the chick,[23] the placental shark,[24] and cephalopods.[25] Aristotle examined chick

[19] G. E. R. Lloyd, *Aristotle: The Growth and Structure of His Thought* (Cambridge: Cambridge University Press, 1968), 306.

[20] Lloyd, ibid., 287.   [21] Aristotle, *History of Animals*, 9.37.620b.10–29.

[22] Aristotle, ibid., 9.40.623b.5–627b.22.   [23] Aristotle, ibid., 6.3.561a.4–562a.21.

[24] Aristotle, ibid., 6.10.565b.2–17.

[25] Aristotle, ibid., 5.6.541b.1–33. For Aristotle as a zoologist, I rely on George Sarton, *A History of Science*, 537–545.

embryos, for example, by breaking open one egg every day and observing the progress of the embryo. On the third day, he observed the beginning of an embryo and noted that "the heart appears, like a speck of blood in the white of the egg";[26] on the tenth day, "the chick and all its parts are distinctly visible."[27] Although Aristotle knew that most fishes produce their young by laying eggs, he recognized an important exception in the placental dogfish, a member of the selachian group, which brought forth its young alive. It does this by laying eggs in the womb, whereupon the eggs become attached to the womb by a navel string. Most were skeptical of Aristotle's claim until 1842, when Johannes Müller, the great German biologist, showed that Aristotle's observations were correct.[28] According to Thomas Lones, Aristotle describes the internal parts of 110 animals, of which he may have dissected as many as forty-nine, perhaps even an elephant.[29] At the very least, Aristotle dissected the eye of a chick, the eye of a mole, the cochlea of the inner ear, and the stomach of ruminants, which he accurately describes.[30]

Aristotle's biological works were still influential and current in the nineteenth century, as two tributes to him reveal. George Henry Lewes (1817–1878), a naturalist who carefully studied Aristotle's biological treatises, was particularly impressed by Aristotle's On the Generation of Animals, of which he said: "It is an extraordinary production. No ancient, and few modern works, equal it in comprehensiveness of detail and profound speculative insight. We there find some of the obscurest problems of Biology treated with a mastery which, when we consider the condition of science at that day, is truly astounding."[31]

The second tribute to Aristotle was from none other than Charles Darwin, who, on receipt from Dr. William Ogle of a copy of the latter's translation of Aristotle's Parts of Animals, sent a letter of thanks to Ogle on February 22, 1882. "You must let me thank you for the pleasure which the introduction to the Aristotle book has given me," began Darwin, who then proclaimed:

I have rarely read anything which has interested me more, though I have not read as yet more than a quarter of the book proper.

---

[26] Aristotle, History of Animals, 6.3.561a.10–11.   [27] Aristotle, ibid., 6.3.561a.26.

[28] See Sarton, A History of Science, 541–542. For Aristotle's discussion, see Cohen and Drabkin, A Source Book in Greek Science, 420.

[29] Thomas E. Lones, Aristotle's Researches in Natural Science (London: West, Newman & Co., 1912), 105–106.

[30] See Aristotle, History of Animals, 6.3.561a.29–31 (chick); 4.8.533a.3–15 (mole); 1.11.15–21 (cochlea of inner ear); 2.17.507a.–507b.11 (stomach of ruminants). For the text and a few notes, see Cohen and Drabkin, A Source Book in Greek Science, 412.

[31] George Henry Lewes, Aristotle: A Chapter from the History of Science, Including Analyses of Aristotle's Scientific Writings (London, 1864), 325. Quoted from Sarton, A History of Science, 540.

From quotations which I had seen, I had a high notion of Aristotle's merits, but I had not the most remote notion what a wonderful man he was. Linnaeus and Cuvier have been my two gods, though in very different ways, but they were mere schoolboys to old Aristotle.[32]

If Aristotle is the founder of biology, he is also the universally acknowledged inventor of formal, syllogistic logic. His logical treatises are known as the Organon – tool or instrument – and consist of six works: the *Categories*, which treats of terms, *On Interpretation*, which is concerned with propositions; *Topics*, in eight books, "contains a study of non-demonstrative reasoning and is effectively a grab-bag of how to conduct a good argument";[33] *Sophistical Refutations*, which describes the different kinds of fallacies and also explains how to resolve them. The last two, the *Prior Analytics* and the *Posterior Analytics* are the most important. The *Prior Analytics* contains Aristotle's greatest contribution to the history of logic, namely, the theory of the syllogism, which is the theory of deductive inference and usually consists of two premises and a conclusion. Not only was Aristotle the first to lay out the formal analysis of the syllogism, but he also "invented the use of schematic letters." As Jonathan Barnes explains:

Logicians are now so familiar with his invention, and employ it so unthinkingly, that they may forget how crucial a device it was: without the use of such letters logic cannot become a general science of argument. The *Prior Analytics* makes constant use of schematic letters,[34]

as is evident early on when Aristotle presents the classic case of the syllogism: "If A is predicated of every B, and B of every C, A must be predicated of every C."[35] Finally, there is the *Posterior Analytics*, wherein Aristotle presents his theory of scientific demonstration, or scientific method, and uses the mathematical sciences as his primary model.

Aristotle's contributions to logic are "particularly comprehensive, very largely original and for the most part eminently lucid. It is, moreover, highly professional, and the specialist will find a great deal that is of interest in the logical treatises apart from those sections of them that contain what still remains an excellent introduction to the study of elementary logic."[36]

---

[32] Quoted by Sarton, ibid., 545 from Francis Darwin, *The Life and Letters of Charles Darwin*, second edition (London, 1887), vol. 3, 251.

[33] See Paul Vincent Spade, *Thoughts, Words, and Things: An Introduction to Late Mediaeval Logic and Semantic Theory*, Version 1.1. August 9, 2002 (Copyright 2002 by Paul Vincent Spade), ch. 2 ("Thumbnail Sketch of the History of Logic to the End of the Middle Ages"), Section B: "Aristotelian Logic," in which Spade gives a brief description of each of Aristotle's logical works. For the *Topics*, see p. 12. For the passage, see Spade's Web address in the Bibliography.

[34] Barnes, *Aristotle*, 30.   [35] Aristotle, *Prior Analytics*, 1.4.26a.1–2.

[36] Lloyd, *Aristotle*, 111.

In addition to his work in biology and logic, Aristotle constructed a system of the cosmos that endured for more than two thousand years in three different civilizations and cultures. His discussions of motion in the *Physics* set the stage for subsequent controversies that resulted in medieval advances and ultimately led to Galileo and Newton. In his brief *Poetics*, he established the categories of drama that are still accepted, namely, "Tragedy, Comedy, Epic, and Lyric."[37] Within the category of Tragedy he distinguished six elements: "plot, character, thought, diction, song, and spectacle."[38] As one author put it, "after twenty-two centuries it remains, the most stimulating and helpful of all analytical works dealing with poetry."[39] His ideas about politics and ethics formed the basis of discussions in those areas until early modern times.

It is difficult to overstate Aristotle's significance for Western thought. His contributions can be summarized in many ways. A brief account by A. E. Taylor will do. "It has not been the lot of philosophers, as it is of great poets," he declares,

that their names should become household words. . . . Yet there are a few philosophers whose influence on thought and language has been so extensive that no one who reads can be ignorant of their names, and that every man who speaks the language of educated Europeans is constantly using their vocabulary. Among this few Aristotle holds not the lowest place. We have all heard of him, as we have all heard of Homer. He has left his impress so firmly on theology that many of the formulae of the Churches are unintelligible without acquaintance with his conception of the universe. If we are interested in the growth of modern science we shall readily discover for ourselves that some knowledge of Aristotelianism is necessary for the understanding of Bacon and Galileo, and the other great anti-Aristotelians who created the "modern scientific" view of Nature. If we turn to the imaginative literature of the modern languages, Dante is a sealed book, and many a passage of Chaucer and Shakespeare and Milton is half unmeaning to us unless we are at home in the outlines of Aristotle's philosophy. And if we turn to ordinary language, we find that many of the familiar turns of modern speech cannot be fully understood without a knowledge of the doctrines they were first forged to express. An Englishman who speaks of the "golden mean" or of "liberal education," or contrasts the "matter" of a work of literature with its "form," or the "essential" features of a situation or a scheme of policy with its "accidents," or "theory" with "practice," is using words which derive their significance from the part they play in the vocabulary of Aristotle.[40]

37 J. A. Oesterle, "Poetics (Aristotelian)," in the *New Catholic Encyclopedia*, vol., XI (Washington, DC: Catholic University of America, 2003), 33. Oesterle cites as his source H. House, *Aristotle's Poetics: A Course of Eight Lectures*, (rev. ed. London: R. Hart-Davis, 1956).

38 Oesterle, ibid. p. 433.

39 L. Cooper, *The Poetics of Aristotle*, rev. ed. (Ithaca, NY: Cornell University Press, 1956), 3; cited by J. Oesterle, *The New Catholic Encyclopedia*, vol., XI, 455.

40 A. E. Taylor, *Aristotle* (New York: Dover Publications, 1955; reprinting of revised edition of 1919), 5–6. Taylor goes on to inform his audience that the aim of his modest book about Aristotle is "to help the English reader to a better understanding of such familiar language and a fuller comprehension of much that he will find in Dante and Shakespeare and Bacon and Milton." Ibid., 6.

Of the numerous themes that one might investigate in the thought of Aristotle, we shall focus our attention on his natural philosophy, which served as the dominant interpretation of nature for approximately two thousand years, encompassing at least three civilizations.

ARISTOTLE'S COSMOS AND NATURAL PHILOSOPHY

Aristotle's cosmos is, of course a product of his natural philosophy. By a variety of means and in a number of treatises, Aristotle identified and described the basic components of our physical world. What he fashioned was destined to serve as the basic conception of the universe for almost two millennia.

For Aristotle, the cosmos was a gigantic spherical plenum that had neither a beginning, nor would it have an end. Everything in existence exists within that sphere; nothing exists, or can possibly exist, outside of it: neither matter, nor empty space, nor time nor place. Aristotle regarded it as nonsensical to inquire about extracosmic existence, consequently rejecting the possibility that other worlds might exist beyond ours. Within the cosmos, Aristotle distinguished two major divisions: the celestial region and the terrestrial. The dividing line between the two regions was the concave surface of the lunar sphere. That surface divided two totally dissimilar regions.

The terrestrial region, which lay below the concave lunar surface, was a region of constant change and transformation. It consisted of four elements: earth, water, air, and fire, arranged by nature in this order from the center of the world to the moon's concave surface. All bodies were compounded of combinations of two or more elements. In the terrestrial region, bodies were always coming into being as elements were compounded into different bodies; and bodies were always passing away because their elements dissociated to combine with other elements and form new compound bodies. At the center of the universe was the earth, surrounded in many of its parts by water and then air and fire. If the motions of the elements were suddenly to cease, the four elements would sort themselves out into four concentric regions, from heaviest to lightest, namely, from earth to water to air to fire. But this cannot happen because it is the nature of all elements to move and thereby to associate and dissociate with other elements. In the upper atmosphere of the terrestrial region, just below the concave surface of the moon, Aristotle assumed that comets, shooting stars, and other similar phenomena occurred. He inferred their existence in this region, because they were changeable phenomena, and therefore could not occur in the celestial region.

For if change and transformation are the characteristic features of the terrestrial area, minimal change is the hallmark of the celestial region, within which lie the planets and stars. The lack of change is attributable to a celestial ether, which Aristotle regarded as a fifth element and which fills the celestial region, leaving no empty spaces. The ether is an incorruptible substance that can suffer no change, other than change of place. Because the planets and stars are composed of the celestial ether, they undergo no change, except change of place, which we can readily observe. Because Aristotle viewed a small degree of change as superior to a greater degree of change, he regarded the celestial region, where the only change was change of place, as nobler than, and vastly superior to, the terrestrial region, where incessant and unremitting change was the most characteristic feature. Because it is nobler and superior to the terrestrial region, Aristotle thought it appropriate that the celestial region should influence terrestrial changes. Future astrologers found this a welcome support to justify their prognostications.

It was Aristotle's understanding of natural philosophy that enabled him to determine the physical nature of the cosmos and to spell out its properties and behavior in his numerous treatises. To grasp the role that natural philosophy played in Aristotle's scheme of things, it is essential to understand the emphasis he placed on human reason, or, which is the same thing, intellect. In the *Nicomachean Ethics* (10.7.1178a.5–8), Aristotle declares: "That which is proper to each thing is by nature best and most pleasant for each thing; for man, therefore, the life according to intellect is best and pleasantest, since intellect more than anything else *is* man." Aristotle frequently emphasizes reasoned discourse and accords it the highest place.[41] Although he did not assume a creation for the world, he did believe in a God, but a rather strange God, one who serves as a final cause for an eternal world, without beginning or end. Indeed, Aristotle's God has no knowledge of our world's existence but is wholly absorbed in thinking about himself, as he alone is worthy of serving as his own object of thought. Even if the world were not the object of God's thoughts, Aristotle regarded it as a rationally structured physical sphere that contained all that exists, with nothing lying beyond.

Aristotle thought it important to classify different kinds of knowledge and actions into appropriate categories. Where, then, did he locate physics, or natural philosophy, or natural science – the three expressions came to be regarded as synonymous – within the all-inclusive domain of knowledge? Aristotle distinguished three broad categories of knowledge that he regarded as scientific: the productive sciences, the practical sciences, and the theoretical sciences. The productive sciences embraced all knowledge concerned with the making of useful objects, whereas the practical sciences were directed toward human conduct. Everything else fell under the jurisdiction of the theoretical sciences, which Aristotle divided into three parts. If we take them in the order of priority, they are (1) metaphysics, or theology, which considers things that are unchangeable and therefore distinct and separable from matter or body, such as God and spiritual substances; (2) mathematics, which also

---

[41] For instances of Aristotle's use of the term reason, see Troy W. Organ, *An Index to Aristotle in English Translation* (New York: Gordian Press, 1966), 138.

considers things that are unchangeable, but, unlike metaphysics, the objects of mathematics have no separate existence because they are abstractions from physical bodies; and (3) finally, there is physics, or as it was often called natural science, or, as it came to be popularly designated, natural philosophy, which is concerned only with things that are changeable, exist separately, and also have within themselves an innate source of movement and rest.[42] From Aristotle's standpoint, physics, or natural philosophy embraces both animate and inanimate bodies and is applicable to the whole physical world, that is, to both the terrestrial and celestial regions.

But how do we derive knowledge about nature? First it is essential to understand that for Aristotle, sense perception was the foundation of human knowledge. He regarded it as "impossible to get an induction without perception."[43] But as basic as he regarded sense perception, Aristotle denied that we could arrive at scientific knowledge by means of sense perception alone. To attain to scientific knowledge, universal propositions are essential. One arrives at universals, from sense perceptions by means of induction.[44] It is by means of universals, not direct perception, that we can generate demonstrations that produce scientific knowledge. As Aristotle explains, perception is of the individual or particular, and from particulars "it is impossible to perceive what is universal and holds in every case." But demonstrations that yield scientific knowledge are based on universal propositions. That is why "if we were on the moon and saw the earth screening it we would not know the explanation of the eclipse. For we would perceive that it is eclipsed and not why at all; for there turned out to be no perception of the universal. Nevertheless, if, from considering this often happening, we hunted the universal, we would have a demonstration; for from several particulars the universal is clear."[45]

Although Aristotle placed great emphasis on demonstrations based on universal propositions arrived at inductively, ultimately from sense perceptions of individual things, he was equally interested in explaining the causes of natural phenomena, even phenomena not directly perceived. In the beginning of the seventh chapter of the first book of his *Meteorology*, Aristotle declares: "We consider a satisfactory explanation of phenomena inaccessible to observation to have been given when our account of them is free from impossibilities."[46] Thus, even if one cannot directly observe a phenomenon, Aristotle feels that an explanation is acceptable if it is compatible with what is

---

[42] For Aristotle's division of the sciences, see his *Metaphysics*, bk. 6, ch. 1.
[43] Aristotle, *Posterior Analytics*, 1.18.81b.6. [44] Ibid., 1.18.81a.37–81b.9.
[45] Ibid., 1.31.87b.38–88a.4. For a helpful discussion of these issues, see A. C. Crombie, *Styles of Scientific Thinking in the European Tradition: The History of Argument and Explanation Especially in the Mathematical and Biomedical Sciences and Arts*, 3 vols. (London: Duckworth, 1994), vol. 1, 245. Crombie devotes chapter 5 to Aristotle.
[46] Aristotle, *Meteorology*, 1.7.344a5–7; translation by E. W. Webster.

possible. His immediately following explanations of the formation of comets and shooting stars exemplify this procedure.[47]

In addition to general methodological considerations, Aristotle lays great emphasis on the role of causes in natural philosophy. Aristotle regarded all bodies as composites of matter and form, the former functioning as a passive principle, the latter as an active principle. How do these bodies change? Aristotle attributed all possible changes to four kinds of causes. The first is the material cause, which is the matter from which something is made, as the bronze is the matter of a bronze statue. The second is the formal cause, which is the essence or inner structure of a thing as expressed in its definition. To pursue the statue example, the sensible aspect of form is the shape the sculptor will give to the statue; the intelligible aspect of form in this case would be the essence of what it is to be a statue. The third type is the efficient cause, which is the agent or producer of the change or action, namely, the sculptor. The fourth is the final cause, which is the end or purpose for which an action is done. In the present case, the final cause of the bronze statue is the original intent of the sculptor to make a statue, for it was what motivated the sculptor to make the statue. Aristotle sometimes reduced the four causes to two. The material cause always remained the material cause. The other three causes can be reduced to a single cause, a formal-final-efficient cause. Thus, if an artist has the formal and final causes of a statue in mind, they will serve as an efficient cause to prompt him or her to make the statue. Or, to use an organic example, an acorn does not yet have the form of an oak tree, although it has the potentiality of becoming an oak tree. Thus, the acorn will try to realize the form of an oak tree. Its ultimate goal of becoming an oak tree also functions as a final cause. The efficient cause operates to enable the acorn to realize its final form as an oak tree.

Aristotle distinguished four kinds of changes that the four causes could produce: "(1) substantial change, where one form supplants another in the underlying matter, as when fire reduces a log to ash; (2) qualitative change, as when the color of a leaf is altered from green to brown in the same underlying matter; (3) change of quantity, as when a body grows or diminishes while otherwise retaining its identity; and finally (4) change of place, when a body suffers change as it moves from one place to another."[48]

Aristotle had other tools of analysis for comprehending nature. In the *Physics*, he contrasts things that exist by nature with those that do not.

---

[47] R. J. Hankinson ("Science," in Jonathan Barnes, ed., *The Cambridge Companion to Aristotle*, 154–155) translates the relevant passage and considers how Aristotle's account of comets and shooting stars (he calls them meteors) exemplifies the methodological passage.

[48] Cited from Edward Grant, *The Foundations of Modern Science in the Middle Ages, Their Religious, Institutional, and Intellectual Contexts* (Cambridge: Cambridge University Press, 1996), 56.

"By nature the animals and their parts exist, and the plants and the simple bodies (earth, fire, air, water) – for we say that these and the like exist by nature."[49] Each of these things "has within itself a principle of motion and of stationariness (in respect of place, or of growth and decrease, or by way of alteration). On the other hand, a bed and a coat and anything else of that sort, *qua* receiving these designations – that is, in so far as they are products of art – have no innate impulse to change."[50] But products of art will undergo change if they are composed of things that do have an impulse to change.

Later, in the *Physics*, Aristotle characterizes nature as "a cause, a cause that operates for a purpose"[51] and then defines it as "a principle of motion and change."[52] Thus, nature operates by causes that produce motions and changes. An investigation of nature by means of physics, or natural philosophy, would involve a study and analysis of those causes and the motions and changes they produce. In the introductory paragraph to his *Meteorology*, Aristotle gives us a good sense of what we should understand by the study of nature by natural philosophy when he says:

We have already discussed the first causes of nature, and all natural motion, also the stars ordered in the motion of the heavens, and the corporeal elements – enumerating and specifying them and showing how they change into one another – and becoming and perishing in general. There remains for consideration a part of this inquiry which all our predecessors called meteorology. It is concerned with events that are natural, though their order is less perfect than that of the first of the elements of bodies. They take place in the region nearest to the motion of the stars. Such are the milky way, and comets, and the movements of meteors. It studies also all the affections we may call common to air and water, and the kinds and parts of the earth and the affections of its parts. These throw light on the causes of winds and earthquakes and all the consequence of their motions. Of these things some puzzle us, while others admit of explanation in some degree. Further, the inquiry is concerned with the falling of thunderbolts and with whirlwinds and fire-winds, and further the recurrent affections produced in these same bodies by concretion. When the inquiry into these matters is concluded let us consider what account we can give, in accordance with the method we have followed, of animals and plants, both generally and in detail. When that has been done we may say that the whole of our original undertaking will have been carried out.[53]

From this it is apparent that for Aristotle, physics, or natural philosophy, embraces the motions of terrestrial and celestial bodies; the motions and transformations of the four elements in the terrestrial region, and the generations and corruptions of the compound bodies they continually produce; it also includes phenomena in the upper region of the atmosphere just below

---

[49] Aristotle, *Physics*, 2.1.192b.9–11; translated by R. P. Hardie and R. K. Gaye.
[50] Ibid., 2.1.192b.14–19.   [51] Aristotle, *Physics*, 2.8.199b.30–31.   [52] Ibid., 3.1.200b.11.
[53] Aristotle, *Meteorology*, 1.1.338a.20–339a.9. Translated by E. W. Webster in *The Complete Works of Aristotle*.

the moon, which was his concern in the *Meteorology*; and, finally, it also includes the study of animals and plants, which Aristotle says he will subsequently present. These, and other topics on natural philosophy, appear in a collection of Aristotle's treatises that came to be known collectively as the "natural books" (*libri naturales*), which include *Physics, On the Heavens (De caelo), On the Soul (De anima), On Generation and Corruption (De generatione et corruptione), Meteorology,* and *The Short Physical Treatises (Parva naturalia)*, which consists of a number of brief treatises.[54]

THE SCOPE OF ARISTOTLE'S NATURAL PHILOSOPHY

Natural philosophy in its many manifestations was practiced, as we have seen, long before Aristotle made his momentous contribution. We saw it in Egyptian civilization and in the pre-Socratic philosophers. But, as far as is known, no one in those places and times sought to define anything resembling what we might regard as natural philosophy. They simply wrote about a variety of topics and it has fallen to modern historians to decide whether what they wrote ought to be categorized as natural philosophy. Since medicine was not excluded in ancient Egypt or in Greece in the sixth and fifth centuries BC, it seems appropriate to include it within the domain of natural philosophy, and, perhaps, even magic as well, although the latter would more properly form part of natural philosophy in ancient Egypt than in the Greece of the Pre-Socratics.

Aristotle's contributions to natural philosophy changed all this. Not only did he leave treatises on almost all aspects of natural philosophy, but he also realized the need to define natural philosophy and delineate its scope, as well as to determine the best methodology for applying it to nature. Aristotle was apparently the first to perform this service. His efforts were destined to have a lasting effect, enduring for nearly two thousand years in three different, major linguistic cultures and civilizations – Greek (Byzantine Empire), Arabic (Islamic Civilization), and Latin (Western Europe).

How did Aristotle define and understand natural philosophy? We have already seen that by defining it, and enumerating the range of subjects to which it applies (in the *Meteorology*), he restricted its scope. This is obvious by his division of the theoretical sciences into metaphysics, mathematics, and natural philosophy, or physics. Clearly, he thought of metaphysics and mathematics as distinct from natural philosophy. Their subject matter was with entities that did not suffer change, while the essence of natural philosophy was to treat wholly of bodies undergoing change and motion. But did

---

[54] They are titled: *Sense and Sensibilia; On Memory; On Sleep; On Dreams; On Divination in Sleep; On Length and Shortness of Life; On Youth, Old Age, Life and Death, and Respiration*. All of these works are printed in *The Complete Works of Aristotle*, cited in the preceding note.

Aristotle really mean *all* bodies subject to change and motion? If so, natural philosophy would embrace virtually every discipline that treats of some aspect of the physical world, every part and subdivision of which undergoes change and motion. Because medicine is concerned with changes in the human body, it seems appropriate to infer that Aristotle included medicine as part of natural philosophy. But this seems unlikely. In the opening passage of his *Meteorology* (cited a few paragraphs earlier), Aristotle intended to mention, or allude to, all the subjects that formed part of his research program. We may infer this from his remark that, when the study of animals and planets has been completed, "we may say that the whole of our original undertaking will have been carried out." Nowhere in that "original undertaking" is medicine mentioned, nor, as far as we know, did Aristotle ever write a treatise on medicine, although he often used examples from medicine, and was the son of a physician. In addition to the exclusion of medicine from natural philosophy, Aristotle also excludes the mathematical, or exact, sciences, which he characterizes as "the more natural of the branches of mathematics, such as optics, harmonics, and astronomy."[55] Some lines earlier, Aristotle explains that when, for example, a mathematician treats of celestial bodies, he does not "treat of them as the limits of a natural body; nor does he consider the attributes indicated [that is, the shapes of celestial bodies] as the attributes of such bodies. That is why he separates them; for in thought they are separable from motion, and it makes no difference, nor does any falsity result, if they are separated."[56] As we saw, Aristotle regards optics, astronomy, and harmonics as "the more natural branches of mathematics," and therefore seemingly more mathematics than natural philosophy. These sciences are "the converse of geometry. While geometry investigates natural lines, but not *qua* natural, optics investigates mathematical lines, but *qua* natural, not *qua* mathematical."[57] For Aristotle, the exact mathematical sciences fell somewhere between natural philosophy and pure mathematics, perhaps closer to the latter than the former. But the exact sciences belong neither wholly to natural philosophy nor to mathematics but are relevant to both. Because they were viewed as lying between the two disciplines, the exact sciences came to be known as *middle sciences (scientiae mediae)* during the Middle Ages.

Aristotle's Approach

If Aristotle furnished the content, scope, and methodology of natural philosophy, he also provided something else of almost equal importance: a positive attitude toward nature and a style of doing natural philosophy. To convey a powerful sense of Aristotle's attitude toward nature, we can do no better

---

55 Aristotle, *Physics*, 2.2.194a.6–7; translated by R. P. Hardie and R. K. Gaye.
56 Ibid., 2.2.193b.32–35. I have inserted the phrase in brackets.   57 Ibid., 2.2.194a.7–10.

than cite a famous passage from his biological treatise, _Parts of Animals_, where he says:

Of substances constituted by nature some are ungenerated, imperishable, and eternal,[58] while others are subject to generation and decay.[59] The former are excellent and divine, but less accessible to knowledge. The evidence that might throw light on them, and on the problems which we long to solve respecting them, is furnished but scantily by sensation; whereas respecting perishable plants and animals we have abundant information, living as we do in their midst, and ample data may be collected concerning all their various kinds, if only we are willing to take sufficient pains. Both departments, however, have their special charm. The scanty conception to which we can attain of celestial things give us, from their excellence, more pleasure than all our knowledge of the world in which we live; just as a half glimpse of persons that we love is more delightful than an accurate view of things, whatever their number and dimensions. On the other hand, in certitude and in completeness our knowledge of terrestrial things has the advantage. Moreover, their greater nearness and affinity to us balances somewhat the loftier interest of the heavenly things that are the objects of the higher philosophy. Having already treated of the celestial world as far as our conjectures could reach, we proceed to treat of animals, without omitting, to the best of our ability any member of the kingdom, however ignoble. For if some have no graces to charm the sense, yet nature, which fashioned them, gives amazing pleasure in their study to all who can trace links of causation, and are inclined to philosophy. Indeed, it would be strange if mimic representations of them were attractive, because they disclose the mimetic skill of the painter or sculptor, and the original realities themselves were not more interesting, to all at any rate who have eyes to discern the causes. We therefore must not recoil with childish aversion from the examination of the humbler animals. Every realm of nature is marvellous: and as Heraclitus, when the strangers who came to visit him found him warming himself at the furnace in the kitchen and hesitated to go in, is reported to have bidden them not to be afraid to enter, as even in that kitchen divinities[60] were present, so we should venture on the study of every kind of animal without distaste; for each and all will reveal to us something natural and something beautiful. Absence of haphazard and conduciveness of everything to an end are to be found in nature's works in the highest degree, and the end for which those works are put together and produced is a form of the beautiful.[61]

If we use hindsight to categorize Aristotle, we would judge that he had an intellectual temperament that was forged from the combined qualities

---

[58] Aristotle is here alluding to the incorruptible celestial ether from which the planets, stars, and orbs are made. All the notes to this passage from Aristotle (this note, and the three following) are drawn from my book, _God and Reason in the Middle Ages_ (Cambridge: Cambridge University Press, 2001), 95–96.

[59] In Aristotle's cosmos, all animate and inanimate entities that exist in the sublunar region are subject to generation and corruption.

[60] That is, gods.

[61] Aristotle, _Parts of Animals_, 1.5.644b.21–645a.25. In this magnificent passage, Aristotle also reveals his bias for a hierarchical universe, where the celestial region is nobler than the terrestrial region, and for a teleological universe, one in which all activities are for an end or purpose.

of a philosopher, scientist, and historian. The historian in Aristotle is manifested in the way he presented problems. In the first book of his *Metaphysics*, Aristotle became the first historian of philosophy when he set forth the opinions of his predecessors, the pre-Socratic philosophers and his teacher, Plato. The problem he was investigating concerned the substance of things. What was their elemental nature? What was the cause of things? "We have studied these causes sufficiently in our work on nature," Aristotle declares, "yet let us call to our aid those who have attacked the investigation of being and philosophized about reality before us. For obviously they too speak of certain principles and causes; to go over their views, then will be of profit to the present inquiry, for we shall either find another kind of cause, or be more convinced of the correctness of those which we now maintain."[62] With this said, Aristotle launches on a discussion of his predecessors for the rest of the first book.

Aristotle again resorts to the opinions of his predecessors in his cosmological treatise *On the Heavens*. After asking "whether the heaven is ungenerated or generated, indestructible or destructible," Aristotle declares:

Let us start with a review of theories of other thinkers; for the proofs of a theory are difficulties for the contrary theory. Besides, those who have first heard the pleas of our adversaries will be more likely to credit the assertions which we are going to make. We shall be less open to the charge of procuring judgement by default. To give a satisfactory decision as to the truth it is necessary to be rather an arbitrator than a party to the dispute.[63]

Aristotle was well aware that natural philosophy had a history, and he appealed to it in numerous places.

But the roles of historian, philosopher, and scientist are not easily distinguished in Aristotle's treatises. When Aristotle invokes history and his predecessors, it is not merely for historical reasons, but rather to set the stage for resolving important scientific and philosophical issues. Thus, in the first chapter of the third book of his *Metaphysics*, he once again cites other opinions:

These include both the other opinions that some have held on certain points, and any points besides these that happen to have been overlooked. For those who wish to get clear of difficulties it is advantageous to state the difficulties well; for the subsequent free play of thought implies the solution of the previous difficulties, and it is not possible to untie a knot which one does not know. But the difficulty of our thinking points to a knot in the object; for in so far as our thought is in difficulties, it is in like case with those who are tied up; for in either case it is impossible to go forward. Therefore one should have surveyed all the difficulties beforehand, both

---

[62] Aristotle, *Metaphysics*, 1.3.983a.32–983b.5; translated by W. D. Ross.
[63] Aristotle, *On the Heavens*, 1.10.279b.6–12; translated by J. L. Stocks.

for the reasons we have stated and because people who inquire without first stating the difficulties are like those who do not know where they have to go; besides, a man does not otherwise know even whether he has found what he is looking for or not; for the end is not clear to such a man, while to him who has first discussed the difficulties it is clear. Further, he who has heard all the contending arguments, as if they were parties to a case, must be in a better position for judging.[64]

Aristotle then lays out various problems and difficulties that he will subsequently consider. He usually sought to identify all the problems relevant to any issue he attempted to resolve.[65]

To see how Aristotle proceeded in his analysis of problems in natural philosophy, it will be helpful to present a few examples. In the fourth book of his *Physics*, Aristotle argued strongly against the possibility that a vacuum could exist. As he so often did, Aristotle thought it best to begin by presenting the conflicting viewpoints, declaring that "We must begin the inquiry by putting down the account given by those who say that it exists, then the account of those who say that it does not exist, and third the common opinions on these questions."[66] After setting out the conflicting opinions, Aristotle, in a typical move, declares that "As a step towards settling which view is true, we must determine the meaning of the word," and then proceeds to present different ways that the term "void," or vacuum, has been defined and conceived.[67] Finally, he offers a series of arguments to demonstrate the impossibility of the existence of void space.[68]

We see the same concern for the meaning of crucial terms in his cosmological work *On the Heavens*. In this treatise, Aristotle asks whether there is only one heaven, or world, and whether it is eternal. In order to answer such questions, Aristotle insists that "we must explain what we mean by 'heaven' and in how many ways we use the word, in order to make clearer the object of our inquiry."[69] He then distinguishes three different usages of the term "heaven" or world. In one sense, world is taken as equivalent to the outermost circumference of the whole cosmos; in another, heaven, or world is conceived to embrace the whole celestial region, including the moon, sun, and the other celestial bodies; and, finally, heaven is taken as equivalent to the entire world. He appears to opt for heaven as the totality of the world, and then argues that "there is no place or void or time outside the heaven."[70]

---

[64] Aristotle, *Metaphysics*, 3.1.995a.24–995b.4.

[65] For more on Aristotle's methodical approach to problems, see Grant, *God and Reason in the Middle Ages*, ch. 3 ("Aristotle's Legacy to the Middle Ages"), 91–97.

[66] Aristotle, *Physics*, 4.6.213a.20–22. For a brief account of Aristotle's views on void space, see Edward Grant, *Much Ado about Nothing: Theories of Space and Vacuum from the Middle Ages to the Scientific Revolution* (Cambridge: Cambridge University Press, 1981), ch. 1 ("Aristotle on Void Space"), 5–8.

[67] He does so in *Physics*, bk. 4, ch. 7.   [68] In ibid., chs. 8, 9.

[69] Aristotle, *On the Heavens*, 1.9.278b.10–11.   [70] Ibid., 1.9.279a.12.

With these distinctions established, Aristotle says "we may now proceed to the question whether the heaven is ungenerated or generated, indestructible or destructible." And then, in order to answer the question properly, he once again feels it essential to review the opinions and theories of others, declaring:

Let us start with a review of the theories of other thinkers; for the proofs of a theory are difficulties for the contrary theory. Besides, those who have first heard the pleas of our adversaries will be more likely to credit the assertions which we are going to make. We shall be less open to the charge of procuring judgement by default. To give a satisfactory decision as to the truth it is necessary to be rather an arbitrator than a party to the dispute.[71]

Aristotle again thought it best to present the opinions of others when, in his *Meteorology*, he asserted: "Let us now explain the origin, cause, and nature of the milky way. And here too let us begin by discussing the statements of others on the subject."[72] Aristotle exhibited the same attitude in *On the Soul*, where he declared that, "For our study of soul it is necessary, while formulating the problems of which in our further advance we are to find the solutions, to call into council the views of those of our predecessors who have declared any opinion on this subject, in order that we may profit by whatever is sound in their suggestions and avoid their errors."[73]

Logic was an exception to Aristotle's usual procedure. He did not begin those treatises, especially the *Prior Analytics*, with a summary of previous opinions because there were no previous opinions, and he owed no debt. As Aristotle explains, "It was not the case that part of the work had been thoroughly done before, while part had not. Nothing existed at all."[74]

It also was characteristic of Aristotle's style in natural philosophy to inform his readers or listeners about the procedure he intended to follow. Thus, at the outset of his *On Interpretation (De interpretatione)*, Aristotle explains that "First we must settle what a name is and what a verb is, and then what a negation, an affirmation, a statement and a sentence are"[75] and then proceeds to consider each of these entities. He begins the *Prior Analytics* by announcing the topics he will consider. "First," he says, "we must state the subject of the enquiry and what it is about: the subject is demonstration, and it is about demonstrative understanding. Next we must determine what a proposition is, what a term is, and what a deduction is (and what sort of

---

[71] Ibid., 1.10.279b.6–12.

[72] Aristotle, *Meteorology*, 1.8.345a.11–12; translated by E. W. Webster. Aristotle considers the opinions of the Pythagoreans, Anaxagoras, and Democritus.

[73] Aristotle, *On the Soul*, 1.2.403b.20–23; translated by J. A. Smith. For other references by Aristotle to his predecessors, see Lloyd, *Aristotle*, 284.

[74] Aristotle, *Sophistical Refutations*, 34.183b.34–35; translated by W. A. Pickard-Cambridge.

[75] Aristotle, *De interpretatione*, 1.16a.1–2; translated by J. L. Ackrill.

deduction is perfect and what imperfect); and after that, what it is for one thing to be or not be in another as a whole, and what we mean by being predicated of every or of no."[76]

From my earlier mention of Aristotle's contributions to biology, it is apparent that he engaged in activities that involved careful observation of the various activities of numerous animals. In some instances, this observational knowledge derived from dissections he did, but often it came from his own desire to observe and report. In his description of the embryological development of the chicks of the common hen, we see Aristotle at his best. As will be evident from the passage quoted later, Aristotle obviously broke open eggs that had been laid at the same time and observed the status of the chick at different stages of its development.[77] He tells us that

With the common hen after three days and three nights there is the first indication of the embryo; with larger birds the interval being longer, with smaller birds shorter. Meanwhile the yoke comes into being, rising towards the sharp end, where the primal element of the egg is situated and where the egg gets hatched; and the heart appears, like a speck of blood, in the white of the egg. This point beats and moves as though endowed with life, and from it, as it grows, two vein ducts with blood in them trend in a convoluted course towards each of the two circumjacent integuments; and a membrane carrying bloody fibres now envelops the white, leading off from the vein-ducts. A little afterwards the body is differentiated, at first very small and white. The head is clearly distinguished, and in it the eyes, swollen out to a great extent. This condition lasts on for a good while, as it is only by degrees that they diminish in size and contract. At the outset the under portion of the body appears insignificant in comparison with the upper portion. Of the two ducts that lead from the heart, the one proceeds towards the circumjacent integument, and the other, like a navel-string, towards the yoke. The origin of the chick is in the white of the egg, and the nutriment comes through the navel-string out of the yolk.

When the egg is now ten days old the chick and all its parts are distinctly visible.... [After providing much more of a detailed description of the tenth day, Aristotle declares:]

About the twentieth day, if you open the egg and touch the chick, it moves inside and chirps; and it is already coming to be covered with down, when, after the twentieth day is past, the chick begins to break the shell. The head is situated over the right leg close to the flank, and the wing is placed over the head; and about this time is plain to be seen the membrane resembling an after-birth that comes next after the

---

[76] Aristotle, *Prior Analytics*, 1.1.24a10–15; translated by A. J. Jenkinson.

[77] Aristotle was not the first to think of this. We find the idea in the Hippocratic treatise *On the Nature of the Child* 29, in which the author says that "if you take twenty or more eggs and put them under two or more hens for hatching, and each day from the second to the last, that is, the day of the hatching, take one egg, break it open and examine it, you will find everything as I have described it, in so far as the nature of a bird may be compared with that of man." This passage is cited in Cohen and Drabkin, *Source Book in Greek Science*, 424–425. Aristotle's detailed account is vastly superior.

outermost membrane of the shell, into which membrane the one of the navel-strings was described as leading (and the chick in its entirety is now within it), and so also is the other membrane resembling an after-birth, namely that surrounding the yolk, into which the second navel-string was described as leading; and both of them were described as being connected with the heart and the big vein. At this time the navel-string that leads to the outer after-birth collapses and becomes detached from the chick, and the membrane that leads into the yolk is fastened on to the thin gut of the creature, and by this time a considerable amount of the yolk is inside the chick and a yellow sediment is inside its stomach. About this time it discharges residuum in the direction of the outer after-birth, and has residuum inside its stomach; and the outer residuum is white and there comes a white substance inside. By and by the yolk, diminishing gradually in size, at length becomes entirely used up and comprehended within the chick (so that, ten days after hatching, if you cut open the chick, a small remnant of the yolk is still left in connexion with the gut), but it is detached from the navel, and there is nothing in the interval between, but it has been used up entirely. During the period above referred to the chick sleeps, but if it is moved it wakes, looks up and chirps; and the heart and the navel together palpitate as though the creature were respiring. So much as to generation from the egg in the case of birds.[78]

In this remarkable description of the embryonic development of a chick, Aristotle shows his masterful powers of observation and his ability to record what he saw in a scientific manner. There is an air of detachment and objectivity worthy of a great scientist and natural philosopher. Virtually all of the experiences and observations Aristotle made exhibit these same qualities. Whether observing and recording the behavior of animals based on direct observation, or reporting observations made by others, or writing about the nature and operation of the terrestrial and celestial regions of the physical world, based on gross observation and many theoretical constructions about its essential features, Aristotle retained the same calm and impersonal mode of presentation. And yet, underlying this impersonal, detached style was a deep love of nature in all its manifestations, as we saw in the famous introductory passage to his biological treatise *On the Parts of Animals* (cited earlier in this chapter).

Aristotle's positive attitude toward nature and his own desire to remain a careful and objective observer are exemplified in all his works on natural philosophy. But his methodological approach to nature did not include the use of experiments. It has been suggested that Aristotle would have had no interest in experiments on natural phenomena, because experiments require one to alter the behavior of nature artificially and arbitrarily. By altering the natural environment of the thing that is being investigated, we would

[78] Aristotle *History of Animals*, 6.3.561a.5–562a.21; translated by d'A. W. Thompson.

not observe its natural behavior, because its natural behavior only occurs under natural conditions. "It is therefore senseless," argues Sarah Waterlow,

to place a substance under artificial conditions for better observation. This cannot enable us to identify the typifying behaviour in a given case; nor, if we believe that we have identified it, can it help us study it better. For if the substance still exhibits the behaviour in question this teaches us no more about its nature than we would have learned through observing it in the natural context. But the artificial conditions are more likely than not to obstruct the typifying behaviour, and in that case we learn nothing at all about the substantial nature, since this could only have been revealed through changes other than those which are taking place. Experiment, in short, opens up no new access to the facts, and may succeed only in suppressing them.[79]

In the chick embryo experiment cited earlier, however, Aristotle did intervene in nature, because he realized that only by breaking eggs on different days could he observe what would otherwise be unobservable, namely, the embryonic development of chicks. In at least this one instance, Aristotle shows that he would interfere with nature. Perhaps, we should assume that Aristotle would have intervened in nature whenever he could see direct benefit from the intervention. That he hardly ever did so, however, tells us that either he rarely ever saw direct benefit from intervening in nature, or that if he did see direct benefit, he was not ingenious enough to conjure up appropriate experiments that might shed light on the natural phenomena in which he was interested. It is not at all clear that Aristotle was reluctant to intervene in nature. A more likely conjecture might be that he rarely ever thought he had to, because he was convinced that he could derive solutions to most problems by contemplating the way things had to be by a priori and deductive means.

With hindsight, we can see that Aristotle was in error in much of what he had to say about the physical world. That is hardly surprising for someone who wrote more than twenty-three hundred years ago. But we cannot judge Aristotle's significance and impact on that basis, for we know, all too well, that much scientific knowledge that appeared in the nineteenth century, and even in the twentieth century, has been shown to be erroneous or misleading, or will be shown to be such. We must, rather, judge Aristotle by the way he approached nature, by the way he organized his research, and on the style and manner in which he presented scientific knowledge. From that standpoint, as we saw, he earns and deserves high praise, which he customarily received, and still receives, from all who have had occasion to judge him in the ancient, medieval, and modern worlds. He taught those who read and studied his works what nature is, and how they ought to appreciate and

---

[79] From Sarah Waterlow, *Nature, Change, and Agency in Aristotle's "Physics"* (Oxford: Clarendon Press, 1982), 34. See also David Lindberg, *The Beginnings of Western Science*, 53.

study it. Aristotle did this through the medium of his many treatises. For it is the phenomenon of Aristotelianism that clustered around Aristotle's works and thoughts that made his name the dominant force in natural philosophy from late antiquity to the seventeenth century. We must now describe this momentous and extraordinary story.

3

Late Antiquity

Aristotle, as we saw, restricted the scope of natural philosophy by defining it as a branch of theoretical knowledge just below metaphysics and mathematics. It was the discipline that studied bodies undergoing change and motion, which included virtually every physical body in the universe and thus seemingly embraced medicine and alchemy within the domain of natural philosophy. But that did not occur, largely, I suspect, because the extant works that came to be identified with the name of Aristotle did not include works on medicine and alchemy and these disciplines were, therefore, not regarded as belonging to natural philosophy.

NEOPLATONISM AND ITS APPROACH TO ARISTOTLE

Aristotle was obviously not the only one who wrote on subjects regarded as part of natural philosophy. But so overwhelming was his influence, and so numerous his works, that Aristotle's works on natural philosophy, or physics, came to be regarded as synonymous with that discipline. To express one's opinions and judgments on natural philosophy as understood by Aristotle, it became customary in antiquity to comment on Aristotle's works. The name of the first commentator, or commentators, is unknown. Indeed, how Aristotle's works fared during the first few centuries after his death is a mystery. But in the second half of the first century BC, Aristotle's fortunes changed dramatically as a result of the efforts of Andronicus of Rhodes, who produced an edition of the works of Aristotle that forms the basis on which subsequent texts were based. Indeed, "the tradition of writing commentaries on Aristotle's works began about the middle of the first century BC."[1] Although there is much that is problematic about the works of Aristotle, as we saw in the preceding chapter, these difficulties were not discussed by those who came to comment on his works, and who formed the tradition of Aristotelianism. In this chapter, I shall attempt to convey something of the commentary tradition that may have begun right after Aristotle's death,

---

[1] Hans B. Gottschalk, "The Earliest Aristotelian Commentators," in Richard Sorabji, ed., *Aristotle Transformed: The Ancient Commentators and Their Influence* (Ithaca, NY: Cornell University Press, 1990), 55.

although no such works have survived. There is evidence that commentaries were written on the *Categories* in the first century, perhaps after the completion of Andronicus's edition. However, none of these commentaries has survived. We must wait until the second century AD for surviving commentaries, when once again the *Categories* was the major interest.[2]

But what about Aristotle's texts on natural philosophy, the focus of this volume. Which of Aristotle's works belong to the category of natural philosophy? As described in the previous chapter, the core treatises, usually included among Aristotle's works on natural philosophy, are his *Physics, On the Heavens, Meteorology, On Generation and Corruption*, and *On the Soul*. Although they were not often commented on, the biological works (*The History of Animals, On the Parts of Animals, On the Generation of Animals, Movement of Animals*, and *Progression of Animals*) also belong to natural philosophy, as do the series of brief treatises subsumed under the title *Parva naturalia*, or *The Small Natural Books*.[3] Although, Aristotle's *Metaphysics* was not part of natural philosophy, it was frequently used and cited in commentaries on natural philosophy.

To appreciate the role of natural philosophy in the late ancient world, it is essential to characterize the status of philosophy in the Roman Empire during the period from the third to sixth centuries. This was a time in which many new religions emerged to compete with traditional religions. Philosophers, and philosophical schools, were strongly affected and fashioned their own religions. The most notable philosophical movement of late antiquity that also served as a religion to its followers is undoubtedly Neoplatonism, founded by Plotinus (ca. 204–269/270), who focused his philosophy on an absolutely transcendent God, which he called the One. Plotinus argued that the One transcends anything and everything we can experience and he ascribed no positive attributes to it.[4] Neoplatonists sought to achieve a mystical union with the One and toward that end employed various theurgic practices, which under the influence of Iamblichus, were called the Chaldaean rites. Indeed, although some Neoplatonists put philosophy first, others regarded theurgy as the most important activity of a philosopher.[5] This strong religious component of Neoplatonist doctrine and dogma had an impact on the way philosophy was taught.

Fortunately for the history of Aristotelianism, Neoplatonists looked favorably on the writings of Aristotle. It was largely through Neoplatonism

---

[2] See Richard Sorabji, "The Ancient Commentators on Aristotle," in Richard Sorabji, ed., *Aristotle Transformed*, 1.

[3] For a list of the works included in the *Parva naturalia*, see Chapter 2, n. 54.

[4] For a brief account of the life of Plotinus, see A. H. Armstrong, "Plotinus" in A. H. Armstrong, ed., *The Cambridge History of Later Greek and Early Medieval Philosophy* (Cambridge: Cambridge University Press, 1970), ch. 12 ("Life: Plotinus and the Religion and Superstition of His Time"), 195–210.

[5] Ibid., 279.

that Aristotelianism was preserved in late antiquity, although Plotinus, the founder of Neoplatonism, complained that in his *Categories*, Aristotle ignored Plato's Ideas or forms.[6] But Porphyry (232–309), the disciple of Plotinus, disagreed, arguing, as Sorabji has put it, that "the *Categories* is not about things, but only about words insofar as they signify things, and words get applied primarily to things in the sensible world, not to the Ideas in the intelligible realm."[7] Porphyry also wrote on how the schools of Plato and Aristotle were really one. Indeed, he wrote a treatise *On the School of Plato and Aristotle Being One* and, perhaps, another independent work *On the Difference between Plato and Aristotle*. The import of Porphyry's ideas about Plato and Aristotle was that Aristotle's works became an integral part of the study of Plato in the Neoplatonic schools of late antiquity. "The harmony of Plato and Aristotle was accepted to a larger or smaller extent by all commentators in the Neoplatonist tradition, and the great bulk of the ancient commentators, Christians included, are in that tradition,"[8] with the notable exception of John Philoponus, a Christian Neoplatonist who, as we shall see, rejected numerous basic ideas held by Aristotle. Philoponus's hostility to Aristotle aroused the ire of the important Neoplatonic author, Simplicius, who countered with his own severe criticisms of Philoponus. Thus did Neoplatonists play an instrumental role in keeping Aristotle's works and ideas alive in the late centuries of antiquity. Indeed, up to the sixth century AD, Neoplatonists wrote most of the extant commentaries on the works of Aristotle,[9] although most of them who studied Aristotle were really more devoted to Plato. Nevertheless, they regarded the study of Aristotle as essential to a proper understanding of Plato. Much more is known about Aristotelianism within the Neoplatonic philosophical tradition than is known of the genuine peripatetic tradition.[10]

Neoplatonists often approached their commentaries on Aristotle and Plato in a spiritual manner. In his important commentary on Aristotle's *On the Heavens (De caelo)*, Simplicius inserted a personal prayer at the end and climax of his commentary. He addressed his appeal to Plato's Demiurge, or creator God. "These reflections, O Lord, Creator of the whole universe and of the simple bodies within it," Simplicius begins,

I offer to you as a hymn, to you and to the beings you have produced, I who have ardently desired to contemplate the greatness of your works and to reveal it to those

---

[6] On this I follow Sorabji, "The Ancient Commentators on Aristotle," in Sorabji, ed., *Aristotle Transformed*, 2.

[7] Ibid.   [8] Ibid., 3.

[9] *Place, Void, and Eternity; Philoponus: Corollaries on Place and Void*, translated by David Furley; *with Simplicius: Against Philoponus on the Eternity of the World*, translated by Christian Wildberg (Ithaca, NY: Cornell University Press, 1991), 146.

[10] See Robert W. Sharples, "The School of Alexander?" in Richard Sorabji, ed., *Aristotle Transformed*, 83.

who are worthy of it, so that conceiving nothing mean or human about you, we may adore you in accordance with your transcendency in relation to all the things you have created.[11]

In composing his hymn, Simplicius changed the prosaic, scholarly style of the commentary to the more uplifting style suitable for a prayer. "Simplicius' commentary is an exercise that derives from the *religio mentis*, the intellectual celebration of divinity."[12] By applying this approach and attitude in his commentary on *De caelo*, Simplicius reveals that "an exegesis of Aristotle's treatise is in itself a religious act."[13] Whether a Neoplatonist philosopher was commenting on Plato or Aristotle, or both, the objective was to place the commentary within a divine, prayerful context, although the textual comments themselves were not religious in character but guided by the issues embedded in the texts, which were usually secular in nature.

There was one other basic feature of Neoplatonic commentaries. It was assumed that Plato and Aristotle could not contradict each other. In his commentary on Aristotle's *Categories*, Simplicius declares that

It is necessary... when Aristotle disagrees with Plato, not merely to look at the letter of the text, and condemn the discord between the philosophers, but to consider the spirit and track down the agreement between them on the majority of points.[14]

After the time of Porphyry, virtually all Neoplatonists accepted the assumption that Plato and Aristotle were in essential doctrinal agreement. Apparent differences between Plato and Aristotle were resolved in essentially Neoplatonic interpretations. "By assuming that any differences were merely superficial, verbal ones, it was perfectly simple to claim a false harmony between the doctrines of the two thinkers."[15]

In coming to grips with apparent disagreements in Plato and Aristotle, Simplicius emphasized the differences in their philosophical methodologies, as we see in the following two passages. In his commentary on Aristotle's *Physics*, Simplicius explains that

the present difference between the philosophers bears not on the matter itself, but on the word, and it is the same in most other cases. In my opinion, the reason is that Aristotle often wants to preserve the customary meaning of the words, and sets

---

[11] See Philippe Hoffmann, "Simplicius' Polemics," in Richard Sorabji, ed., *Philoponus and the Rejection of Aristotelian Science* (Ithaca, NY: Cornell University Press, 1987), 72.

[12] Ibid., 74. Hoffmann (ibid.) also translates (from French) a brief statement from an article by H. D. Saffrey, in which Saffrey declares that "as philosophy in Greece was never a purely intellectual activity, but also a life-style, the spiritual life of these philosophers became an unbroken succession of prayer and liturgy."

[13] Ibid. For more on Simplicius's religious attitude toward Aristotle's thought, see Hoffmann, ibid., 75–76.

[14] See Hoffmann, ibid., 77.

[15] Hoffmann, ibid. 78. Hoffmann translates the quotation from I. Hadot, *Le Problème du néoplatonisme alexandrin: Hiéroclès et Simplicius* (Paris, 1978), 195.

out, in constructing his argument, from what is manifest to the senses, whereas Plato frequently displays contempt for that kind of evidence, and deliberately rises to the level of intellectual contemplation.[16]

And in his commentary on the *Categories*, Simplicius offers a similar distinction between the two, when he declares:

Convincing evidence can be of two sorts: one is based on intellectual intuition, the other on sensation. Because he speaks to beings who live with sensation, Aristotle prefers the evidence that is based on sensation. He is always unwilling to stray from nature, and even objects which are above nature he studies in their relation to nature. The divine Plato, on the other hand, goes the other way and, in conformity with Pythagorean practice, examines physical objects in so far as they participate in what lies above nature.[17]

Interpretations of Aristotle's works by Neoplatonists were thus made from the standpoint of Plato's thought, a procedure that would obviously color their judgments about Aristotle. But Neoplatonists also devised a ten-point approach to the study of Aristotle's treatises.[18] Thus commentators were expected to classify Aristotle's writings (point 2) and to emphasize that courses on his works should begin with logic (point 3). The final goal of the study of Aristotle's works (point 4) "was knowledge of God, the First Principle." The way toward this goal (point 5) was to use Aristotle's works and ideas about ethics, physics, mathematics, and theology (presumably metaphysics). Commentators, or exegetes, were expected to be critical (point 7) and to be familiar with the thought of both Plato and Aristotle in order to show their essential agreement.

In commenting on, and discussing, Aristotle's texts, it was customary to read them in a certain order. Aristotle's *Categories* and *Metaphysics* were read before the works on natural philosophy, which were usually taken in the order of *Physics*, *On the Heavens*, followed by *On Generation and Corruption* and the *Meteorology*. Indeed, the late Greek commentators were here following Aristotle's own order of presentation, given at the very outset of the *Meteorology*, where Aristotle explains:

We have already discussed the first causes of nature, and all natural motion [i.e., in the *Physics*], also the stars ordered in the motion of the heavens [i.e., in his *On the Heavens*], and the corporeal elements – enumerating and specifying them and showing how they change into one another – and becoming and perishing in general

---

[16] Hoffmann, ibid., 78. In the two passages I cite here, I have omitted transliterations of Greek phrases and words that Hoffman interpolates into his translation.

[17] Hoffmann, ibid., 78–79.

[18] The ten points are presented in synoptic form by L. G. Westerink, "The Alexandrian Commentators and the Introductions to Their Commentaries," in Richard Sorabji, ed., *Aristotle Transformed*, 342–343. The points mentioned here are drawn from Westerink's article.

[i.e., *On Generation and Corruption*] There remains for consideration a part of this inquiry which all our predecessors called meteorology [i.e., *Meteorology*].[19]

Of the twenty-five "Principal Greek commentators" on Aristotle mentioned by Richard Sorabji,[20] the most significant for the subsequent history of natural philosophy were Alexander of Aphrodisias (fl ca. 198–209 AD), Themistius (fl. late 340s to 384 or 385 AD), John Philoponus (ca. 490–570s AD), and Simplicius (wrote after 529 AD).

"Alexander of Aphrodisias was appointed by the emperors as a public teacher of Aristotelian philosophy at some time between 198 and 209 AD."[21] He wrote numerous commentaries and independent works. Two of his commentaries on Aristotle's natural philosophy have been wholly preserved, namely his *Meteorology* and *On Sense and Sensibilia*.[22] Part of his commentary on the *Metaphysics* also survives.

Alexander is regarded as a genuine peripatetic philosopher who was one of the last of a line of peripatetic thinkers. He was, however, followed more than a century later by Themistius, who is also regarded as a periaptetic philosopher. Son of a philosopher, Themistius was a teacher of philosophy who focused on ethics but also wrote on natural philosophy. Indeed, his Aristotelian writings are in the form of paraphrases, which include paraphrases on Aristotle's *On the Soul* (*De anima*), *On the Heavens* (*De caelo*), and *Physics*.[23] Among the students of Aristotle's thought, genuine peripatetics, such as Alexander and Themistius, were relatively rare.

As we saw, however, most Aristotelian commentators in late antiquity were Neoplatonists. One of the most important and influential Neoplatonic Aristotelian commentators was John Philoponus, a Christian, who, although a commentator on some of Aristotle's works, was hostile to him. Philoponus had occasion to air his disagreements in at least four commentaries on Aristotle's natural philosophy: on the *Physics* (only the first four books are extant, along with fragments from books 5 to 8), *On the Soul*, *On Generation and Corruption*, and the *Meteorology*. He also treated themes in natural philosophy in other works, such as *On the Creation of the World* (*De opificio mundi*), *On the Eternity of the World against Proclus* (*De*

---

[19] Aristotle, *Meteorology*, bk. 1, ch. 1 (trans. E. W. Webster). I have added Aristotle's book titles in square brackets. Aristotle goes on to explain what he understands by meteorology.

[20] Sorabji, "The Ancient Commentators on Aristotle," *Aristotle Transformed*, 29–30. I have omitted transliterations of interpolated Greek phrases and words.

[21] Sharples, "The School of Alexander?" in Richard Sorabji, ed., *Aristotle Transformed*, 83. Sharples emphasizes that Alexander was not the head of the Lyceum, "which had probably ceased to exist in 86 BC." Ibid.

[22] See Philip Merlan, "Alexander of Aphrodisias," in Charles C. Gillispie, ed., *Dictionary of Scientific Biography*, 16 vols. (New York: Charles Scribner's Sons, 1970–1980), vol. 1 (1970), 117.

[23] See G. Verbeke, "Themistius," in *Dictionary of Scientific Biography*, vol. 13 (1976), 308.

aeternitate mundi contra Proclum),[24] and a treatise titled *Against Aristo-tle (contra Aristotelem)*, which is extant only in fragments, most of them preserved by Simplicius in the latter's commentaries on Aristotle's *On the Heavens* and *Physics*.[25]

Philoponus and Simplicius

Among Greek commentators on Aristotle, Simplicius, the last of the four commentators mentioned earlier, ranks in importance with Philoponus. In addition to his commentaries on Aristotle's *On the Heavens* and *Physics*, there is also his commentary on *On the Soul*.[26] Moreover, as a severe critic of Philoponus, Simplicius is linked to Philoponus.

As a Christian, Philoponus had seen fit to criticize Aristotle's views on the eternity of the world. He did so in two treatises, *On the Eternity of the World Against Proclus (De aeternitate mundi contra Proclum)*, which appeared in 529 AD, and *On the Eternity of the World against Aristotle*. Indeed, Philoponus disapproved of much in Aristotle's natural philosophy and attacked Aristotle's worldview on many points.[27] Unlike Aristotle, Philoponus, as a Christian, regarded the world as something with a beginning and an end. He also denied Aristotle's rigid division of the cosmos into two distinct parts, one celestial and composed of a fifth, unchanging, ethereal element, the other terrestrial, composed of four incessantly changing elements (earth, water, air, and fire). Philoponus rejected Aristotle's fifth element, or celestial ether, and regarded the heavens as "composed of a mixture of the purest parts of the four elements, with fire predominating."[28] He also assumed that bodies were not moved from one place to another by air, as Aristotle had argued, but by an impressed force, or impetus, which played a sig-nificant role in the history of dynamics, influencing natural philosophers in Islam and the Latin West all the way to Galileo. Philoponus believed that the medium through which a body moved served to resist the motion of that body, but not to move it, as Aristotle argued. He also rejected Aristotle's denial of the possibility of motion in a vacuum. In the absence of a resistant medium, Aristotle believed that bodies would fall with infinite speed. Philoponus disagreed, insisting that motion in a vacuum would be

[24] See S. Sambursky, "John Philoponus," in *Dictionary of Scientific Biography*, vol. 7 (1973), 138–139.

[25] For a bibliography of the works of Philoponus, see Richard Sorabji, ed., *Philoponus and the Rejection of Aristotelian Science* (Ithaca, NY: Cornell University Press, 1987), 231–235.

[26] See G. Verbeke, "Simplicius," in *Dictionary of Scientific Biography*, vol. 12 (1975), 442.

[27] Philoponus's anti-Aristotelian sentiments are described by Richard Sorabji, ed., *Philoponus and the Rejection of Aristotelian Science* (Ithaca, NY: Cornell University Press, 1987). See ch. 1: "John Philoponus," 1–31.

[28] Sorabji, "John Philoponus," in Sorabji, ed., *Philoponus and the Rejection of Aristotelian Science*, 25.

finite. Contrary to Aristotle's doctrine of place as "the boundary of the container," Philoponus regarded place or space as "a certain extension in three dimensions."[29] In effect, he regarded place or space as a three-dimensional void, which, however, is never empty of body.[30] Philoponus disagreed with Aristotle on other important points, including the nature of matter and light.[31]

As a staunch defender of Aristotle, Simplicius mounted a wide-ranging attack against Philoponus, focusing on the latter's treatise, *Against Aristotle* (*Contra Aristotelem*). As the vehicle for his criticisms of Philoponus, Simplicius used his own commentaries on Aristotle's *On the Heavens* and *Physics*. In the process, Simplicius quoted extensively from *Against Aristotle*, thus preserving significant parts of Philoponus's attack on Aristotle, which would otherwise have been lost. Indeed, he was apparently quite historically minded, preserving many important bits and pieces of information from the Pre-Socratics, in the sixth and fifth centuries BC, to John Philoponus in the sixth century AD Simplicius's interest in his predecessors is manifested in his commentary on Aristotle's *On the Heavens*, where Aristotle had declared that air would have weight in its own place. "Of this we have evidence," Aristotle argued, "in the fact that a bladder when inflated weighs more than when empty."[32] In commenting on this passage, Simplicius explains that Ptolemy, the famous astronomer, used the same inflated bladder experiment to argue the opposite, namely, that air has no weight in air. Indeed, Simplicius relates that Ptolemy "not only contradicts Aristotle's view that the inflated bladder is heavier than when uninflated, but he maintains that the inflated bladder actually becomes lighter."[33] Simplicus then also appeals to empirical evidence, when he explains: "Having tried this out myself with the greatest possible accuracy I found that the weight of the bladder when uninflated and inflated was the same."[34] Thus did Aristotle, Ptolemy, and Simplicius arrive at three different empirically derived conclusions about the weight of an inflated bladder as compared to the weight of an uninflated bladder. Aristotle declares that the inflated bladder weighs more than an empty one; Ptolemy that it weighs less; and Simplicius that they are of equal weight. As

[29] On all these points, see David Furley's translation of *Philoponus: Corollaries on Place and Void* in *Place, Void, and Eternity*, 15–73; also see Richard Sorabji, "John Philoponus," in Richard Sorabji, ed., *Philoponus and the Rejection of Aristotelian Science*, 1–40.

[30] See David Furley's translation of *Philoponus: Corollaries on Place and Void*, in *Place, Void, and Eternity*, 29–30.

[31] On matter and light, see Sorabji, "John Philoponus," in Sorabji, ed., *Philoponus and the Rejection of Aristotelian Science*, 18–23, 26–30. For a brief account of Philoponus's disagreements with Aristotle's ideas about dynamics and cosmology, see G. E. R. Lloyd, *Greek Science after Aristotle* (New York: W. W. Norton, 1973), 158–162.

[32] Aristotle, *On the Heavens* 4.4.311b.9–10; trans. J. L. Stocks. In this, I am indebted to Lloyd, *Greek Science after Aristotle*, 157–158.

[33] Lloyd's translation in *Greek Science after Aristotle*, 157.   [34] Lloyd's translation, ibid.

Lloyd observes, this experiment, like many others cited in the ancient world, produced no decisive results.[35]

In their commentaries on Aristotle's *Physics*, both Philoponus and Simplicius included lengthy discussions and arguments on different topics. As I have already mentioned, Philoponus disagreed with Aristotle on numerous themes. One of his most basic attacks on Aristotle involved the latter's views on the impossibility of the existence of void space as a three-dimensional extension and the impossibility of motion in a hypothetical vacuum. Philoponus posed cogent arguments in favor of void space as a three-dimensional extension in which motions could occur in finite times. He concluded: "I believe, then, that it has been shown well enough that even if there were a void, nothing would prevent the occurrence of motion – motion in time, since there is no timeless motion."[36] Indeed, Philoponus insists that "it is impossible for motion to occur without void,"[37] although he went on to emphasize that "the void, although having its own existence, is never without body."[38]

Just as Philoponus attacked Aristotle, Simplicius defended Aristotle and attacked Philoponus, sometimes referring to the latter as the Grammarian, or simply as "this man."[39] Because Philoponus had made the eternity of the world his major critique against Aristotle, Simplicius devoted a considerable part of his commentary on Aristotle's *Physics* to a defense of Aristotle's position and to a severe attack against Philoponus.[40]

We see that Aristotelian commentaries in late antiquity sometimes included lengthy discussions on specific topics that might be an attack on, or defense of, Aristotle. Commentaries were not solely efforts to explain the meaning of Aristotle's text but often went well beyond that. Commentators occasionally presented paraphrases of Aristotle's texts. But whatever form late ancient Greek Aristotelian commentaries took, their content was destined to play a significant role in shaping the subsequent development of natural philosophy.

---

[35] In a note (ibid., 157, n. 1), Lloyd explains that "The different results obtained by Aristotle, Ptolemy and Simplicius may not be due simply to negligence in carrying out the test. The weight of the inflated bladder depends on, among other things, the proportion of carbon dioxide it contains, and this will vary according to whether it is filled with exhaled breath or atmospheric air."

[36] *Place, Void, and Eternity; Philoponus: Corollaries on Place and Void*, translated by David Furley, 72.

[37] Furley, tr., ibid.   [38] Furley, tr., ibid., 73.

[39] See *Place, Void, and Eternity; Simplicius: Against Philoponus on the Eternity of the World*, translated by Christian Wildberg (Ithaca, NY: Cornell University Press, 1991), 107, n. 1.

[40] See ibid., 107–128, for Wildberg's translation of Simplicius's arguments.

4

Islam and the Eastward Shift of Aristotelian Natural Philosophy

Although Neoplatonism and Neoplatonic interpretations of Aristotle's works shaped the understanding of Aristotle's ideas in late antiquity in the Greek world of the Byzantine Empire, Greek science, medicine, and natural philosophy, especially Aristotle's natural philosophy, were disseminated eastward into Syria and Persia, largely by way of translations, first from Greek into the Syriac language, and then Syriac and Greek into Arabic. Religious tensions and animosities were the major catalyst for this eastward thrust of Greek science and natural philosophy. To strike a blow against paganism, the Roman Emperor Justinian closed the Neoplatonic philosophical school in Athens in 529 AD. A number of the philosophers at the school, including Simplicius, opted to move to Persia and continue their philosophizing under the aegis of the Persian king, Chosroes. In time, however, all of these philosophers chose to return to the Byzantine Empire. King Chosroes made an arrangement with Emperor Justinian to allow their return on condition that Justinian would not coerce them into embracing the Christian faith. Justinian honored his commitment and the returning philosophers lived in peace.[1]

THE TRANSLATIONS

But the major problem for the Byzantine Empire derived from two schismatic sects, the Nestorians and Monophysites. The Nestorians took their name from Nestorius, who had been a monk in Antioch but was made Patriarch of Constantinople in 428. Anastasius, who also had been a monk in Antioch, insisted that the Blessed Virgin Mary was the mother only of the human body of Christ, and not of Christ's divinity. When bishop Nestorius sided with Anastasius, the former was excommunicated and those who shared his belief became known as Nestorians.[2] Although persecuted within the Byzantine Empire, Nestorians were numerous in Syria and Persia, where they were protected.

---

[1] See A. A. Vasiliev, *History of the Byzantine Empire 324–1453* (Madison: University of Wisconsin Press, 1952), 150.

[2] For a brief account, see De Lacy O'Leary, *How Greek Science Passed to the Arabs* (London: Routledge & Kegan Paul, 1949), 52–53.

The Monophysite heresy was condemned at the Council of Chalcedon in 451. The two natures of Christ, divine and human, were enunciated at Chalcedon but rejected by those who came to be called Monophysites, because they believed in the unity – that is, one nature – of the divine and human in the living, physical Christ.

The Beginnings: Nestorian and Monophysite Christian Translators

The Nestorians and Monophysites each formed a Syrian Christian Church: "the Nestorian Church of Persia, now doctrinally beyond the pale and administratively beyond the reach of the Emperor. And the so-called Jacobite Church of the Monophysites, subject to constant persecutions from the Byzantine State, yet vital enough to propagate itself even in the Nestorian preserves in Persia."[3] The theological quarrels and disputes that wracked the Byzantine Empire took on a scholastic aspect. Similar theological battles occurred among Nestorians and Monophysites, especially one against the other. In these battles, Aristotle's logical treatises were found essential. In their schools at Edessa (in northern Mesopotamia) and Nisibis in Persia, Nestorians translated a number of Aristotle's elementary logical treatises – *Categories*, *On Interpretation*, and probably the *Prior Analytics* – from Greek into Syriac.[4] By the sixth century, Nestorians also had developed an intellectual center in Jundi-Shapur, where some translating was done but probably not as much as once believed.[5] In all these places, translations were made from Greek into Syriac and treatises were also written directly in Syraic. In addition to Aristotle's logical treatises, translations were made of Greek medical treatises by Hippocrates and Galen, and probably also of some mathematical and astronomical treatises.

During the same period, the Monophysites were also actively translating Greek texts into Syriac. As with the Nestorians, the Monophysites translated Aristotle's elementary logical works and numerous medical treatises, as did Sergius of Reshaina (d. ca. 536), a priest and physician. In the seventh century, Severus Sebokht, a Monophysite monk, translated logical and astronomical works, perhaps including Ptolemy's *Almagest*.[6]

Missing from these early translations are Aristotle's treatises on natural philosophy. They were perhaps ignored in favor of Aristotle's logical texts and medical treatises, because those were of immediate practical use. But it is even more likely that the Nestorians and Monophysites were not primarily interested in philosophy or natural philosophy, but in scriptural exegesis,

[3] F. E. Peters, *Aristotle and the Arabs: The Aristotelian Tradition in Islam* (New York: New York University Press, 1968), 37–38. In *How Greek Science Passed to the Arabs*, O'Leary devotes chapter 5 to the Nestorians (pp. 47–72) and chapter 6 to the Monophysites (pp. 73–95).

[4] See Marshall Clagett, *Greek Science in Antiquity* (London: Abelard-Schuman, 1957), 179.

[5] See David C. Lindberg, *The Beginnings of Western Science*, 164–165; for the earlier view, see O'Leary, *How Greek Science Passed to the Arabs*, 71–72.

[6] Clagett, *Greek Science in Antiquity*, 180–181.

for which Aristotle's logic was far more important than his natural philosophy.[7] It was not until the foundation of Baghdad in 762 and the ascent to the Caliphate of the Abbasid dynasty, beginning with Hārūn al-Rashīd in 786, that Aristotle's treatises were translated, along with numerous works in Greek science and medicine. The rulers in Baghdad utilized the translation skills and services of Syriac Christians in Jundi-Shapur, which by the sixth century had become a center of Hellenism.[8] Syriac-speaking Christians sought out Greek texts to translate into Syriac. Although the early emphasis, from the third to the early fifth centuries, was on Greek theological works, from the fifth century onward, many Greek secular works also were translated.[9]

In the reign of Hārūn al-Rashīd, Muslims at his court became interested in the traditional learning of the Greeks, which in medieval Islam came to be called "the foreign sciences," or "pre-Islamic sciences," as opposed to the Islamic sciences based on the Koran and Islamic law and traditions. A major catalyst was Ja'far ibn Barmak, Hārūn al-Rashīd's minister, who sought to introduce Greek science into the Arabic language and for that purpose induced numerous Nestorian translators to come to Baghdad from Jundi-Shapur. "Thus the Nestorian heritage of Greek scholarship passed from Edessa and Nisibis, through Jundi-Shapur, to Baghdad."[10] Most of the translator's were Syriac Christians.

It was not until the reign of the Caliph al-Ma'mūn (813–833) that translations of Aristotle's nonlogical works appeared. These translations were associated with the House of Wisdom (*bayt al-ḥikmah*), which emerged under the sponsorship of al-Ma'mūn. The House of Wisdom was more a research center and library than an academic institution. It was the place in Baghdad where the translations from Greek to Arabic were made.[11]

Al-Nadīm's "Fihrist": A Catalogue of Translations

Many translators contributed to the conversion of Aristotle's natural philosophy from Greek into Arabic, and Syriac into Arabic. We glean some sense of the process from an extraordinary treatise titled the *Fihrist*, or "catalogue," written in the tenth century by Muḥammad ibn Ishaq al-Nadīm (ca. 935–990).[12] The *Fihrist* "is largely an annotated list of works and authors in all branches of knowledge, reaching over a thousand entries in number and including translations from Persian, Greek, Coptic, Syriac, Hebrew,

---

[7] See Peters, *Aristotle and the Arabs*, 58. [8] See Clagett, *Greek Science in Antiquity*, 179.

[9] See Scott Montgomery, *Science in Translation: Movements of Knowledge through Culture and Time* (Chicago: University of Chicago Press, 2000), 68.

[10] O'Leary, *How Greek Science Passed to the Arabs*, 72.

[11] See Peters, *Aristotle and the Arabs*, 73–74.

[12] For a brief biography of al-Nadīm, see Bayard Dodge, ed. and trans., *The Fihrist of al-Nadīm: A Tenth-Century Survey of Muslim Culture*, 2 vols. (New York: Columbia University Press, 1970), xv–xxiii. Although the work is in two volumes, the pagination is continuous.

and Hindu."[13] On the translation of Aristotle's *On the Heavens*, al-Nadīm writes:

> This is composed of four books. Ibn al-Biṭrīq translated this work and Ḥunayn emended the translation. And Abū Bishr Mattā translated a portion of the first book. There is a commentary by Alexander of Aphrodisias on part of the first book of this work, and from Themistius a commentary on the entire work. Yaḥya ibn ʿAdi translated or (al-Qifṭi and) emended it. Ḥunayn also produced something on this work, namely sixteen questions. Abū Zayd al-Balkhi wrote a commentary on the introduction of this work for Abū Jaʿfar al-Khazan.[14]

Not only does al-Nadīm mention the translators he knows for the work in question, but he also provides information about Greek commentaries on that work in late antiquity and about subsequent Arabic commentaries. In his comments on the translations of Aristotle's *On Generation and Corruption* (*De generatione et corruptione*), al-Nadīm explains:

> Ḥunayn translated it into Syraic and Isḥāq into Arabic, and al-Dimashqi [translated it into Arabic].[15] It is related that Ibn Bakūsh translated it. Alexander commented the entire work and Mattā translated it. Qusṭā translated the first book. There is a commentary by Olympiodorus in the version of Asṭāt; Abū Bishr Mattā translated it and this, namely the version of Mattā, was emended by Abū Zakarīyā when he studied it. Recently there came to light the commentary of Themistius on the *De generatione et corruptione*; there are really two commentaries, a great one and a small one. There is a complete commentary by Yaḥyā the Grammarian on the *De generatione et corruptione*. The Arabic version is worse than the Syriac version.[16]

Here again, al-Nadīm tells most, or all, of what he knows about translations of Aristotle's *Meteorology*. Ḥunayn ibn Isḥāq had translated it into Syriac and others had rendered it into Arabic, including Ḥunayn's son, Isḥāq, although it is not clear whether Isḥāq used his father's Syriac version for

---

[13] Montgomery, *Science in Translation*, 106. For biographical and bibliographical information about al-Nadīm, and a table of contents for section 7 ("On Philosophers and the Ancient Sciences and Books on These Subjects"), part 1 of the *Fihrist*, see Peters, *Aristotle and the Arabs*, 277–280. This part includes what al-Nadīm knew about the translations of Aristotle's works.

[14] Translated in F. E. Peters, *Aristoteles Arabus: The Oriental Translations and Commentaries on the Aristotelian "Corpus"* (Leiden: E. J. Brill, 1968), 35. Al-Qifṭi, who added material to al-Nadīm's text has the following immediately after the quotation above: "And there is a discourse by Abū Hāshim al-Jubbāʾi on it and a refutation which is called 'An Investigation in which the Foundations of Aristotle are Abrogated.' He attacked the meaning of Aristotle thus shaking the foundations." The entire *Fihrist* has been translated in Bayard Dodge, ed. and trans., *The Fihrist of al-Nadim: A Tenth-Century Survey of Muslim Culture*, cited earlier. Al-Nadīm's coverage of Aristotle and his works appears in vol. 2, 594–606. For his discussion of *On the Heavens*, see 603.

[15] Peters adds the following to his translation of al-Nadīm's text: (*add.* Al-Qifṭi: translated it into Arabic)." I have omitted "(add. Al-Qifṭi:"

[16] Peters, *Aristoteles Arabus*, 37. For Dodge's translation of this section, see *The Fihrist of al-Nadim*, 604.

his Arabic translation. Once again, al-Nadīm mentions Greek and Arabic commentaries and concludes with a judgment that the Arabic translation was inferior to the Syriac version. This may well be a tribute to the translator of the Syriac version, Ḥunayn ibn Isḥāq, perhaps the greatest and most reliable translator of the Islamic and Latin Middle Ages.

Ḥunayn ibn Isḥāq (808–873)

Ḥunayn was the son of a Nestorian druggist in the city of Hira. Because he asked too many questions in the school he attended in Jundi-Shapur, his teacher expelled him. Ḥunayn then departed for "the land of the Greeks," where he learned Greek and acquired knowledge about textual criticism. Somewhat later, he went to Basra and studied Arabic. Thus, Ḥunayn was familiar with the Syriac, Greek and Arabic languages. He was probably a member of a mission to Byzantium to acquire Greek manuscripts. Ḥunayn translated works of Plato and Aristotle and their commentators. But he is far better known as a translator of the works of the three great Greek medical writers: Hippocrates, Galen, and Dioscorides.[17] Indeed, Ḥunayn was himself a physician and the author of at least fourteen original medical treatises.[18]

The transmission of Greek science and natural philosophy depended on reasonably sound translations that were intelligible to those who lived in a society and culture that was far removed from Greek antiquity in both custom and language. Ḥunayn ibn Isḥāq had a reputation as the best and most reliable translator of Greek into Syriac and of Greek and Syraic into Arabic. The fourteenth-century Muslim biographer, al-Ṣafadi, paid tribute to Ḥunayn, and also provided illuminating insight into the art of translation, when he declared:

There are two methods of translations used by the translators. One is the method of Yuḥannā ibn Biṭrīq, Ibn an-Nā'ima al-Himsi, and others. According to this method the translator renders each Greek word by a single Arabic word of an exactly corresponding meaning, thus establishing the translation of one word after the other, until the whole has been translated. This method is bad on two counts. (1) there are no corresponding Arabic words for all Greek words; therefore, in this kind of translation many Greek expressions remain as they are. (2) Syntactic peculiarities and constructions are not the same in one language as in the other. Mistakes are also caused by the use of metaphors which are frequently used in all languages.

The other method of translating into Arabic is that of Ḥunayn ibn Isḥāq, al-Jawhari, and others. According to this method the translator grasps in his mind the meaning of the whole sentence and then renders it by a corresponding sentence in

---

[17] See G. C. Anawati and Albert Z. Iskander, "Ḥunayn ibn Isḥāq al-'Ibādī Abū Zayd, known in the Latin West as Johannitius," in Charles C. Gillispie, ed., *Dictionary of Scientific Biography*, vol. 15, supplement 1 (1978), 230–249. The first part (230–234) is by Anawati; the second part, which includes "Hunayn the Translator" and "Hunayn the Physician" is by Iskander (234–249).

[18] Iskander, ibid., 247–248.

Arabic, regardless of the congruence or lack of congruence of the individual words. This method is the better. Therefore Ḥunayn's books need no revision, except in the field of mathematics which he did not completely master.[19]

From al-Nadīm's comments, it appears that Aristotle's treatises in natural philosophy were translated in stages. Part of a treatise might be translated and then another part by a later translator; or one translator might render the Greek into Syriac and then another translator would convert the latter to Arabic, as happened with *On Generation and Corruption*, with Ḥunayn translating into Syriac and his son, Isḥāq, into Arabic.[20] As Walzer explains, the translators at the House of Wisdom "produced a great number of partially improved and partially first translations of Aristotle. The translators sometimes worked from the Greek original, sometimes from older or recent intermediate Syriac translations."[21]

Ḥunayn ibn Isḥāq established a school of translators, which consisted primarily of himself, his son, Isḥāq (d. 911), his nephew Ḥubaysh ibn al-Ḥasan, and a number of disciples, including 'Isa ibn Yaḥyā, Yaḥyā b. Harun, Stephanus son of Basilius and Musa b. Khalid. All were Nestorian Christians. Because of his unsurpassed mastery of Greek, Ḥunayn usually made a primary translation from Greek into Syriac, and occasionally from Greek into Arabic.[22] Others translated the Syriac versions into Arabic. Isḥāq, Ḥubaysh, and 'Isa translated philosophic and mathematical works, and did most of Aristotle's treatises. Isḥāq is said to have translated *On Generation and Corruption*, and "to Ḥunayn and Isḥāq are ascribed translations, paraphrases, elucidations and abridgements of Plato's *Republic*, Aristotle's *Categories*, *De Interpretatione*, *Analytica*, *Topica*, *Sophistica*, *Rhetorica*, *Physica*, *De anima*, *Metaphysica*, *De caelo....*"[23] and other works.

The Nestorian and Monophysite translators were undoubtedly favorably disposed toward the Greek learning they translated. They translated not only because they regarded certain Greek treatises in logic, and perhaps also natural philosophy and metaphysics, as essential for the study of Christian theology, but probably because they loved the pursuit of learning and respected Greek secular thinkers. A northern Syrian scholar, David bar Paulos, who

---

[19] Peters, *Aristotle and the Arabs*, 63–64.

[20] Neither of these two translations survives, although a later translation from Arabic to Latin by Gerard of Cremona, and an Arabic to Hebrew translation by Zecharia ben Isaac, were both made from Isḥāq's Arabic translation. See Peters, *Aristoteles Arabus*, 38.

[21] See R. Walzer, "Aristutalis," in *The Encyclopaedia of Islam*, New Edition, Prepared by a Number of Leading Orientalists, edited by H. A. R. Gibb, J. H. Kramers, E. Lévi-Provençal, and J. Schacht, vol. 1 (A-B) (Leiden: E. J. Brill, 1986), 631.

[22] See L. E. Goodman, "The Translation of Greek Materials into Arabic," in M. J. L. Young, J. D. Latham, and R. B. Serjeant, eds., *Religion, Learning and Science in the 'Abbasid Period* (Cambridge: Cambridge University Press, 1990), 487.

[23] Goodman, ibid., 489.

lived in the late eighth or early ninth centuries, wrote a poem that reveals his love and admiration of Greek thought. He declared that:

> Above all the Greeks is the wise Porphyry held in honor,
> The master of all sciences, after the likeness of the godhead.
> In all fields of knowledge did the great Plato too shine out,
> And likewise subtle Democritus and the glorious Socrates,
> The astute Epicurus and Pythagoras the wise;
> So too Hippocrates the great, and the wise Galen,
> But exalted above these all is Aristotle,
> surpassing all in his knowledge, both predecessors and successors.[24]

David Bar Paulos's sentiments probably reflect the attitudes of the Syriac translators, for whom Greek learning and culture formed the basis of intellectual life.

Translations in Baghdad

The final stage of Aristotelian translations, or the post-Ḥunayn phase, occurred in Baghdad between 900 and 1020 AD, a period that parallels, but extends some thirty years beyond, the life of al-Nadīm. In this final phase, the translators usually revised older translations.[25] In this group were Abū 'Uthmān al-Dimashqī, Qusṭa b. Lūqā, Abū Bishr Mattā b. Yunus (d. AD 940) and his disciple, Abū Zakariyyā' Yaḥyā b. 'Adī, Abū 'Alī 'Īsā b. Isḥāq b. Zur'ah (942–1008), al-Ḥasan b. Suwar, known as Ibn al-Khammār (942–1017), and Abū'l-Faraj 'Abdullāh b. al-Ṭayyib (d. 1043).[26] When the process was completed, there were Arabic versions of all of Aristotle's works on logic, natural philosophy, and metaphysics,[27] as well as numerous works falsely attributed to Aristotle,[28] especially the *Theology of Aristotle* (*Theologia Aristotelis*) and the *Book on Causes* (*Liber de causis*). Although attributed to Aristotle, both of these pseudo-Aristotelian treatises were far removed from his real philosophical thought.[29] Both were, in fact, Neoplatonic treatises. The *Theology of Aristotle* was really a paraphrase of Books IV to VI of the *Enneads* of Plotinus, the founder of Neoplatonism, while the *Book on Causes* was based on Proclus's *Elements of Theology* (*Elementatio theologica*). In these two treatises, "the doctrine of emanation, which served as the cornerstone of almost the whole of Arab philosophical thought, is fully expounded

---

[24] Quoted by Montgomery, *Science in Translation*, 76.

[25] Peters cites seven translators who worked in Baghdad in this period. See Peters, *Aristotle and the Arabs*, 60–61.

[26] See Goodman, "The Translation of Greek Materials into Arabic," 491–494.

[27] For a list of these works and those that have been edited and published, see Walzer, "Aristutalis," in *The Encyclopaedia of Islam*, vol. 1 (A-B), 631–632.

[28] Walzer, ibid., 632–633.

[29] See Majid Fakhry, *A History of Islamic Philosophy* (New York: Columbia University Press, 1970), 33.

and discussed."[30] Thus was the Neoplatonic doctrine of emanation falsely attributed to Aristotle and made of him a full blown Neoplatonist. Because of this, as we shall see, the Aristotle of Islamic thought was radically different from the Aristotle of the Latin West.

Some of the translators mentioned earlier in this paragraph also rendered into Arabic works by the Late Greek commentators. Thus by around 1050 AD, virtually all that the Muslims would obtain from Greek thought was translated into Arabic, which, perhaps, explains why the translating movement was largely over, although some activity would continue for two more centuries.[31] It is for this reason, L. E. Goodman explains, that "in Arabic letters from the time of Ibn Sīnā [980–1037; known in the West as Avicenna] we do not find a thirst for new materials but an endeavour to assimilate, synthesize, and – not only in al-Ghazālī [1058–1111] but in Ibn Sīnā himself – to overcome the influence of the Greeks."[32] This suggests that once Muslim thinkers absorbed Greek thought from available translations, they sought to reduce Greek influence by replacing it with something more Islamic, or more in keeping with Islamic attitudes and values. But the Muslim response to Greek thought, especially to that of Aristotle, was complex and is difficult to judge and assess. It is, however, obvious that some important Muslim scholars did not wish "to overcome the influence of the Greeks" but, rather, to defend and advance those ideals. There is, however, no doubt that a major battleground in Islam was over the status of natural philosophy. Was it friend or foe?

THE FATE OF NATURAL PHILOSOPHY IN ISLAM

Anyone who attempts to grapple with this question must keep in mind the decentralized nature of the Islamic religion. There is no "pope" of Islam and no hierarchy of religious officials. Religious authority is largely a local matter, with prestigious mullahs exercising wider authority than those less prestigious. Religious leaders who saw dangers to the Islamic faith in the secular learning of Greek natural philosophy could deter its use by public attacks against it. If the reputation of that leader extended well beyond his home locale, he could subvert natural philosophy and secular learning over a much wider region. In a similar manner, when ruling caliphs chose to support and encourage secular learning at their courts, or by supporting one or more

---

[30] Fakhry, ibid.

[31] Fakhry, ibid., 16–31, gives a good, brief account of philosophical translations into Arabic, including Aristotle's works.

[32] Goodman rejects the argument that a religious reaction against the influx of rationalistic Greek thought brought translating activity to a halt. For his ideas, see Goodman, "The Translation of Greek Materials into Arabic," 495–497. I have added the bracketed information.

schools, scholars devoted to the study of natural philosophy could flourish in a friendly environment.

In the earlier discussion about the late Neoplatonic Greek commentators on Aristotle, I had occasion to mention that they approached philosophy, and indeed natural philosophy, with a religious and spiritual attitude. Thus, religion and natural philosophy were not regarded as conflicting areas of thought and belief. In Islam, by contrast, historical circumstances produced a radically different view of natural philosophy.

When, after the death of Muḥammad in 632, Muslim armies burst out of Arabia, the religion of Islam became the dominant religion over a vast area stretching from the Straits of Gibraltar in the West to India in the East. In contrast to Christianity, which spread slowly and by proselytizing, Islam was disseminated rapidly by conquest. In areas such as Egypt and Syria, Greek learning had been entrenched for many centuries. Converts to Islam in these lands were frequently Christians or pagans who had been exposed to Greek learning. The Muslim invaders, however, often viewed Greek learning as alien to Islam. As evidence of this, Muslims distinguished two kinds of sciences: the Islamic sciences, based on the Qur'an and Islamic law and traditions, and the foreign sciences, or "pre-Islamic" sciences, which encompassed Greek science and natural philosophy. We might say that the slow spread of Christianity provided Christians an opportunity to adjust to Greek secular learning, whereas Islam's rapid dissemination made its relations with Greek learning much more problematic. Not only was Greek philosophy regarded as a foreign science, but the term *philosopher* (*failasuf;* plural: *falasifa*) was often employed pejoratively.

Three Levels of the Intellectual Hierarchy

In the intellectual hierarchy of medieval Islamic society, scholars distinguish three levels.[33] Because Islam was a nomocracy, the first level comprised legal scholars. The religious law and traditions were valued above all else, and, therefore, valued even more than theology. Next in order came the *mutakallimun*, scholars who used Greek philosophy to interpret and defend the Muslim religion. The *mutakallimun* emphasized rational discourse, to which they added the authority of revelation. And, finally, at the bottom, were the *falasifa*, the rationalistic Islamic philosophers, who followed Greek thought, especially the thought of Aristotle. Not surprisingly, the philosophers placed greatest reliance on reasoned argument while downplaying revelation. The philosophers sought to develop natural philosophy in an Islamic environment, and, as A. I. Sabra has put it, did so,

[33] Toby Huff, *The Rise of Early Modern Science*, second edition (Cambridge: Cambridge University Press, 2003), 71.

"often in the face of suspicion and opposition from certain quarters in Islamic society."[34]

Of the three Islamic groups just distinguished, namely, (1) legal scholars, who were almost always traditionalists, (2) the *mutakallimun*, and (3) philosophers, the legal scholars, or traditionalists, made little use of Greek philosophy, largely because they found it a threat to revealed truth and the Islamic faith. In their bitter struggle with each other and with the traditionalists, the *mutakallimun* and the philosophers made much use of Greek philosophy. The *mutakallimun* were primarily concerned with the kalam, which, according to Sabra, is "an inquiry into God, and into the world as God's creation, and into man as the special creature placed by God in the world under obligation to his creator."[35] Thus, Kalam is a theology that used Greek philosophical ideas to explicate and defend the Islamic faith.

The philosophers, or natural philosophers, were the least popular, because they usually tended to use logic and natural philosophy to seek truth for its own sake, which was taken as a sign that they were ignoring religion and revelation. Although philosophers may have placed much less emphasis on religion than on reason, they rarely opposed their philosophical opinions to basic Qur'anic beliefs. That would have been rash and foolhardy. But there were numerous points of tension between the philosophers and the theological guardians of the sanctity of Islam. Tensions arose, for example, over the problem of creation in time and on the resurrection of the body.[36] They also arose over the claims, implicit and explicit, of philosophers and natural philosophers on the powers of reason to explicate, almost demonstratively, various tenets of revealed religion. To convey a sense of Islamic natural philosophy, I shall now briefly describe the attitudes, approaches, and contributions of five of the most significant natural philosophers in medieval Islam.

Muslim Natural Philosophers

Following the translations of Greek natural philosophy and science in the early centuries of Islam, a number of independent Islamic scholars of Greek philosophy – primarily Aristotle's Neoplatonized philosophy – emerged during the ninth to twelfth centuries. Foremost in this group, in chronological order, were al-Kindī (ca. 800–870), al-Rāzī (ca. 854–925 or 935), al-Fārābī (d. 950), ibn Sīnā (Avicenna) (980–1037); and ibn Rushd (Averroes) (1126–1198), all of whom made an impact on the Latin West. Of the five, all but

---

[34] A. I. Sabra, "Science and Philosophy in Medieval Islamic Theology," in *Zeitschrift für Geschichte der Arabisch-Islamischen Wissenschaften*, vol. 9 (1994), 3.

[35] Sabra, ibid., 5.

[36] See Georges C. Anawati, "Philosophy, Theology, and Mysticism," in the Late Joseph Schacht with C. E. Bosworth, eds., *The Legacy of Islam* (Oxford: Clarendon Press, 1974), 356.

one, al-Rāzī, sought "to assimilate the data of revelation by fitting them into the framework of Greek philosophy," a procedure that "did not fail to arouse the suspicion of traditional believers, and still more their reprobation."[37] Al-Rāzī is the most radical of the five, because he was essentially hostile to revealed religion and showed this by exalting reason over revelation. He was hardly typical of Islamic natural philosophers and will therefore be considered last. I shall take up the other four in chronological order.

Al-Kindī (ca. 800–870)

Yaqʿūb ibn Isḥāq al-Kindī, or simply al-Kindī, is regarded as the first philosopher of Islam.[38] As the first philosopher, al-Kindī saw as his primary task the need to introduce philosophical analysis into Islamic life. He distinguished philosophy from theology, the latter based on revelation, the former not.[39] At the very outset, philosophy, including natural philosophy, was made independent of revelation. Although philosophy was not based on revelation, al-Kindī, perhaps to mollify the theologians, allowed that "the knowledge due to revelation and communicated to men by divinely inspired prophets is fundamentally different from any knowledge acquired through philosophical training and unambiguously superior to it."[40] Indeed, this is the reason al-Kindī, who greatly admired Aristotle, sought to demonstrate that the world was created, despite Aristotle's unequivocal belief in the eternality of the world.[41] Revelation had declared the world created, therefore it must be true, and al-Kindī sought to use philosophy to demonstrate this truth. Thus, al-Kindī was using philosophy as the handmaid to theology, which, as Walzer indicates, "is more in keeping with the true Islamic way of life than the attempts of Al-Fārābī and Ibn Sīnā and Ibn Rushd to understand prophecy and revelation in exclusively philosophical terms."[42] Not only did al-Kindī defend the doctrine of the creation of the world from nothing (*ex nihilo*), but he also defended "the resurrection of the body, the possibility of miracles, the validity of prophetic revelation, and the origination and destruction of the world by God."[43] Al-Kindī was a firm supporter of Qur'anic doctrine and upheld the Islamic theological tradition. Nevertheless, he believed that philosophy was independent of theology, and

---

[37] Anawati, "Philosophy, Theology, and Mysticism," 359.

[38] See Kevin Staley, "Al-Kindi on Creation: Aristotle's Challenge to Islam," in *The Journal of the History of Ideas* 50, no. 3 (July–Sept. 1989), 355.

[39] Staley, ibid., 357.

[40] This is Richard Walzer's description of al-Kindī's attitude. See Richard Walzer, "New Studies on Al-Kindi," in Richard Walzer, *Greek into Arabic: Essays on Islamic Philosophy* (Cambridge, MA: Harvard University Press, 1962), 179.

[41] For a summary of al-Kindī's demonstration, see Staley, "Al-Kindi on Creation: Aristotle's Challenge to Islam," 355–370.

[42] Walzer, "New Studies on Al-Kindi," 180.

[43] Fakhry, *A History of Islamic Philosophy*, 85.

that philosophers would eventually find demonstrative evidence for revealed truths.

Al-Kindī did not write commentaries on Aristotle's treatises, but chose to write epitomes and paraphrases of themes developed by Aristotle. For example, he wrote a treatise titled "*Five Principles or Essences*, i.e., matter, space, form, motion, and time."[44] Another of his works was titled *Nature of the Sphere is Different from That of the Four Elements*. In this treatise, he compared the celestial bodies, composed of celestial ether, with the four terrestrial elements.[45] Al-Kindī also wrote treatises giving definitions of technical philosophical terms, such as his *Definitions and Descriptions of Things*.[46]

With al-Kindī, Islamic philosophy was launched. He accorded to philosophy a role that did not arouse the ire of the theologians. Majid Fakhry has contrasted al-Kindī's role in the development of Islamic philosophy with that of the Neoplatonists and peripatetics who followed him:

Being committed to the fundamental tenets of Islamic belief in the manner of the Mu'tazilite theologians, he ran a far less grave risk of public disapproval than the tenth- and eleventh-century Neo-Platonists who sought artfully to effect the impossible marriage of philosophy to Islamic belief. At the root of the difficulty was the reluctance of these Neo-Platonists to surrender any aspect of the former, or to attribute any mark of privilege or distinction to the latter by virtue of its supernatural or divine origin. For al-Kindī, however, the true vocation of philosophy was not to contest the truth of revelation, or make impudent claims of superiority, or even parity, with it. Philosophy, he believed, should simply surrender its claims to be the highest pathway to truth and be willing to subordinate itself as an ancillary to revelation.[47]

Al-Kindī was convinced that although revelation and human reason took different paths, they would eventually reach the same conclusions. He viewed philosophy as the handmaid of theology and revelation.[48] It was later philosophers, especially al-Rāzī, al-Fārābī, Ibn Sīnā, and Ibn Rushd, who, as we shall see, used the power of philosophy much more for its own sake than for upholding revealed truth. Indeed, their views were sometimes in opposition to revealed truth.

Among the Islamic philosophers considered here, al-Kindī's opinions and interpretations were easily the most compatible with traditional theological viewpoints. Al-Kindī, it would appear, gave theologians little cause for concern. But in his *Fihrist*, al-Nadīm tells a story about al-Kindī that reveals a hostile reaction to the latter's thought by Abu Ma'shar (787–886), who also was known as Albumasar in the Latin West. Al-Nadīm explains that "at first," Abu Ma'shar was "a scholar of the Hadith"; that is, he was a

---

[44] Fakhry, ibid., 106.   [45] Ibid.   [46] Ibid., 107.   [47] Ibid., 108–109.
[48] See R. Walzer, "Early Islamic Philosophy," in A. H. Armstrong, ed., *The Cambridge History of Later Greek and Early Medieval Philosophy* (Cambridge: Cambridge University Press, 1970), 650.

student of the sayings attributed to Muḥammad and his companions. Aware of al-Kindī's interests in Aristotelian natural philosophy, he became antagonistic to al-Kindī and "stirred up the populace against him, accusing him because of his philosophical sciences."[49] Eventually, Abu Maʿshar himself became interested in the same subjects as al-Kindī and, al-Nadīm informs us, "ended his ill-will for al-Kindī."[50] We see that even a rather conservative natural philosopher such as al-Kindī could draw a hostile reaction from a conservative theologian.

But the philosophers who followed al-Kindī approached philosophy and natural philosophy more radically, some even exalting the teachings of philosophy over those of religion. Some, like al-Rāzī, opposed Aristotle's physics and cosmology, as well as opposing revelation and religious authority. But most chose to follow a philosophical path, seeking to adhere to their philosophical beliefs about the nature of the world and hoping to reconcile them with theology and also avoid antagonizing the theologians. The philosophical path that became dominant in Islamic philosophy was that which followed the tradition inherited from late Greek antiquity (described earlier) that intermingled the teachings of Plato, Aristotle, and Neoplatonism, seeking always to reconcile Plato and Aristotle.[51]

Al-Fārābī (d. 950)

In the Islamic world, Al-Fārābī is regarded as the founder of Neoplatonism,[52] and a deep student of Aristotle. More than one hundred works were attributed to him, which included introductions to philosophical topics, monographs on special questions, and commentaries on Aristotle's treatises, including Aristotle's logical works, as well as on the latter's *Physics, On the Heavens*, and the *Meteorology*.[53] Although it was not a commentary, al-Fārābī also explicated Aristotle's *Metaphysics* in a treatise titled *Intentions of Aristotle's Metaphysics*.[54] He also wrote what may be described as supercommentaries; that is, he wrote commentaries on Aristotelian commentaries written by Late Greek commentators.[55] Although al-Fārābī did not oppose

[49] Bayard Dodge, ed. and tr., *The Fihrist of al-Nadim*, vol. 2, 656. See also David Pingree, "Abū Maʿshar al-Balkhī," in *Dictionary of Scientific Biography*, 18 vols. (New York: Charles Scribner's Sons, 1970–1990), vol. 1, 33.

[50] *Fihrist*, ibid.

[51] See Arthur Hyman and James J. Walsh, eds., *Philosophy in the Middle Ages: The Christian, Islamic, and Jewish Traditions* (Indianapolis: Hackett Publishing Company, 1973), 206.

[52] Fakhry, *A History of Islamic Philosophy*, 147.

[53] In his *Fihrist*, al-Nadīm mentions al-Fārābī's commentaries on Aristotle's logical works; see *The Fihrist of al-Nadim*, 600, 601, and 629. For mention of al-Fārābī's commentaries on the *Physics, On the Heavens, and Meteorology*, see R. Walzer, "al-Fārābī," 780.

[54] Fakhry, *A History of Islamic Philosophy*, 149. See also R. Walzer, "Al-Fārābī," in *The Encyclopaedia of Islam*, Vol. 2 (C–G) (Leiden: E. J. Brill, 1991), 780, where the title is given as *About the scope of Aristotle's Metaphysics*.

[55] R. Walzer, "Al-Fārābī," in *The Encyclopaedia of Islam*, vol. 2 (C–G), 780.

philosophy directly to the Islamic religion, he regarded human reason as superior to religious faith. Philosophical truth is universally valid, whereas religious concepts vary among the different peoples. Al-Fārābī categorized Muḥammad as a philosopher-prophet.

Al-Fārābī's judgment about the relation of reason and revelation is left somewhat unclear. In a discussion about scholastic theology, al-Fārābī argued that when scriptural statements conflict with reason, one should interpret Scripture allegorically. Where this proves unfeasible, Scripture must be vindicated. However, Fakhry explains that "considering the role of reason in his general conception of the scheme of things ... the inference is inescapable that it devolved upon reason, rather than revelation, to arbitrate in the conflict."[56]

In one of his polemical writings, he argued against John Philoponus's attack against Aristotle's views on the eternity of the world, thereby defending a position hostile to the Islamic religion.[57] Al-Fārābī did this because he adopted the typical Neoplatonic view of the world's origin; that is, he based it on the Neoplatonic doctrine of emanation, which begins with God thinking about Himself, an act that causes an intellect to emanate as the first emanation. Nine other intellects emanate from the first intellect, each associated with a celestial body. The last emanation is an Agent Intellect, which governs the perpetually changing sublunar realm.[58] If al-Fārābī were following the traditional emanationist theory derived from Plotinus, and that was repeated in the two widely used Neoplatonic treatises, *The Book on Causes* (*Liber de causis*) and *Theology of Aristotle*, he would have assumed that God does not create the world by free choice but from necessity and that He has been emanating the world through all eternity without any consequent diminution of His being. By following the Neoplatonic mode of explanation, al-Fārābī moved away from the Muslim belief in a world that had a beginning in time and was created from nothing.[59] By assuming immortality only to the intellectual part of the soul, al-Fārābī also departed from the Islamic concept of the immortality of the whole soul and of all souls. He also seems to have denied another vital feature of Islam: bodily resurrection.[60]

Many Muslims came to regard al-Fārābī as the greatest authority in philosophy after Aristotle.[61] He was even called "the second teacher," Aristotle being the first.[62] He presented the philosophies of Plato and Aristotle in

[56] Fakhry, *A History of Islamic Philosophy*, 135.
[57] Muhsin Mahdi, "Al-Fārābī," in *Dictionary of Scientific Biography*, vol. 4 (1971), 525.
[58] See Hyman and Walsh, eds., *Philosophy in the Middle Ages*, 212.
[59] See Frederick Copleston, S. J., *A History of Philosophy*, vol. 1: *Ancient Philosophy: The Bellarmine Series;* (no place or date), 466–467; also Fakhry.
[60] Fakhry, *A History of Islamic Philosophy*, 147.
[61] Mahdi, "Al-Fārābī," *Dictionary of Scientific Biography*, vol. 4, 524.
[62] See George Sarton, *Introduction to the History of Science*, 3 vols. (Baltimore: Williams & Wilkins, 1927–1948), vol. 1, 628.

separate treatises. In his *Philosophy of Plato*, he reveals a thorough and profound knowledge of Plato's works and in the *Philosophy of Aristotle*, he surveys the whole range of Aristotle's philosophy and concludes that "'scientific knowledge' is a necessary part of the good life."[63] Al-Fārābī's most significant work may have been *The Enumeration of the Sciences*, in which he groups the sciences of his day into eight categories: linguistic, logical, mathematical, physical, metaphysical, political, juridical, and theological.[64] This treatise, as Fakhry remarks, "is perhaps the most crucial for the understanding of al-Fārābī's conception of philosophy in relation to the other sciences, as indeed of the conception of the whole Islamic philosophical school of the nature and interrelation of the Greek and the Islamic syllabus of the sciences, echoes of which still ring four centuries later in the writings of the anti-Hellenic encyclopedist and historiographer Ibn Khaldūn of Tunis (d. 1406)."[65] The *Enumeration of the Sciences* was translated into Latin by Gerard of Cremona in the twelfth century and exerted an influence on the West.[66]

As a philosopher, Al-Fārābī seems to have emphasized reason, but it is not obvious that he advocated it over revelation, although, as we saw, he seems to have held some opinions that were in conflict with Islamic doctrine. There seems little doubt that his interpretations of the relationship between philosophy and revelation were more daring than those given by al-Kindī.

Ibn Sīnā (Avicenna) (980–1037)

Ibn Sīnā (Abū 'Alī al-Ḥusayn Ibn Sina ibn 'Abdallāh), is regarded, like al-Fārābī before him, as a Neoplatonist. In an unusual autobiography he wrote covering his first thirty years, Ibn Sīnā depicts himself as a genius, possessed of an extraordinary memory and intellect. He informs us that his father hired a philosopher, Abū 'Abd Allāh al-Nātilī, to live with his family and to teach him logic. So impressed was al-Natili with his pupil that, as Ibn Sīnā tells us, he "warned my father that I should not engage in any other occupation but learning."[67] Ibn Sīnā soon knew more than his teacher and continued studying by himself until he had mastered all of logic. He then studied and mastered Euclid's *Elements* and then Ptolemy's *Almagest*, and also began the study of medicine, undertaking to treat the sick soon after. While engaged in

---

[63] Fakhry, *A History of Islamic Philosophy*, 128–129.

[64] I follow Fakhry, ibid., 131.   [65] Fakhry, ibid., 130.

[66] For the list of Gerard of Cremona's translations, see Edward Grant, ed., *A Source Book in Medieval Science* (Cambridge, MA: Harvard University Press, 1974), 36–38. Al-Fārābī's treatise is number 42 on p. 37. The Latin and Arabic texts of al-Fārābī's work have been published by Angel González Palencia, *Al-Fārābī, Catalogo de las ciencias*, second edition (Madrid: Instituto Miguel Asin, 1953).

[67] Ibn Sīnā's autobiography to age thirty, and its continuation to Ibn Sīnā's death in 1037, by his disciple, Abū 'Ubaid al-Jūzjānī, appear in Arthur J. Arberry, trans., *Avicenna on Theology* (London: John Murray, 1951), 9–14 for the autobiography and 15–24 for al-Jūzjānī's continuation.

all this, he also studied and disputed the law. Ibn Sīnā informs us that he had done all of this by the time he was sixteen years of age![68] "I continued," he tells us, "until I had made myself master of all the sciences; I now comprehended them to the limits of human possibility. All that I learned during that time is exactly as I know it now; I have added nothing more to my knowledge to this day."[69] In his study of Aristotle's thought, Ibn Sīnā confesses that Aristotle's *Metaphysics* defied his understanding, even after he read it forty times. Only after purchasing and reading al-Fārābī's explication of it, in the latter's *Intentions of Aristotle's Metaphysics*, did he gain mastery of it.[70]

Ibn Sīnā was a prolific author – 276 works are attributed to him – and unlike those of most of his colleagues, many of his works have been preserved. Ibn Sīnā wrote no commentaries, preferring to treat various themes of Greek philosophy in separate treatises and also to include numerous themes in comprehensive philosophical encyclopedias.[71] The most important of his treatises is *Kitab al-Shifā*, or *Book of Healing*, which was translated into Latin and known by the title *Sufficientia*. The *Kitab al-Shifā* "is an encyclopedia of Islamic-Greek learning in the eleventh century, ranging from logic to mathematics,"[72] and also including a summary of Aristotelian natural philosophy, with Ibn Sīnā's elaborations of it, as well as a section on metaphysics.

In turning to Ibn Sīnā's substantive views about the world, we observe first that he interpreted the creation in purely Neoplatonic terms. God created the world of necessity – that is, it was not a free act. Nor did God create the world directly. In the emanationist interpretation, the Neoplatonic One, who is God, produces an Intelligence who, in turn, creates another Intelligence, and so on. Avicenna's theory is quite similar to al-Fārābī's described earlier.[73]

With such a cosmic picture, Avicenna, like al-Fārābī before him, depicts a world that God chose not to create directly by Himself, but achieved the same end by creating intermediaries to carry out the task. Moreover, following the Neoplatonic tradition, God is seen to create from necessity, rather than from free choice; and if He is contemporaneous with the world, that implies a world that is not created in time. Indeed, Avicenna held that if God had chosen to create our world that would have caused a change in Him, which cannot occur. "Hence creation is eternal (IX,1)," Gilson concludes and "our sole problem then is to know how becoming is eternally flowing from the first and necessary being."[74]

---

[68] Arberry, ibid., 10–11.   [69] Ibid., 11.

[70] As his version of the title of al-Fārābī's treatise, Arberry gives *On the Objects of the Metaphysica* (ibid., 12). I have used Fakhry's title, as it appears in *A History of Islamic Philosophy*, 147.

[71] See R. Walzer, "Early Islamic Philosophy," 652–653.

[72] Fakhry, *A History of Islamic Philosophy*, 150.

[73] See also Hyman and Walsh, *Philosophy in the Middle Ages*, 234–235.

[74] Etienne Gilson, *A History of Christian Philosophy in the Middle Ages* (London, Sheed and Ward, 1955), 213. The reference IX,1 is to the Latin translation of Avicenna's *Metaphysics*.

In one of his final works, Avicenna argued against the resurrection of the body alone, as well as of body and soul together (in other works, he conceded bodily resurrection, as stated in the Qur'an). He opted for the immortality of the human soul and held that after death, the soul would endure eternal happiness or eternal misery. "This is the boldest and most persuasive argument in favour of a spiritual and against a physical survival to be found in Arabic literature."[75] Avicenna espouses the virtues of spiritual happiness over physical happiness in a treatise called the *Book of Salvation* (*Kitab al-Najāt*), which was his own abridgment of his lengthy *Kitab al-Shifā*. He declares that:

the true religion brought into this world by our Prophet Muhammad has described in detail the state of happiness or misery awaiting us hereafter so far as the body is concerned. Some further support for the idea of a hereafter is attainable through reason and logical demonstration – and this is confirmed by prophetic teaching – namely, that happiness or misery posited by spiritual appraisement. . . . Metaphysicians have a greater desire to achieve this spiritual happiness than the happiness which is purely physical; indeed they scarcely heed the latter, and were they granted it would not consider it of great moment in comparison with the former kind, which is proximity to the First Truth, in a manner to be described presently. Let us therefore consider this state of happiness, and of contrasting misery: the physical sort is fully dealt with in the teachings of religion.[76]

Avicenna argues that "the peculiar perfection towards which the rational soul strives is that it should become as it were an intellectual microcosm, impressed with the form of the All, the order intelligible in the All, and the good pervading the All." He then declares most emphatically that:

When this state is compared with those other perfections so ardently beloved of the other faculties, it will be found to be of an order so exalted as to make it seem monstrous to describe it as more complete or more excellent than they; indeed, there is no relation between it and them whatsoever, whether it be of excellence, completeness, abundance, or any other of the respects wherein delight in sensual attainment is consummated.[77]

In Avicenna's view, the body corrupts the soul "causing it to forget its proper yearning and its quest for perfection."[78]

Thus did Avicenna add a spiritual and philosophical dimension to resurrection as presented in the Qur'an. Indeed, he diminishes the importance of bodily resurrection while extolling the glories of the resurrection of the rational soul. Avicenna apparently believed that the Prophet had to preach

[75] A. J. Arberry, *Revelation and Reason in Islam* (London: George Allen & Unwin Ltd., 1957), 55. Arberry cites the work in which Avicenna proclaims his doctrine as *al-Risalat al-adhawiya fi amr al-ma'ad*. See also A. M. Goichon, "Ibn Sīnā," in *The Encyclopaedia of Islam*, new edition, vol. III, H-IRAM, 944, col. 1.

[76] Translated by Arthur J. Arberry, *Avicenna on Theology*, 64.

[77] Translated by Arberry, ibid., 67.   [78] Translated by Arberry, ibid., 73.

a physical resurrection because the masses could not comprehend a spiritual resurrection. Arberry argues that for Avicenna, "It proves the superiority of Mohammed to all other prophets, that he painted for men the most realistic and emotive picture of heaven and hell."[79] Avicenna regarded it as far more pleasurable than the sterile Christian concept of a physical resurrection.

The positions just described involved Avicenna in philosophy, natural philosophy, and theology. Much, if not most, of what he wrote did not impinge on religion or theology. Except for his section on metaphysics, the *Book of Healing* (*Kitab al-Shifā*) was otherwise concerned with logic, natural philosophy, and mathematics. But, as we shall see, it was as much Avicenna's great emphasis on, and exaltation of, reason and reasoned arguments that could, and sometimes did, arouse the ire of theologians and traditional Muslim thinkers.

Ibn Rushd (Averroes) (1126–1198)

Ibn Rushd has been called "the first and last great Aristotelian in Islam."[80] Abu'l-Walīd Muḥammad Ibn Aḥmad, or Ibn Rushd, or Averroes as he was known in the Latin West, was born in Cordova, in what was then Muslim Spain. He was born into an eminent family that had a distinguished history as scholars and jurists. His early education was largely traditional, focused on linguistic and legal studies and scholastic theology. He also learned medicine in a professional sense and became a physician, composing a major medical treatise in 1169 called *al-Kulliyat*, which was translated from Arabic into Latin in the thirteenth century, bearing the title *Colliget*.[81]

Two major Muslim philosophers in Spain – Ibn Bājjah and Ibn Ṭufayl – served as Ibn Rushd's teachers. Around 1169, Ibn Ṭufayl introduced Ibn Rushd to the caliph Abū Ya'qūb Yūsuf, who had a strong interest in philosophy and science. Two significant events derived from this meeting. The caliph appointed Ibn Rushd to serve as the religious judge, or *qadi*, of Seville. And, more significantly for the history of natural philosophy, the caliph had complained to Ibn Ṭufayl that the texts of Aristotle were too obscure and requested that Ibn Ṭufayl write explanatory commentaries on those texts. Ibn Ṭufayl considered himself too busy and too old to do so. He advised the caliph to request Ibn Rushd to perform this task. And so it was that Ibn Rushd came to write his numerous and monumentally important Aristotelian commentaries.[82]

---

[79] Arberry, *Revelation and Reason in Islam*, 53.
[80] Fakhry, *A History of Islamic Philosophy*, 302.
[81] The Jewish translator, Bonacosa, made the translation in 1255 in Padua. See Sarton, *Introduction to the History of Science*, vol. 2, part 1, 360.
[82] Before he was presented to the caliph, Ibn Rushd had already written some commentaries on Aristotle's treatises, including commentaries on the logical works, the *Physics* and *Metaphysics*. The request from the caliph not only enlarged his task, but gave it the aura of a command performance. See R. Arnaldez, "Ibn Rushd," in *The Encyclopaedia of Islam*, New Edition, Vol. III (H–IRAM), 910, col. 2.

Abū Yaʿqūb Yūsuf's successor, his son, al-Manṣūr Yaʿqūb ibn Yūsuf (r. 1184–1199), found reason to persecute Ibn Rushd in 1195. The persecution was motivated by al-Manṣur's desire to appease orthodox allies, who took a dim view of Ibn Rushd's philosophy. The caliph ordered Ibn Rushd's books to be burned and exiled him from Cordova. Shortly after, the caliph relented and reinstated Ibn Rushd, who then retired to Marrakesh, where he died soon after in 1198.

Like Ibn Sīnā, before him, Ibn Rushd wrote numerous treatises on a wide range of topics, including medicine and law. But there can be little doubt that his most significant contributions were in philosophy and natural philosophy and foremost among these were his famous commentaries on the works of Aristotle. Ibn Rushd composed three types of commentary:

1. a major commentary in which he included the original text and explained it section by section;
2. a middle commentary in which Ibn Rushd often expanded on the original text with his own interpretations;
3. and a short commentary, which is really a paraphrase, or epitome, of the text.

All told, Averroes completed thirty-eight commentaries of the three kinds. "Of these, only twenty-eight are still extant in the original Arabic, thirty-six in Hebrew translations, and thirty-four in Latin translations dating from various times."[83] On some of Aristotle's works, Ibn Rushd completed versions of all three commentaries, as he did for the *Physics, Metaphysics, On the Soul (De anima), On the Heavens (De caelo),* and *Posterior Analytics.*[84]

Commentaries on Aristotle's works represent only a part of Ibn Rushd's literary output. I have already mentioned his medical and legal works. He also produced treatises that were concerned with the relationship between religion and philosophy. Thus, he wrote an important, lengthy refutation of al-Ghazālī's attack on al-Fārābī and Ibn Sīnā, which Ghazālī called *The Incoherence of the Philosophers.* In his rebuttal of al-Ghazālī, Ibn Rushd titled his work *The Incoherence of the Incoherence (Tahāfut al-Tahāfut).* Another work in this category is *A Determination of What There Is of Connection between Religion and Philosophy.*[85] George Sarton presents the following categories, and the numbers within each category, of the works written by Ibn Rushd: "(1) Philosophy, 28 items; (2) Theology, 5 items;

---

[83] Hyman and Walsh, eds., *Philosophy in the Middle Ages,* 286. For a more detailed analysis of Averroes's works, see Richard C. Taylor, "Averroes" in Jorge J. E. Gracia and Timothy B. Noone, eds., *A Companion to Medieval Philosophy in the Middle Ages* (Oxford: Blackwell Publishing Ltd., 2003), 182–195; see especially 182–185.

[84] For more on Averroes's commentaries on Aristotle, see Chapter 6.

[85] In his translation of the work, George Hourani gives as the full English title: *The book of the decision (or distinction) of the discourse, and a determination of what there is of connection between religion and philosophy.* See George F. Hourani, *Averroes On the Harmony of Religion and Philosophy* (London: Luzac & Co, 1976), 1.

(3) Law, 8 items; (4) Astronomy, 4 items; (5) Grammar, 2 items; (6) Medicine, 20 items."[86]

No one in the Middle Ages, whether in Islam or the Latin West, produced a more thorough and extensive analysis of Aristotle's thought than did Ibn Rushd. He was a model for the West on how to analyze Aristotle's ideas and concepts. It is commonplace to observe that Ibn Rushd's greatest influence on the history of natural philosophy occurred in the West, while his influence within Islam was minimal. But it is necessary to add that Averroes was a controversial figure in the West from the time Michael Scot began to translate his works into Latin around 1220 and Hermann the German in 1240.[87] As a dedicated Aristotelian, who rejected the Neoplatonic intrusions that had long been a part of the Islamic interpretation of Aristotle's thought, Ibn Rushd held doctrines that were offensive to both Muslims and Christians. So threatening were Averroes's ideas in the West that the bishop of Paris condemned some of them in 1277. The history of Averroes's thought in the West, which gave rise to a movement designated as "Latin Averroism," was tempestuous. But that is another story. Now, it is his place in the history of Islamic natural philosophy that is of immediate concern. What was it about Ibn Rushd's ideas that made him so controversial, and in the end, deprived him of any significant influence in the history of Islamic thought? Much, indeed most, of what Ibn Rushd had to say in his numerous Aristotelian commentaries was noncontroversial and would have been of little consequence to the wider world. But aspects of Aristotle's natural philosophy impinged on theological matters and inevitably aroused the ire of orthodox theologians.

As do all Muslims, Ibn Rushd regarded the Qur'an as infallible. But he also believed in one truth, which, in his judgment, was derived from two primary sources: philosophy and Scripture, or reason and revelation.[88] In arriving at truth, they were equal. Even if they agreed, few Islamic natural philosophers would have enunciated such beliefs as blatantly as did Ibn Rushd.[89] But Ibn Rushd based his judgment on a passage in the Qur'an in which it is said that some Qur'anic verses are unambiguous and others ambiguous, but "only God and those confirmed in knowledge know its interpretation [ta'wil]."[90]

---

[86] Sarton, Introduction to the History of Science, vol. 2, part 1, 359.   [87] Sarton, ibid.

[88] See Fakhry, A History of Islamic Philosophy, 308–309.

[89] Al-Kindi, as we saw, held that revelation is superior to reason and philosophy.

[90] Cited by Fakhry, A History of Islamic Philosophy, 310. The passage is from Qur'an 3, 5 and seems to have been translated by Fakhry. To give the proper context, I shall quote the relevant passage from another translation, which, however, occurs in Qur'an 3, 7. "He it is Who has revealed the Book to you; some of its verses are decisive, they are the basis of the Book, and others are allegorical; then as for those in whose hearts there is perversity, they follow the part of it which is allegorical, seeking to mislead, and seeking to give it (their own) interpretation, but none knows its interpretation except Allah, and those who are firmly rooted in knowledge say: We believe in it, it is all from our Lord; and none do mind except those having understanding." From The Qur'an, translated by M. H. Shakir; 11

For Ibn Rushd, "those confirmed in knowledge" are none other than the philosophers.

Philosophers usually regarded philosophy as the most important discipline, and, because they pursued that discipline, they considered themselves the most worthy of all beings. Al-Fārābī, for example, viewed the philosopher as the perfect man, one who could provide the proper judgments about the good life.[91] Ibn Rushd, however, provided the most elaborate argument for the high status of philosophers. In a treatise titled *A Determination of What There Is of Connection between Religion and Philosophy*, Ibn Rushd declared that philosophers, who had mastered Aristotelian demonstrative science, were the proper judges of the true inner meaning of the Qur'an. With respect to Scripture, Ibn Rushd divided all people into three classes. The first group constitutes "the rhetorical class," which includes the overwhelming mass of believers. The second class is made up of dialecticians, who would, presumably, argue on the basis of probable premises. The third class "is the people of certain interpretation: these are the demonstrative class, by nature and training, i.e. in the art of philosophy. This interpretation ought not to be expressed to the dialectical class, let alone to the masses."[92] Ibn Rushd left no doubt philosophers were the proper interpreters of Scripture, not theologians or lawyers, who could, at best, approach the Qur'an only dialectically. Not only are the philosophers the proper interpreters of Scripture, but the Law derived from Scripture makes it obligatory to study philosophy. Believing that he had shown this, Ibn Rushd declares:

Since it has now been established that the Law has rendered obligatory the study of beings by the intellect, and reflection on them, and since reflection is nothing more than inference and drawing out of the unknown from the known, and since this is reasoning or at any rate done by reasoning, therefore we are under an obligation to carry on our study of beings by intellectual reasoning. It is further evident that this manner of study, to which the Law summons and urges, is the most perfect kind of study using the most perfect kind of reasoning; and this is the kind called "demonstration."[93]

Thus did Ibn Rushd exalt the status of philosophy and philosophers. He was not alone. Al-Fārābī and others had done it before. It was an attitude that was not shared by the numerous opponents of the philosophers.

Ibn Rushd employed philosophy and natural philosophy to resolve issues that were relevant to both theology and philosophy. His interpretations

---

U.S. edition (Elmhurst, NY: Tahrike Tarsile Qur'an, Inc., 1999), 30–31. The phrase "rooted in knowledge" is equivalent to Fakhry's "confirmed in knowledge."

[91] See R. Walzer, "Al- Fārābī," in *The Encyclopaedia of Islam*, new edition, vol. 2, C–G, 779, col. 1.

[92] *Averroes On the Harmony of Religion and Philosophy*, 65. See also Gilson, *History of Christian Philosophy in the Middle Ages*, 218.

[93] Hourani, *Averroes On the Harmony of Religion and Philosophy*, 45.

would have elicited strong opposition from theologians, as well as from Neoplatonic philosophers, as Ibn Rushd was convinced that Neoplatonist philosophers like al-Fārābī and Ibn Sīnā had distorted Aristotle's thought.[94] In his *Long Commentary on the Metaphysics*, he opposed the doctrine of creation from nothing (*ex nihilo*) and upheld Aristotle, or at least found Aristotle's explanation "the least doubtful and the most congruent with the nature of being," namely, the idea that creation would have to be made from preexistent matter. He also opposed the Neoplatonic emanation theory espoused by Ibn Sīnā and al-Fārābī, primarily because it assumed that God could only create one thing, which in turn created another thing, and so on. But Ibn Rushd believed that God could produce a multiplicity of beings. However, the creation he attributed to God is an eternal one, because empty time could not have existed before the world.[95] Ibn Rushd saw nothing in the Qur'an that indicated God created the world after He had existed for a time without the world. Indeed, he saw reason to believe that the Qur'an suggested that matter and time are uncreated.[96]

Ibn Rushd briefly confronted the problem of the resurrection of the body near the end of his lengthy *The Incoherence of the Incoherence* (*Tahāfut al-Tahāfut*), which was a rebuttal of al-Ghazālī's major assault on the philosophers, titled the *Incoherence of the Philosophers*. Al-Ghazālī charged that the philosophers denied the resurrection of the body, which he regarded as blatant heresy. Ibn Rushd declares that "the philosophers in particular, as is only natural, regard this doctrine [of the bodily resurrection] as most important and believe in it most, and the reason is that it is conducive to an order amongst men on which man's being, as man, depends and through which he can attain the greatest happiness proper to him. ... [97] Ibn Rushd's idea of bodily resurrection, however, could not have satisfied al-Ghazālī or the theologians. For although Ibn Rushd agrees with al-Ghazālī that the soul is immortal, "what rises from the dead," Ibn Rushd explains,

is simulacra of these earthly bodies, not these bodies themselves, for that which has perished does not return individually and a thing can only return as an image of that which has perished, not as a being identical with what has perished, as Ghazali declares. Therefore the doctrine of resurrection of those theologians who believe that the soul is an accident and that the bodies which arise are identical with those that perished cannot be true. For what perished and became anew can only be specifically, not numerically, one. ... [98]

---

[94] See R. Arnaldez, "Ibn Rushd," 915, col. 2.   [95] Arnaldez, ibid.

[96] See Fakhry, *A History of Islamic Philosophy*, 314–315.

[97] See *Averroes' Tahāfut al-Tahāfut (The Incoherence of the Incoherence)*, translated from the Arabic with Introduction and Notes by Simon van den Bergh, 2 vols. (Oxford: University Press, 1954), vol. 1, 359.

[98] *Averroes' Tahāfut al-Tahāfut*, vol. 1, 362.

Finally, I mention one more important issue on which Ibn Rushd took issue with al-Ghazālī. He cites Ghazālī's lengthy argument against the assumption of necessary causal connections in nature. To Ghazālī, necessary causal connections were a restriction on God's power. God can cause any effect He chooses to create, and He need not utilize the same cause to produce the same effect consistently. To emphasize this point, Ibn Rushd quotes lengthy passages from Ghazālī's *Incoherence of the Philosophers*. Thus, Ghazālī declares that:

A man who had left a book at home might find it on his return changed into a youth, handsome, intelligent, and efficient, or into an animal; or if he left a youth at home, he might find him turned into a dog; or he might leave ashes and find them changed into musk; or a stone changed into gold, and gold changed into a stone. And if he were asked about any of these things, he would answer: "I do not know what there is at present in my house; I only know that I left a book in my house, but perhaps by now it is a horse which has soiled the library with its urine and excrement, and I left in my house a piece of bread which has perhaps changed into an apple tree." For God is able to do all these things, and It does not belong to the necessity of a horse that it should be created from a sperm, nor is it of the necessity of a tree that it should be created from a seed; no, there is no necessity that it should be created out of anything at all. And perhaps God creates things which never existed before; indeed, when one sees a man one never saw before and is asked whether this man has been generated, one should answer hesitantly: "It may be that he was one of the fruits in the market which has been changed into a man, and that this is that man." For God can do any possible thing, and this is possible, and one cannot avoid being perplexed by it; and to this kind of fancy one may yield *ad infinitum*, but these examples will do.[99]

Ibn Rushd simply denies Ghazālī's arguments and examples. To accept them would be to deny that "every act must have an agent."[100] Moreover,

---

[99] Ibid., vol. 1, 323–324. In a note to these remarks by al-Ghazālī, Simon van den Bergh compares Ghazālī's deliberately chosen bizarre examples with similar remarks by Sextus Empiricus (fl. 250 AD), the Greek sceptic. For these remarks, Van de Bergh refers to Sextus's *Adv. Phys.* I. 202–204 (that is, *Against the Physicists*), "where it is said that if there were no causes anything might come from anything at any time and place; a horse might come from a man, a plant from a horse, snow might congeal in Egypt, there might be a drought in Pontus, things happening in summer might occur in winter and vice versa; and again, *Hyp. Pyrrh.* iii, 18, where we have again as an example the horse which might come from mice, or, as another example, elephants that might come from ants." For the translation of the passage from *Against the Physicists* in the Loeb Classical Library, see *Sextus Empiricus with an English Translation* by the Rev. R. G. Bury: *Against the Physicists; Against the Ethicists*, 103. The translation is substantially the same as that of Van den Bergh. The second passage to which van den Bergh refers occurs in Sextus's *Outlines of Pyrrhonism*, III, 18, where in Bury's translation, we read: "Moreover, if cause were non-existent everything would have been produced by everything and at random. Horses, for instance, might be born, perchance, of flies, and elephants of ants." See *Sextus Empiricus: Outlines of Pyrrhonism*, with an English Translation by R. G. Bury (Cambridge, MA: Harvard University Press, 1933), 337.

[100] Ibid., vol. 1, 318.

God would not alter his course of action to perform the kinds of acts that Ghazālī invented.[101] Ibn Rushd viewed al-Ghazālī's assault on causal connections as subversive of all knowledge. In his counterattack, he declares:

Logic implies the existence of causes and effects, and knowledge of these effects can only be rendered perfect through knowledge of their causes. Denial of cause implies the denial of knowledge, and denial of knowledge implies that nothing in this world can be really known, and that what is supposed to be known is nothing but opinion, that neither proof nor definition exist, and that the essential attributes which compose definitions are void. The man who denies the necessity of any item of knowledge must admit that even this, his own affirmation, is not necessary knowledge.[102]

Ibn Rushd's reply to Ghazālī is largely based on Aristotle's natural philosophy in which causal connections are essential for nature's regular operation.

Ibn Rushd was the last of the great natural philosophers we have thus far considered. The philosophical opinions and interpretations of al-Kindī, al-Fārābī, Ibn Sīnā, and Ibn Rushd, were relatively mild by comparison to that of the physician and philosopher, al-Rāzī. Before describing the views of al-Rāzī, however, brief mention must be made of al-Rāzī's genuine predecessor, Ibn al-Rāwandī (d. ca. 910).

Ibn al-Rāwandī and al-Rāzī

Early in his life, Ibn al-Rāwandī was a Mutazilite scholar, who, like all Mutazilite scholars, sought to apply Greek philosophy to explicate Islamic theology. After rejecting Mutazilism, he turned for a while to Shi'ism. At some point, however, and for reasons that are apparently unknown, al-Rāwandī became a free thinker and repudiated Islam and revealed religion. Ibn al-Rāwandī's works have not survived, perhaps because of his antireligious views. Some of his opinions, however, have been preserved. Drawing

---

[101] See ibid., vol. 1, 333.

[102] Ibid., vol. 1, 319. In a note to Ibn Rushd's remarks, Simon van den Bergh explains (vol. 2, 179) that "This is a well known dictum... Sextus Empiricus says (Adv. Phys. I.204 and Hyp. Pyrrh. ii 19; 23) that the man who denies cause does so either without a cause or with a cause–but in the former case his assertion is worthless." In Bury's translation in the passage in Against the Physicists, Sextus says: "Also, he who says that cause does not exist says so either without a cause or with some cause. And if he does so without any cause, he is untrustworthy, besides the consequence he incurs of not maintaining this position any more than its opposite, as there pre-exists no reasonable cause which makes him say that cause is non-existent. But if he says so with some cause, he is self-refuted, and in the act of saying that no cause exists he is affirming the existence of some cause." See Sextus Empiricus: Against the Physicists; Against the Ethicists, 103–105. The passages in Bury's translation of the relevant texts in Outlines of Pyrrhonism are too lengthy to include here. They occur in Sextus Empiricus: Outlines of Pyrrhonism. With an English translation by R. G. Bury (Cambridge, MA: Harvard University Press, 1933), 163–165, 167.

on a variety of sources, Majid Fakhry presents the following as al-Rāwandī's judgments about religion and Islam:

> Ibn al-Rāwandī denounced the whole fabric of revelation as superfluous. He is reported to have argued that human reason was sufficient to determine the knowledge of God and the distinction between good and evil, a view in keeping with the teachings of the majority of the Mu'tazilah to whom he was originally affiliated; revelation therefore was altogether unnecessary, and miracles, upon which the claims of prophecy are alleged to rest, were altogether absurd. The most important miracle from the Islamic point of view, that of the inimitable literary perfection of the Koran, is quite untenable, according to him, since it is not beyond reason that an Arab (i.e., Muhammad) should so excel all other Arabs in literary proficiency that his work would be unquestionably the best. Yet this excellence would not necessarily involve any extraordinary or miraculous character in his output. Nor can it be denied that this alleged literary miraculousness is hardly relevant, as probative evidence, in regard to foreigners to whom Arabic is an alien tongue.[103]

Despite his unusual beliefs, the fact that his works have not survived means that the little that is known about al-Rāwandī comes to us indirectly, probably from the writings of his critics and enemies.

It is quite otherwise with another extraordinary free thinker, Abū Bakr Muḥammad b. Zakariyā' al-Rāzī, known as al-Rāzī and as Rhazes in the Latin West. He has been described as "the greatest nonconformist in the whole history of Islam and undoubtedly the most celebrated medical authority in the tenth century."[104] A prolific author, al-Rāzī claimed to have written more than two hundered works, very few of which have been preserved, probably for reasons similar to those that may have caused al-Rāwandī's works to disappear. Indeed, in one year, he claimed to have written the staggering number of twenty thousand pages.[105] He wrote a great compendium of medicine called *al-Ḥāwī*, which was translated into Latin in 1279 under the title *Liber continens*, a work widely used in Western Europe until the sixteenth century. Among his other medical works is the first treatise on smallpox, called *Smallpox and Measles*. Al-Rāzī's undogmatic approach and his splendid clinical observations mark his medical works as among the best in the Middle Ages.

He wrote numerous works on natural philosophy, including a commentary on Plato's *Timaeus*. Although he does not seem to have written commentaries on Aristotle's works, he did compose treatises that were concerned with major themes in Aristotle's natural philosophy, as, for example, his *Absolute and Particular Matter, Plenum and Vacuum, Fire and Space, That the World Has a Wise Creator*, and *On the Eternity and Noneternity of Bodies*. Al-Rāzī is noteworthy, because he disagreed with Aristotle on major points about the

---

[103] Fakhry, *A History of Islamic Philosophy*, 114–115.   [104] Ibid., 115.
[105] See L. E. Goodman, "al-Rāzī," in *The Encyclopaedia of Islam*, vol. 8 (NED-SAM), 474, col. 2.

nature of the cosmos. He boldly adopted the Greek atomistic theory identified with the names of Democritus and Leucippus, a theory that Aristotle had
gone to great lengths to refute. Thus, although Aristotle denied the existence
of void, al-Rāzī assumed that bodies consisted of atoms and void space.[106]
He assumed that the space in which the world and all bodies are located
is infinite. Infinite, universal space thus includes both atomic matter and
void.[107]

Although he accepted Aristotle's theory of the four elements – earth, water,
air, and fire – al-Rāzī departed from Aristotle on the nature and motion of
those elements. Motion is not determined by the natural places of the four
elements, either up or down. Motion was, rather, an attribute of a body, with
all bodies tending to move naturally downward toward the center of the
world.[108] For al-Rāzī "the qualities of the four elements ... as well as those
of heaven – that is their lightness, heaviness, brightness, darkness, softness,
hardness, etc. – are determined by their greater or lesser density, that is, by
the number and size of the vacant gaps."[109] By adopting a cosmic system
built on eternal atoms and void, al-Rāzī was inevitably led to a rejection of
most of Aristotle's physics and cosmology.

By adopting atomistic physics, al-Rāzī was led to disagree with some
of the most important theological and religious ideas in the Islamic faith.
Although he accepted the creation of the world, al-Rāzī rejected the idea
of creation from nothing and argued that God did not have the power to
create from nothing.[110] God created from preexistent, eternal atoms moving
in the void. Al-Rāzī found other reasons to depart from the Islamic faith. He
believed that philosophers should reject religious messages from the prophets
of the three monotheistic religions – Moses, Jesus, and Muḥammad – arguing that a philosopher should not accept prophecies made by great leaders.
"How can anyone think philosophically," he declared, "while committed to
those old wives' tales, founded on contradictions, obdurate ignorance, and
dogmatism."[111] Because al-Rāzī included Muḥammad in his blanket condemnation of prophets, one can only wonder how, if his opinions were
known, he survived within a Muslim society. If he were a medieval Christian
with analogous opinions, it is not likely that he would have survived the
wrath of the Church.

---

[106] See Shlomo Pines, *Studies in Islamic Atomism*, translated from German by Michael Schwarz,
edited by Tzvi Langermann (Jerusalem: The Magnes Press, The Hebrew University, 1997),
48.

[107] Pines, ibid., 55.

[108] See Fakhry, *A History of Islamic Philosophy*, 118–119; and Pines, *Studies in Islamic Atomism*, 50.

[109] Pines, ibid., 50.   [110] Ibid., 48–49.

[111] Goodman, "al-Rāzī," 476, col. 1. Goodman gives a transliteration of an Arabic title. Perhaps
the passage cited here is from a lost treatise attributed to al-Rāzī, titled *The Tricks of the
Prophets*. See Shlomo Pines, "Al-Rāzī, Abū Bakr Muḥammad ibn Zakariyyā," in *Dictionary
of Scientific Biography*, vol. 11 (1975), 323, and also R. Walzer, "Early Islamic Philosophy,"
651.

Al-Rāzī firmly believed that a philosopher should formulate his own ideas, and not follow the ideas and opinions of a master. He regarded controversy and disagreement on philosophical issues as natural and healthy for the advance of the subject. By learning and retaining knowledge that our predecessors knew, a philosopher could surpass them. "For inquiry, thought and originality," al-Rāzī insisted, "make progress and improvement inevitable."[112] He refused to accept authority in either religion or science and believed that the sciences continually progressed because scientists build upon the knowledge they inherit from their predecessors.

There were, of course, other lesser natural philosophers than the five major figures discussed in this chapter. They undoubtedly shared many basic intellectual features with their more famous colleagues. In presenting brief descriptions of the attitudes and opinions of the five great natural philosophers, it becomes apparent why more traditional minded theologians were disturbed by the way Aristotelian and Neoplatonic natural philosophers viewed the relations between reason and revelation. Natural philosophers seemed to interpret basic Islamic doctrines, such as the creation of the world and the resurrection of the body, in ways that seemed contrary to traditional interpretations. As a consequence, it was not uncommon for religious-minded traditionalists to attack natural philosophy and natural philosophers.

The Assault against the Philosophers: al-Ghazālī (1058–1111); Ibn aṣ-Ṣalah ash-Shahrazūrī (d. 1245); and Ibn Khaldūn (1332–1406)

The most prominent critic of the philosophers was al-Ghazālī, who, in addition to his specific attacks against al-Fārābī and Ibn Sīnā (mentioned earlier), formulated a more general assault on philosophy in his famous quasi-autobiographical treatise, *Deliverance from Error*. Al-Ghazālī was fearful of the detrimental effects on the Islamic religion of subjects like natural philosophy, metaphysics (he calls it theology), logic, and mathematics. In *Deliverance from Error*, he explains that religion does not require the rejection of natural philosophy, but that there are serious objections to it because nature is completely subject to God, and no part of it can act from its own essence. The implication is obvious: Aristotelian natural philosophy is unacceptable because it assumes that natural objects can act by virtue of their own essences and natures. That is, Aristotle believed in secondary causation – namely, that physical objects are capable of causing effects in other physical objects. For al-Ghazālī, as we saw, and for many other Muslim philosophers and theologians, God is assumed the direct cause of all effects. There is no secondary causation.

Al-Ghazālī found mathematics dangerous because it uses clear demonstrations, thus leading the innocent to think that all the philosophical sciences are

---

[112] Goodman, "al-Rāzī," 476, col. 1.

equally lucid. A man will say to himself, declares al-Ghazālī that "if religion were true, it would not have escaped the notice of these men [that is, the mathematicians] since they are so precise in this science."[113] Ghazālī explains further that such a man will be so impressed with what he hears about the techniques and demonstrations of the mathematicians that "he draws the conclusion that the truth is the denial and rejection of religion. How many have I seen," al-Ghazālī continues, "who err from the truth because of this high opinion of the philosophers and without any other basis."[114] Although al-Ghazālī allowed that the subject matter of mathematics is not directly relevant to religion, he included the mathematical sciences within the class of philosophical sciences (these are mathematics, logic, natural science, theology or metaphysics, politics, and ethics) and concluded that a student who studied these sciences would be "infected with the evil and corruption of the philosophers. Few there are who devote themselves to this study without being stripped of religion and having the bridle of godly fear removed from their heads."[115]

In his great philosophical work, *The Incoherence of the Philosophers*, al-Ghazālī attacked ancient philosophy, especially the views of Aristotle. He did this, as we saw, by describing and criticizing the ideas of al-Fārābī and Ibn Sīnā, two of the most important Islamic philosophical commentators on Aristotle. After criticizing their opinions on twenty philosophical problems, including these three: eternality of the world; that God knows only universals and not particulars; and that bodies will not be resurrected after death, al-Ghazālī declares:

All these three theories are in violent opposition to Islam. To believe in them is to accuse the prophets of falsehood, and to consider their teachings as a hypocritical misrepresentation designed to appeal to the masses. And this is blatant blasphemy to which no Muslim sect would subscribe.[116]

Al-Ghazālī regarded theology and natural philosophy as dangerous to the faith. He had an abiding distrust of philosophers and praised the "unsophisticated masses of men," who "have an instinctive aversion to following the example of misguided genius." Indeed, "their simplicity is nearer to salvation than sterile genius can be."[117] As one of the greatest and most respected thinkers in the history of Islam, al-Ghazālī's opinions were not taken lightly.

In view of al-Ghazālī's attack on the philosophers, it is not surprising to learn that philosophers were often subject to persecution by religious leaders. Many religious scholars regarded philosophy, logic and the foreign

---

[113] Translated in M. Montgomery Watt, *The Faith and Practice of al-Ghazali* (London: George Allen and Unwin Ltd, 1953), 33.

[114] Ibid.   [115] Ibid., 34.

[116] From *Al-Ghazali's Tahafut al-Falasifah [Incoherence of the Philosophers]*, translated into English by Sabih Ahmad Kamali (Pakistan Philosophical Congress Publication, No. 3, 1963), 249.

[117] Ibid., 3.

Greek sciences generally, as useless, and even ungodly, because they were not directly useful to religion. Indeed, they might even make one disrespectful of religion,[118] as al-Ghazālī argued. In the thirteenth century, Ibn aṣ-Ṣalāḥ ash-Shahrazūrī (d. 1245), a religious leader in the field of tradition (hadith), declared in a *fatwa* that "He who studies or teaches philosophy will be abandoned by God's favor, and Satan will overpower him. What field of learning could be more despicable than one that blinds those who cultivate it and darkens their hearts against the prophetic teaching of Muhammad...."[119] Logic also was targeted, because, as Ibn aṣ-Ṣalāḥ, put it, "it is a means of access to philosophy. Now the means of access to something bad is also bad."[120] Ibn aṣ-Ṣalāḥ was not content to confine his hostility to words. In a rather chilling passage, he urges vigorous action against students and teachers of philosophy and logic, because

Those who think they can occupy themselves with philosophy and logic merely out of personal interest or through belief in its usefulness are betrayed and duped by Satan. It is the duty of the civil authorities to protect Muslims against the evil that such people can cause. Persons of this sort must be removed from the schools and punished for their cultivation of these fields. All those who give evidence of pursuing the teachings of philosophy must be confronted with the following alternatives: either (execution) by the sword or (conversion to) Islam, so that the land may be protected and the traces of those people and their sciences may be eradicated. May God support and expedite it. However, the most important concern at the moment is to identify all of those who pursue philosophy, those who have written about it, have taught it, and to remove them from their positions insofar as they are employed as teachers in schools.[121]

Although numerous others shared the attitude of Ibn aṣ-Ṣalāḥ, logic continued to be used as an ancillary subject in scholastic theology (Kalam) and in many orthodox religious schools. But there was enough hostility toward philosophy and logic in Islam to prompt philosophers to keep a low profile. Ibn Khaldūn (1332–1406), perhaps the greatest intellect produced by medieval Islam, was convinced, as were al-Ghazālī and Ibn aṣ-Ṣalāḥ before him, that philosophy and logic were great potential dangers to the Islamic religion. In his *Muqaddimah* (*Introduction to History*), a treatise that has been rightly described as "the first large-scale attempt to analyze the group relationships that govern human political and social organization on the basis of environmental and psychological factors,"[122] ibn Khaldūn

---

[118] Huff, *The Rise of Early Modern Science*, 70.
[119] Ignaz Goldziher, "The Attitude of Orthodox Islam Toward the 'Ancient Sciences,'" in Merlin L. Swartz, ed. and tr., *Studies on Islam* (New York/Oxford: Oxford University Press, 1981), 205.
[120] Goldziher, ibid.   [121] Goldziher, ibid., 206.
[122] Franz Rosenthal, "Ibn Khaldūn," in *Dictionary of Scientific Biography*, vol. 7 (1973), 321. Arnold Toynbee spoke in superlative terms about Ibn Khaldūn, declaring that in his *Muqaddima* "he has conceived and formulated a philosophy of history which is undoubtedly the greatest work of its kind that has ever yet been created by any mind in any time or place."

enunciated his sentiments in a chapter titled "A Refutation of Philosophy: The Corruption of the Students of Philosophy."[123] With philosophy as the subject of this chapter, followed by chapters on astrology and alchemy, respectively, ibn Khaldūn begins the chapter on philosophy with the proclamation that

This and the following (two) sections are important. The sciences (of philosophy, astrology, and alchemy) occur in civilization. They are much cultivated in the cities. The harm they (can) do to religion is great. Therefore, it is necessary that we make it clear what they are about and that we reveal what the right attitude concerning them (should be).[124]

After identifying al-Fārābī and Ibn Sīnā as the most famous of Aristotelian scholars, ibn Khaldūn declares that "it should be known that the (opinion) the (philosophers) hold is wrong in all its aspects."[125] It is best, in ibn Khaldūn's judgment, for Muslims to refrain from studying natural philosophy or physics, because "The problems of physics are of no importance for us in our religious affairs or our livelihoods. Therefore we must leave them alone."[126] The study of logic poses even greater problems, because it is inherently dangerous. "Therefore, the student of it should beware of its pernicious aspects as much as he can. Whoever studies it should do so (only) after he is saturated with the religious law and has studied the interpretation of the Qur'an and jurisprudence. No one who has no knowledge of the Muslim religious sciences should apply himself to it. Without that knowledge, he can hardly remain safe from its pernicious aspects."[127]

Higher Education: The Madrasas

The hostile attitudes toward natural philosophy that have been described here were rather common. Aristotle's logic and natural philosophy were often viewed with suspicion, because they were often concerned with themes that were regarded as subversive of religion, as, for example, the themes that al-Ghazālī regarded as heretical, namely, the eternality of the world; that God knows only universals and not particulars; and that bodies will not be resurrected after death. It is therefore not surprising that the foreign sciences, and especially Aristotle's natural philosophy, were not included as a regular feature of Muslim higher education in the madrasas, as the schools of higher education were called. A madrasa was a charitable trust, which was established freely by an individual Muslim, known as a *waqf*, who endowed the

See Arnold Toynbee, *A Study of History*, 12 vols. (Oxford: Oxford University Press, 1934–1960), vol. 3 (1934), 322.

[123] See Ibn Khaldūn, *The Muqaddimah: An Introduction to History*, translated from the Arabic by Franz Rosenthal, 3 vols. (Princeton, NJ: Princeton University Press, 1958; corrected, 1967), vol. 3, ch. 6, sec. 30, 246–258.

[124] Ibn Khaldūn, *The Muqaddimah*, vol. 3, ch. VI, sec. 30, 246.　　[125] Ibid., 250.

[126] Ibid., 251–252.　　[127] Ibid., 257–258.

trust with substantial funds to be used for a public purpose. The founder had great latitude in determining the conditions for the operation of the madrasa he had founded with his own property. "The legal status of the *madrasa* allowed the founder to retain complete control over the administrative and instructional staff of the institution."[128] But the founder of a madrasa had to accept one condition: the terms of the foundation could not violate the tenets of Islam.[129]

The madrasa was essentially a school for the study of the religious sciences and subordinate and related subjects. Excluded from its curriculum were the "foreign sciences," that is the philosophical and natural sciences.[130] The exclusion of the foreign sciences resulted from the fact that

The Islamic waqf, upon which rested the whole edifice of institutions of learning, excluded any and all things that were considered to be inimical to the tenets of Islam. Hence the exclusion of the godless "sciences of the Ancients" from the curriculum. Philosophical doctrines clashed with such monotheistic doctrines as the existence of a personal, provident, almighty God, the non-eternity of the world, and the resurrection of the body.[131]

Those who wished to study natural philosophy or the sciences for their own sakes had to either teach themselves, or make arrangements for private instruction with someone knowledgeable in such matters.[132] Makdisi explains that "there was nothing to stop the subsidized student from studying the foreign sciences unaided, or learning in secret from masters teaching in the privacy of their homes, or in the waqf institutions, outside of the regular curriculum."[133] Occasionally, nonreligious courses were taught in the madrasas on an optional basis. In his splendid book, *The Mantle of the Prophet*, Roy Mottahedeh explains that "*Madreseh* learning had formerly been a conspectus of higher learning, with its optional courses in Ptolemaic astronomy, Avicennian medicine, and the algebra of Omar Khayyam. But . . . even the mullahs recognized that their learning really was 'religious' learning, and only a few enthusiasts studied the traditional nonreligious sciences such as the old astronomy in private."[134] However, only those subjects were taught that illuminated the Qur'an or the religious law. One such subject was logic, which was found useful in semantics and in avoiding "simple errors of inference," although philosophical logic, popular in the West, was usually avoided.[135] The primary function of the madrasa, however, was "to preserve

[128] Article "Madrasa" by J. Pedersen and G. Makdisi in *Encyclopedia of Islam*, new edition, vol. 5, 1128, col. 2.

[129] See George Makdisi, *The Rise of Colleges: Institutions of Learning in Islam and the West* (Edinburgh: Edinburgh University Press, 1981), 36.

[130] Makdisi, ibid., 77.   [131] Ibid., 77–78.   [132] Ibid., 78.   [133] Ibid.

[134] See Roy Mottahedeh, *The Mantle of the Prophet* (New York: Pantheon Books, 1985), 237.

[135] On the subject of logic in Islam, see John Walbridge's excellent article, "Logic in the Islamic Intellectual Tradition: The Recent Centuries," in *Islamic Studies*, 39, Nr 1 (Spring 2000), 55–75. On attitudes toward philosophical logic, see page 68.

learning and defend orthodoxy."[136] In Iran, the madrasas existed into the twentieth century, limping on until the end of World War II.

It is difficult to imagine that the hostility exhibited toward natural philosophers by al-Ghazālī, Ibn aṣ-Ṣalāḥ, Ibn Khaldūn, and others, did not discourage scholars from studying and writing about natural philosophy. It certainly put prospective students of the subject on the defensive. Nevertheless, logic continued to be used as an ancillary subject in scholastic theology (Kalam) and in many orthodox religious schools. But there was enough hostility toward philosophy and logic in Islam to prompt philosophers to keep a low profile. Because natural philosophy in Islam, was never a regular part of the curriculum in the madrasas, those who wished to study natural philosophy found suitable teachers and studied privately with them. As a consequence, schools and traditions of philosophy failed to develop in any meaningful manner. Dedication to the discipline of natural philosophy was, for its devotees, mostly a private, independent matter.

The Marginal Existence of Islamic Natural Philosophy

From the ninth to twelfth centuries, Islam produced some great natural philosophers. There is no other way to characterize al-Kindī, al-Fārābī, Ibn Sīnā, Ibn Rushd, and al-Rāzī. They were equal to the best that was yet to come in the Latin West. And yet natural philosophy never took root in Islam. By comparison to jurisprudence and theology, natural philosophy remained a peripheral activity. This may be partially explicable by the fact that natural philosophy was always regarded as a "foreign science," an alien intrusion into Islam from the pagan Greeks, in contrast to the Islamic sciences associated with the Islamic religion itself. Moreover, as we saw, Aristotelian and Neoplatonic natural philosophy contained a number of elements – the eternity of the world and denial of bodily resurrection, to name two – that were in direct conflict with Islamic religious doctrine.

But al-Ghazālī may have hit on the most worrisome aspect of natural philosophy: its great emphasis on reason and logical argument, which made followers of traditional theology seem old fashioned and unsure of themselves. Indeed, as Ibn Rushd explained, philosophical interpretations of passages in the Qur'an that involved demonstrations could only be presented to other philosophers, or those who were trained in philosophical argumentation, but not to theologians or the mass of common people. As A. J. Arberry has put it,

The Moslem philosophers cheerfully advocated the expediency of permitting one truth for the masses, and another truth for the elect. The theologians saw clearly enough where that kind of double-talk might also lead; the only safe course, as they

---

[136] Mottahedeh, *The Mantle of the Prophet*, 91.

thought, and perhaps rightly in a world menaced by political disruption and beset by growing doubt, was to uphold the pure tradition of one truth sufficient for all men, the truth of the Koran. That was God's undoubted speech, as communicated to His chosen Messenger; and the plain words of the Almighty were a surer guide for perplexed humanity than all the airy theorisings of Plato and Aristotle and their latter day exponents.[137]

Philosophers not only rested their case on philosophy and its methods of analysis, but they emphasized the inferior intellectual status of theologians and the multitudes that accepted all things on faith. In the light of this ever-present hostility between religious traditionalists and philosophers, it is not surprising that natural philosophy never became an integral part of the educational curriculum of the madrasas. It was never institutionalized within Islamic higher education, as it would be during the thirteenth to fifteenth centuries in the Latin West. As a consequence, natural philosophy remained a peripheral and ephemeral activity in Islam. The spirit of reasoned inquiry that developed within natural philosophy in the West established a culture of probing and investigating nature that spilled over into the physical sciences, where it became a regular feature of Western thought. The neglect of natural philosophy in Islam, and frequently open hostility toward it, eventually led to stagnation in both natural philosophy and science.

With the death of Ibn Rushd in 1198, Islamic natural philosophy went on a downward course. But the quality of Islamic natural philosophy from the ninth to the end of the twelfth century would have led us to expect further significant, and even unusual, accomplishments. There can be little doubt that Islamic natural philosophy as exhibited in the works of al-Kindī, al-Fārābī, Ibn Sīnā, Ibn Rushd, and al-Rāzī was far more daring than anything produced in the West between the thirteenth and fifteenth centuries. There was no challenge to the theologians in the West as we find it in the works of Ibn Rushd. There were no defenses of the eternity of the world in the West, as there were in Islam; no one in the West denied the resurrection of the body as was done by Ibn Sīnā and Ibn Rushd; natural philosophers in the West never dared to challenge the theologians on interpretations of Scripture, as Ibn Rushd did within Islam; and, finally, no natural philosophers in the West would have dared to reject the prophets and repudiate revelation, as did al-Rāzī.

And yet, for reasons already given, natural philosophy in Islam stagnated and was little studied. It never became part of the main stream of Islamic education, and therefore never found a significant place in Islamic thought. Natural philosophy was most actively pursued from the ninth to twelfth centuries and thereafter was largely ignored; perhaps because it was often viewed with suspicion as a potential threat to revealed truth. The emphasis on reason, and the implication, or outright proclamation by some natural

---

[137] Arberry, *Revelation and Reason in Islam*, 56.

philosophers, that reason was equal, or superior, to revelation gave traditionalists and theologians constant cause for concern about a discipline that was, in any event, a "foreign science," alien to fundamental Islamic traditions.

This might explain the decline of interest in natural philosophy in the civilization of Islam. But there is yet another kind of explanation as to why Islamic intellectuals turned away from the reason and logical argumentation that was characteristic of Aristotelian natural philosophers, and, in conjunction with the neglect of natural philosophy, also turned away from the kind of science that was developed subsequently in the West. Let me conjecture one explanation out of perhaps many possibilities.

A recent investigation into cultural differences may offer support to those who would distinguish between Western and Islamic science along cultural lines. Dr. Richard Nesbitt, a social psychologist at the University of Michigan, and his colleagues challenge the widely held view among Western philosophers and psychologists that "the same basic processes underlie all human thought, whether in the mountains of Tibet or the grasslands of the Serengeti."[138] The basic processes that all humans followed were alleged to embrace "a devotion to logical reasoning, a penchant for categorization and an urge to understand situations and events in linear terms of cause and effect." However, in comparing East Asians and European Americans, Dr. Nesbitt and his colleagues arrived at a radically different assessment. They "found that people who grow up in different cultures do not just think about different things: they think differently." Easterners, they discovered, "appear to think more 'holistically,' paying greater attention to context and relationship, relying more on experience-based knowledge than abstract logic and showing more tolerance for contradiction. Westerners are more 'analytic' in their thinking, tending to detach objects from their context, to avoid contradictions and to rely more heavily on formal logic." The cultural differences that Dr. Nesbitt found between East Asians and European Americans also may apply to a comparison between medieval Islam and Western Europe in the late Middle Ages. If so, this might account for the lack of enthusiasm and support for natural philosophy in Islamic culture. But even if the comparison is inappropriate, Dr. Nesbitt's characterization of "European Americans" is wholly applicable and appropriate for the late Middle Ages in Western Europe. Let me now show why this is true.

[138] I rely here on the article "How Culture Molds Habits of Thought" by Erica Goode in the Science Times section of the *New York Times* for August 8, 2000. All the quotations in this paragraph are from Ms. Goode's article.

5

Natural Philosophy before the Latin Translations

Aristotle's natural philosophy and Greek science generally did not begin to enter Western Europe until the middle of the twelfth century. Before that time, only a minuscule part of Greek science and natural philosophy was available in the West. The Romans had not been sufficiently interested in such subjects to translate relevant Greek texts into Latin. But some Roman authors wrote treatises that modern scholars would regard as primarily concerned with natural philosophy. Among those in this group who have left extant treatises, the most famous are Lucretius (ca. 95–ca. 55 BC), Seneca (ca. 4 BC–AD 65), and Pliny the Elder (ca. AD 23–79).

ROMAN AUTHORS

The fame of Lucretius derives from his great poem, *On the Nature of Things*, which presents a picture of nature based on the atomic theory of Epicurus (341–270 BC), a Greek philosopher. Lucretius dealt with many topics, but the most important was his cosmic vision based on an assumption of an infinity of worlds, each composed of atoms moving in an infinite void space. He assumed that each world comes into being by a chance coming-together of atoms in the void; eventually each world passes away when its atoms dissociate and move into the void to form parts of other worlds.[1] Lucretius was largely ignored by medieval Christianity, because of his attacks on religion and his denial of a created world, as we see in his statement that "our starting-point will be this principle: *Nothing can ever be created by divine power out of nothing.*"[2]

---

[1] For a fine, brief account of Lucretius's *On the Nature of Things*, see Marshall Clagett, *Greek Science in Antiquity* (London: Abelard-Schuman Ltd, 1957), 101–104. Also in brief compass, but equally informative, is David C. Lindberg's account of Epicurus's atomic theory in his *The Beginnings of Western Science: The European Scientific Tradition in Philosophical, Religious, and Institutional Context, 600 A.C. to A.D. 1450* (Chicago: University of Chicago Press, 1992), 77–80.

[2] *Lucretius, "On the Nature of the Universe,"* translated and introduced by R. E. Latham (Harmondsworth, Middlesex, England: Penguin Books, 1968; first published 1951), Book I, 31.

Although Lucretius's treatise illustrates a Roman interest in Greek philosophy, it had little subsequent influence in the Middle Ages. Seneca and Pliny the Elder exercised considerable influence on the discussions of natural philosophy that followed in the early Middle Ages. In his *Natural Questions*, Seneca, a Roman Stoic, seems to have indirectly drawn on Aristotle's *Meteorology* (through Posidonius's book on meteorology)[3] and *On the Heavens*, along with a few other Greek sources. In a rather disorganized style, Seneca covered topics that Aristotle had included in the two treatises just mentioned. Thus he considered topics in the upper atmosphere that Aristotle discussed, such as comets, meteors, rainbows, thunder, and lightning; and also described earthquakes here below. Seneca was given to drawing morals from natural phenomena, much as Christian authors did in later centuries.[4] The apparent purpose of his treatise was to use natural philosophy to underwrite religion and morality.

Pliny the Elder wrote a lengthy treatise titled *Natural History* in thirty-seven books. The *Natural History* is a vast encyclopedia of knowledge about the natural world crammed with thousands of tidbits of information and misinformation. Most remarkable is the fact that so many manuscript copies of the work exist, for as one author has remarked "the usual fate of voluminous – and popular – works has been to become fragmented or to undergo abridgments and epitomes of abridgments; in later ages when scribes did not have the means or fortitude for long copying nor readers the stamina or motivation for reading, the original work would disappear."[5] But Pliny's work survived in full, along with various abbreviated versions.

Among authors in the ancient world, Pliny was unusual, because he scrupulously cited his numerous sources. Among the topics he included were meteorology, astronomy, geography, zoology, botany, pharmacology, medicine, and mineralogy. The arrangement of the books is as follows: Book I is dedicatory and includes a table of contents; Book II is astronomical; Books III–VI are on geography; Book VII is on man and his inventions; Books VIII–XI are devoted to zoology; Books XII–XIX treat botany; Books XX–XXVII describe medicines made from plants; Books XXVIII–XXXII treat of materia medica made from animal sources; and Books XXXIII–XXXVII are devoted to metals and stones used in medicine, architecture, and art.[6]

In his preface to the work, Pliny comments on the novelty of his encyclopedia, which no Greek or Roman had ever done. It would influence many authors in the centuries to come. It was a natural philosophy of facts and

[3] See William H. Stahl, *Roman Science: Origins, Development and Influence to the Later Middle Ages* (Madison: University of Wisconsin Press, 1962), 46.
[4] See Clagett, *Greek Science in Antiquity*, 109.   [5] Stahl, *Roman Science*, 101.
[6] See Clagett, *Greek Science in Antiquity*, 110–111; and David E. Eichholz, "Pliny (Gaius Plinius Secundus)," in Charles C. Gillispie, ed., *Dictionary of Scientific Biography*, 18 vols. (New York: Charles Scribner's Sons, 1970–1990), vol. 11 (1975), 39.

information, with relatively little theoretical structure. It was worthy of the concrete, practical Roman approach to the world.

Although Christianity was already in existence when Seneca and Pliny lived and wrote, neither of them was a Christian. Of the authors who wrote on natural philosophy in Latin after the advent of Christianity, but prior to the age of translation in the twelfth century, some were Christians, many were not, at least before the sixth century. And some of those who may have been Christians show little or no religious influence in their written treatises. But whether or not they were Christians, it is important to determine what authors in the early Middle Ages – before the influx of Greek and Arabic science and natural philosophy reshaped learning in Western Europe – regarded as natural philosophy. This is a troublesome problem, one that can only be answered by examining works that seem to have been regarded by their authors as in the domain of natural philosophy. The task is made difficult because there is no term for *natural philosophy*, or *physics*, in the Latin West before the translating activity of the twelfth century.[7] To facilitate matters, I shall attempt to judge the extent to which an author seems intent on describing the operations of the physical universe. This will be partly determined by inspecting the topics included.

THE LATIN ENCYCLOPEDISTS: EUROPEAN LEARNING TO THE NINTH CENTURY

An important group of authors in the Latin West during the fourth to eight centuries is known collectively as the Latin Encyclopedists, because they wrote encyclopedic treatises in Latin. In this group are Calcidius (fl. fourth or fifth century AD), Macrobius (fl. Early Fifth Century AD), and Martianus Capella (fl. ca. 365–440), who are all regarded as Neoplatonists. Their writings were the vehicle for the transmission of Platonic cosmography, Calcidius being the most significant because he not only translated two-thirds of Plato's *Timaeus*, which is Plato's major contribution to natural philosophy, but also wrote a widely used Latin commentary on it.[8] To these three, we should add Boethius (ca. 480–525), Cassiodorus (ca. 480–ca. 575), Isidore of Seville (ca. 560–636), and Venerable Bede (672–735). These authors, more than any others, provided the early Middle Ages with its intellectual content, a content in its physical aspects that was largely Platonic and Neoplatonic, with little influence from the works of Aristotle.

A word of caution is necessary, however. Although individual Latin Encyclopedists mention the great philosophers, especially Plato and Aristotle, they were not directly familiar with their works on natural philosophy. They derived their meager knowledge of Plato and Aristotle from a handbook

---

[7] As we saw in Chapter 2, Aristotle used the term *physics* to signify what later came to be called *natural philosophy*.

[8] See Stahl, *Roman Science*, 142.

tradition that had been developed since Hellenistic times. Latin Encyclopedists were more likely to have been influenced by the Platonic and Neoplatonic traditions as these were formed in late antiquity from the writings
of Plotinus and Porphyry. But even the thoughts of Plotinus and Porphyry
probably reached the Latin Encyclopedists after filtering through numerous
summarizers and compilers.

Let us now see what may have passed for natural philosophy in the early
Middle Ages, before the term "natural philosophy" was introduced. To do
this, I shall briefly describe the contents of a few encyclopedic works and seek
to determine whether part or all of them were devoted to what we might
want to characterize as natural philosophy.

Macrobius (fl. Early Fifth Century AD)

Macrobius's *Commentary on the Dream of Scipio* was one of the most important and influential treatises written in the early Middle Ages. From his opening remarks, Macrobius indicates that he did not intend his *Commentary* as
a treatise on natural philosophy. This is made evident when he informs his
readers that his intent is to present a commentary on Plato's *Republic* and
Cicero's *Republic*, of which two treatises "the former drafted plans for the
organization of a state, the latter described one already in existence; the
one discussed an ideal state, the other the government established by his
forefathers."[9] In Plato's *Republic*, Plato reveals "the conditions of souls lib-
erated from their bodies, introducing as well an interesting description of
the spheres and constellations." Macrobius tells us that Cicero covered the
same subjects in his treatise, except that Cicero used the device of revelations
that came to Scipio in a dream. Macrobius then declares that:

The reason for including such a fiction and dream in books dealing with governmental
problems, and the justification for introducing a description of celestial circles, orbits,
and spheres, the movements of planets, and the revolutions of the heavens into a
discussion of the regulations governing commonwealths seemed to me to be worth
investigating; and the reader, too, will perhaps be curious. Otherwise we may be
led to believe that men of surpassing wisdom, whose habit it was to regard the
search for truth as nothing if not divine, have padded their treatises, nowhere else
prolix, with something superfluous, so that the reader may clearly comprehend what
follows.[10]

Macrobius sees his own commentary as essentially concerned with treatises about government by Plato and Cicero. He wishes to investigate why
they included cosmological information about orbits, spheres, and planets.
They did so, apparently, because immortal souls and numbers determine the

---

[9] Macrobius, book 1, ch. 1 in *Macrobius: Commentary on the Dream of Scipio*, translated
with an Introduction and Notes by William Harris Stahl (New York: Columbia University
Press, 1952), 81.

[10] *Macrobius: Commentary on the Dream of Scipio*, book 1, ch. 1, 81.

nature of our universe and the latter is but a manifestation of the nature of soul and numbers. Thus Macrobius joins Pythagorean number lore to natural philosophy. He speaks of the combination of two and five. The dyad refers to the corporeal sphere below the celestial sphere, which represents matter, change, and corruption. The number five "embraces all things that are and seem to be. . . . Consequently, this number designates at once all things in the higher and lower realms." Macrobius then gives the five things which the number five designates:

There is the Supreme God; then Mind sprung from him, in which the patterns of things are contained; there is the World-Soul, which is the fount of all souls; there are the celestial realms extending down to us; and last, the terrestrial realm; thus the number five marks the sum total of the universe.[11]

The number seven has enormous powers. "It was by this number first of all, indeed," Macrobius informs us, "that the World-Soul was begotten, as Plato's *Timaeus* has shown."[12] Moreover, "the Creator, in his constructive foresight, arranged seven errant spheres, beneath the star-bearing celestial sphere, which embraces the universe, so that they might counteract the swift motion of the sphere above and govern everything beneath."[13] Macrobius also links numbers with the four elements and their qualities, ideas that he draws ultimately from Plato's *Timaeus*.[14] Thus did Macrobius intertwine soul and number with the physical universe and, as Plato did, link them intimately to natural philosophy.

Macrobius had a great impact on the Middle Ages. Like most of the encyclopedists, he was a compiler and drew information from a great variety of sources, usually without acknowledgment. As the vehicle for his thoughts, Macrobius wrote a *Commentary on the Dream of Scipio*. The *Dream of Scipio* was the concluding part of Cicero's *Republic*. Macrobius used it to convey special aspects of his Neoplatonic philosophy. William Stahl explains that Macrobius's *Commentary*:

is seventeen times as long as the *Dream of Scipio*. However, because it is a lucid and compendious exposition of Neoplatonic doctrine and contains lengthy excursuses on such popular topics as dreams, Pythagorean number lore, cosmography, and world geography, it was a fascinating book to medieval readers. Most important of all, it was responsible for the preservation of Cicero's *Dream*.[15]

It is an emphasis on cosmography and world geography that makes Macrobius's Commentary relevant to natural philosophy. Nearly half of the treatise – seventeen chapters – is devoted to cosmography. These chapters sometimes circulated as a separate treatise, or were bound with other astronomical bits and pieces to form a seemingly separate work.[16]

---

[11] *Macrobius*, ibid., 104.   [12] *Macrobius*, ibid., 109.   [13] Ibid.   [14] Ibid., 104–107.
[15] Stahl, ibid., 155.   [16] Ibid., 156.

Macrobius probably drew his knowledge of cosmology, or cosmography, from Porphyry's lost commentary on Plato's *Timaeus*.[17] But Macrobius included much information about the heavens, asserting such frequently mentioned ideas as "Astronomers have shown us that the earth occupies the space of a point in comparison with the size of the orbit in which the sun revolves."[18] He also presented the general Greek model of the celestial part of the cosmos.[19] Macrobius included much nontechnical astronomical and cosmological data in his commentary, much of it wrong and confused.[20] Following the cosmological part of his commentary, Macrobius concludes his treatise (book 2, chapters 14 to 16) with an attack on Aristotle's idea that the soul is not self-moved. In the process of refuting Aristotle's opinion, and defending Plato, Macrobius discusses and rejects Aristotle's ideas about motion, especially Aristotle's major thesis that the first cause of motion is itself stationary. Macrobius's arguments "are the clichés of Platonists and Aristotelians worn threadbare through centuries of wrangling." The direct quotations allegedly drawn from Aristotle "prove to be oversimplified statements of doctrines of his found mainly in the *Physics* and *De anima*, removed from their context and sequence, and set up in such a way as to be vulnerable to Macrobius' attacks which follow in the fifteenth and sixteenth chapters."[21] By his vigorous attack on Aristotle, Macrobius appears to have abandoned the usual Neoplatonic goal of reconciling the writings of Plato and Aristotle.[22]

Calcidius (fl. Fourth or Fifth Century AD)

In translating two-thirds of Plato's *Timaeus*, Calcidius made available the most important treatise on cosmology that would be known in the West until the twelfth century. He added a commentary that was almost five times as long as his translation of the *Timaeus*.[23] As in the *Timaeus*, Calcidius has much about numbers and geometry and how these shape the world.[24] Chapter 5 of Calcidius's commentary is a concise treatise on astronomy lifted largely from Theon of Smyrna's *A Manual of Mathematical Knowledge for an Understanding of Plato*. The latter was probably written in the first half of the second century AD and incorporated much Hellenistic astronomy, which Theon borrowed from Adrastus, who had written an earlier commentary on

---

[17] Ibid.   [18] Bk. I, ch. 16 (*Macrobius*, 154).

[19] For a good brief account of Macrobius's astronomical sections, see Stephen C. McCluskey, *Astronomies and Cultures in Early Medieval Europe* (Cambridge: Cambridge University Press, 1998), 117–119.

[20] See Stahl, *Roman Science*, 157–164.

[21] These are Stahl's assessments on page 22 of the Introduction to his translation of Macrobius's *Commentary on the Dream of Scipio*.

[22] See Chapter 3.   [23] See Stahl, *Roman Science*, 144.

[24] For a brief summary account, see Stahl, ibid., 142–150.

the *Timaeus*.[25] Calcidius included commonly repeated statements about the world, such as the earth is a mere point compared with the magnitude of the universe and that the celestial sphere rotates around an axis drawn through the center of the earth.[26] Calcidius also supplied his readers with information about the motions of the planets and stars.

Martianus Capella (fl. ca. 365–440)

The third of the major Neoplatonists who influenced the early Middle Ages is Martianus Capella, the author of a popular treatise titled *On the Marriage of Mercury and Philology*. The latter is a treatise on the seven liberal arts – dialectic, rhetoric, and grammar, constituting what came to be known as the trivium; and, arithmetic, music, geometry, and astronomy, the four scientific subjects that were known as the quadrivium – which Martianus represents as bridesmaids at a marriage between Mercury and Philology. In the eighth book of that treatise, Martianus presents astronomy in a nonmathematical and nontechnical manner. Martianus was not very knowledgeable about astronomy, but his chapter on astronomy served to introduce many scholars to that subject between the sixth and early thirteenth centuries.[27] To this aspect of natural philosophy, Martianus, in his chapter on geometry (Book VI), added very little on geometry, but devoted much of that book to geography, giving Eratosthenes's famous measurement of the earth's circumference as 252,000 stades. He copied much of the geography he presented from Pliny's *Natural History*.

Isidore of Seville (ca. 560–636)

To the literary formats of the three Neoplatonist encyclopedists, Isidore of Seville added a more practical method of presentation. In *On the Nature of Things (De natura rerum)*, a largely cosmological treatise, Isidore organized subjects and themes in ways that proved more useful than that of his earlier fellow encyclopedists. The work is divided into forty-eight chapters all of which treat astronomical, cosmological, geographical, or meteorological themes.[28] Isidore begins with chapters on days, nights, the week, months, years, and the seasons (chapters 1–7). There are chapters on the solstice and equinox (chapter 8), on the world (chapter 9), on the parts of the world (chapter 11), which is really about the four elements, fire, air, water, and

---

[25] Stahl, ibid., 56–57.

[26] Stahl, ibid., 146. See also McCluskey, *Astronomies and Cultures in Early Medieval Europe*, 119–120.

[27] See McCluskey, *Astronomies and Cultures*, 120–122.

[28] See Jacques Fontaine, *Isidore de Seville: Traité de la Nature* (Bordeaux: Féret et fils, 1960). The list of chapter titles appears on pages 168–173.

earth. He includes a sequence of cosmological and astronomical chapters with such titles as "On the celestial planets" (chapter 13), "On the celestial waters," a biblical theme (chapter 14); and seven chapters about the sun and moon (chapters 15–21) followed by six chapters (chapters 22–27) on the stars, including their motions; their positions; how they derive their light; on falling stars; on the names of stars; and whether they have a soul. Isidore then moves on to a series of chapters about meteorological phenomena, to use Aristotle's terminology, in the upper atmosphere. He includes chapters on thunder, lightning, rainbows, clouds, rain, snow, hail, winds, pestilence, oceans, seas, and rivers, and concludes his work with four chapters on the earth, devoting one chapter each to the position of the earth, its motion, Mount Etna, and the parts of the earth.

In his *Etymologies* (*Etymologiae*), a much later and lengthier work, Isidore was ostensibly concerned with the etymological derivations of terms. But "the less said about Isidore's word derivations, the better," is William Stahl's blunt assessment of Isidore's etymological achievement.[29] The treatise is actually an encyclopedia, in which Isidore included much that we would categorize as natural philosophy. He devoted the first three books to the seven liberal arts, and therefore found occasion to include a discussion of astronomy, a subject that he also considers in other parts of his treatise.[30] Indeed, he seems to have copied much of the astronomical material from his earlier work, *On the Nature of Things*, which had a more thorough treatment of the subject.[31] In Books XIII and XIV, Isidore treated cosmography and physical geography. His views on cosmological matters were influenced by Church Fathers such as St. Basil, St. Augustine, and St. Ambrose, whose views he sought to reconcile with secular authorities. Not surprisingly, Isidore is inconsistent and in error on numerous occasions. He drew most of his ideas about the world from other authors, although he exercised some judgment as to what to include or exclude. He was, however, no better than his sources, and occasionally worse. Nevertheless, "in assigning two books of his *Etymologies* to physical geography, Isidore established himself as the great authority on the subject in Western Europe during the Middle Ages."[32]

Pseudo-Bede (ca. Second Half of Eleventh Century)

An anonymous treatise is also noteworthy for early medieval natural philosophy. The work is attributed to Pseudo-Bede, because it was first falsely attributed to Venerable Bede. It bears the title *The Book on the Constitution of the Heavenly and Earthly World* (*De mundi celestis terrestrisque*

---

[29] Stahl, *Roman Science*, 216.
[30] See McCluskey, *Astronomies and Cultures in Early Medieval Europe*, 125.
[31] See Stahl, *Roman Science*, 220.   [32] Stahl, ibid., 221.

constitutione liber).[33] Although it might have been composed anywhere from the ninth to the twelfth century, the editor and translator of the treatise regards the second half of the eleventh century as the most plausible date of origin.[34] The treatise consists of two parts. In the first, and lengthier, part, the author considers numerous topics about the physical universe and then adds a second part devoted to the human soul. Some of the most significant authors in the early Middle Ages – Macrobius, Calcidius, and Martianus Capella – had similarly joined cosmographical and cosmological topics on the physical world with discussions of the human soul.[35]

The author begins the first part at the center of the world and moves upward through the concentric spheres to the farthest reaches of the universe. Many of the topics correspond to themes that Isidore of Seville included in his *On the Nature of Things*. In an introductory segment, Pseudo-Bede considers the elements; humors; transmutation of the elements; and "the position of the elementary spheres and the other spheres of the universe."[36] He next considers the earth, including its shape, zones, climates, and earthquakes. In the next section on earth and water, Pseudo-Bede takes up a range of topics, including ocean current, tides, mists and rain, snow, salt and fresh water, and hail. Moving upward to air, the author treats of winds, lightning, dew, shooting stars, and the rainbow. Under the rubric of "ether," Pseudo-Bede includes discussions of the planets, taking up: the duskiness, phases, and eclipses of the Moon, as well as the eclipse of the Sun, the order of the planets, the intersecting orbits of Venus and Mercury, the order of the upper planets, and the latitudinal and longitudinal motions of the planets. The author also was interested in astrology and included sections on the planetary aspects, planetary houses, and the causes of the coldness of Saturn. After a brief description of the supercelestial waters, the author provides a sequence of random information, treating such themes as the periods of the planetary orbits, the celestial circles (that is, the parallels, colures, the Milky Way, and the Zodiac), the meridian circle, the horizon, and why we see stars at night. One manuscript contains glosses and additions not found in the other manuscripts. Whoever wrote this seems to have reconsidered some topics treated in the work itself. He concludes the additions, however, with a brief response to the query: "Whether an atom is corporeal or incorporeal."[37]

---

[33] The Latin text and English translation appear in *Pseudo-Bede: "De mundi celestis terrestrisque constitutione," A Treatise on the Universe and the Soul*, edited and translated by Charles Burnett, with the collaboration of members of a seminar group at the Warburg Institute, London (London: The Warburg Institute, University of London, 1985).

[34] Burnett, *Pseudo-Bede*, 3.   [35] Burnett, ibid., 5.

[36] Charles Burnett, not the author of the treatise, presents an outline of the topics considered (see pp. 11–14). I follow his outline.

[37] Burnett, *Pseudo-Bede*, 14. In the section on the soul, which follows, Pseudo-Bede includes material in the philosophical tradition from Plato's *Timaeus* and Macrobius's *Commentary on the Dream of Scipio*, as well as ideas from the Christian faith.

The level of natural philosophy in this treatise is low indeed. Inconsistencies abound. Topics are discussed with no sense of relationship. Thus, we find a paragraph on air, followed by a discussion of lightning, followed by one about dew, followed by a paragraph on the rainbow. The next paragraph is on the ether and begins as if it were the immediate successor to the paragraph on air. It begins with:

The ether has its place next to this air and in its bounds the Moon is next in place, which the natural philosophers have called the ethereal earth, perhaps that just as the earth is so called from the word from "rubbing" (*terendum*), so whatever is rubbed off by purgation of the heavenly bodies is received into the Moon, and this is faced by the Sun so that it is as in a mirror. There are also lunar inhabitants which are souls, ascending or descending, which, whatever envelope they have picked up in the various spheres, they leave or take up there; these are <also called> good spirits. And whatever duskiness it is seen to have in the middle is ascribable to the aforementioned cause.[38]

What Pseudo-Bede may have intended as the function of his ether is a complete mystery. He places it after – that is, above – the air. But in the four-element theory fire is next after air, not ether. Earlier, Pseudo-Bede included fire among the four elements and even explained how the elements transmute into one another. Now ether seems to have displaced fire. Is it the same as fire? Virtually nothing has been said about the ether that would enable a reader to understand its role or function, or its relation to fire; or if it isn't fire, its relation to the four elements. Isidore of Seville was not much more helpful in relating fire and ether. In *On the Nature of Things*, Isidore rejects the idea that stars fall at night. This does not happen, because "we know that particles of fire fall from the ether, traverse the sky and are carried by the winds, imitating the light of a star."[39] Once again our Latin Encyclopedist obscures his meaning.

It is overwhelmingly likely that Pseudo-Bede and Isidore of Seville knew Macrobius's *Commentary on the Dream of Scipio*. They therefore should have known that Macrobius equated the ether with fire. In Book I, chapter 21, Macrobius explains that "all things that lie between the topmost border and the moon are holy, imperishable, and divine because they always have in them the same ether and are never subject to the vacillations of change."[40] He goes on to announce that "below the moon are earth, water, and air; but a body capable of living cannot be made from these alone; it requires the aid of heavenly fire to enable the terrestrial limbs to sustain life and breath, and to instill and keep vital heat."[41] It is apparent that for Macrobius, ether is fire and occupies all the space between the outermost sphere of the world and the sphere of the moon. Moreover the

---

[38] Burnett, ibid., 33.

[39] My translation from Isidore of Seville, *De natura rerum*, XXV, 1–3 in Fontaine, ed., *Isidore of Seville "Traité de la Nature,"* 263.

[40] Stahl, tr., *Macrobius, Commentary on the Dream of Scipio*, 180. [41] Stahl, ibid., 181.

heavenly fire is imperishable and things composed of it are also imperishable. Below the moon are the three elements air, water, and earth, in descending order. Although his discussion of ether is brief, we can see what Macrobius intended. But Isidore of Seville and Pseudo-Bede failed to equate fire with ether and to convey the picture that Macrobius had of the universe, although it is very likely that they intended to do so.

Pseudo-Bede's treatise illustrates the low level of communication and understanding that had become characteristic of much of Western thought in the early Middle Ages to the end of the eleventh, and beginning of the twelfth centuries. Not only was much of Pseudo-Bede's treatise on the universe largely incomprehensible, but it was also devoid of sustained discussions on any topic. And yet, Western Civilization owes a large debt to the small band of scholars who kept alive a tradition of learning in astronomy, cosmology, and physical geography. All this was achieved on the basis of a modicum of Platonic and Neoplatonic learning transmitted down the centuries, often in garbled form. But a dramatic change of attitude toward learning is apparent in the twelfth century, and even before. It preceded the great wave of translations of Greek and Arabic science and natural philosophy that transformed Western Europe. With very little that was new, medieval scholars of the twelfth century reveal a more critical attitude toward the learning they had inherited. Scholars such as Adelard of Bath, William of Conches, and John of Salisbury exhibited the new spirit.

THE TWELFTH CENTURY AND ITS IMMEDIATE ANTECEDENTS

The early Latin Encyclopedists, the ones I have discussed, and others, left an important legacy to Western Europe. What scholars in the West knew, they came to know by reading this or that work by one of the early Latin Encyclopedists, who kept alive what learning was available. What is noteworthy about the Latin Encyclopedists, however, is the absence of issues. They rarely discussed any genuine issue in natural philosophy. They merely present bits and pieces of information as if they were all true. They do not raise, or consider, the kinds of questions posed by Aristotle and later medieval natural philosophers. It apparently did not occur to them to pose broad questions about nature and its operations. They merely regurgitated information, which they organized as best they could. But they offered no analyses of issues, because they were not issue oriented. It was simply not the way they did things. But beginning in the eleventh century, and gaining momentum in the twelfth century, a century that many historians have characterized as a period of renaissance,[42] a change in attitude occurred that

---

[42] It is frequently cited as "The Renaissance of the Twelfth Century." See the book by the same name by Charles Homer Haskins.

made scholars perceive their inheritance of learning in ways that differed radically from that of their predecessors in the sixth to eleventh centuries. To appreciate this change of attitude in the late eleventh and twelfth centuries, the final years of the pre-Aristotelian era in Western Europe, it is essential to realize that the context of these changes was a Europe that was itself undergoing dramatic developments and transformations following centuries of barbarian invasions that had reduced the Europe of the Roman Empire to a ruinous condition. Social, economic, political, and educational activities had been seriously diminished and curtailed. But by the eleventh and twelfth centuries, this had changed as Western Europe was transformed into a new, and vigorous civilization.[43]

Once relative peace settled over Western Europe, agricultural advances – the heavy plough, nailed horseshoe, horse-collar, and improved crop rotation – allowed for an increase in the food supply, which made possible a considerable growth in population. With an increased population, Europeans built new towns and also began to colonize unpopulated and underpopulated lands. Germans moved east of the Elbe River and contested with Slavs for the lands in that vast region. With the advent of many new towns and the considerable growth of older cities, new wealth was created by the commerce and manufacturing that were almost inevitable effects of the reurbanization of Europe. European technological ingenuity made possible labor-saving devices that prompted Lynn White to declare that:

the chief glory of the later Middle Ages was not its cathedrals or its epics or its scholasticism: it was the building for the first time in history of a complex civilization which rested not on the backs of sweating slaves or coolies but primarily on non-human power.[44]

By about 1500, when the Middle Ages came to a close,

Europe's technology and political and economic organization had given it a decisive edge over all other civilizations on earth. Columbus had discovered America; the Portuguese had sailed around Africa to India; Europe had developed the cannon, the printing press, the mechanical clock, eyeglasses, distilled liquor, and numerous other ingredients of modern civilization.[45]

To these achievements, we should add one of the truly great contributions of the Middle Ages to the advance of civilization: the use of human

---

[43] I have briefly discussed this in my book, *God and Reason in the Middle Ages* (Cambridge: Cambridge University Press, 2001), ch. 2: "The Emergence of a Transformed Europe in the Twelfth Century," 17–30.

[44] Lynn White, Jr., "Technology and Invention in the Middle Ages," in *Medieval Religion and Technology: Collected Essays* (Berkeley: University of California Press, 1978), 22. Reprinted from *Speculum* 15 (1940), 141–159.

[45] C. Warren Hollister, *Medieval Europe: A Short History*, seventh edition (New York: McGraw-Hill, 1994), 1.

dissection in postmortems in Italy in the late thirteenth century, which led to the introduction of human dissection into medieval medical schools where it was used to study human anatomy.[46] Although human dissection in the late Middle Ages was done largely for teaching purposes, rather than for research to advance knowledge of the human body, subsequent advances in human anatomy would have been impossible without this monumental breakthrough.

Advances in Education and Learning

In conjunction with advances in the social, economic, and political realms, education and learning also emerged from centuries of neglect. Charlemagne marks a turning point, when in 789 he issued a decree ordering the establishment of schools in monasteries and cathedrals. At first, during the ninth century, schools were established in monasteries, the most famous located at Fulda, Corbey, and St. Gallen. But from the late tenth to twelfth centuries, a number of schools were founded in cathedrals. This marked a major shift from rural monastic schools to urban schools, as cathedrals were located in major cities. Among cathedral schools that became famous are those at Paris, Liège, Rheims, Orleans, and Chartres. Many famous teachers were associated with cathedral schools, including Gerbert of Aurillac at Rheims, a school he founded in the late tenth century, and Peter Abelard, who was associated with the cathedral school of Paris in the first half of the twelfth century.

Cathedral schools were initially intended to teach Latin grammar and rhetoric. In time they came to teach the logic derived from the Roman scholar, Boethius, in the fifth century, as well as some natural philosophy and science derived from the Latin Encyclopedists to better understand the physical world, much of which was taught by way of the seven liberal arts. And yet, although geometry was one of the seven liberal arts, Euclid's *Elements* was virtually unknown. At some schools, theology, medicine, and civil and canon law were taught. No single school would have taught all these disciplines, but students knew where to go for the particular subjects in which they were interested. During the period 1050 to 1150, when cathedral schools exerted their greatest influence, the student population of Europe was on the move, seeking the best teachers and schools.

In the ferment that affected Europe in the eleventh and twelfth centuries before the introduction of Greek and Islamic science and natural philosophy – especially Aristotle's logic and natural philosophy – students and scholars had to rest content with the same old intellectual fare on which Europe had subsisted for centuries. But it would be a serious error to infer from this that Europe was in a stagnant condition. Dramatic changes were already under way.

[46] I have here drawn upon my account in *God and Reason in the Middle Ages*, 110–113.

Although the intellectual fare remained fairly constant during the centuries before the twelfth, the attitude toward that body of knowledge began to change as early as the ninth century. A more critical approach began to take hold, an approach that emphasized the centrality of reason.[47] "At the courts of Charlemagne and Charles the Bald, and in the monasteries of Corbie and Auxerre," John Marenbon explains, "men of the early Middle Ages made their first attempts to grapple with abstract problems by the exercise of reason."[48] John Scotus Eriugena (b. ca. 800–d. ca. 877) brilliantly exemplifies the new attitude as early as the ninth century when he declared that:

authority proceeds from true reason, but reason certainly does not proceed from authority. For every authority which is not upheld by true reason is seen to be weak, whereas true reason is kept firm and immutable by her own powers and does not require to be confirmed by the assent of any authority. For it seems to me that true authority is nothing else but the truth that has been discovered by the power of reason and set down in writing by the Holy Fathers for the use of posterity.[49]

Undoubtedly aiding the new emphasis on reason, and perhaps even causative of it, were the logical treatises that came to be called the "old logic" (*logica vetus*), which was largely the work of the great Roman scholar, Boethius (ca. 480–524/525).[50] The old logic consisted of Boethius's translations from Greek to Latin of Aristotle's *Categories* and *On Interpretation*,[51] along with a translation of Porphyry's *Introduction to Aristotle's Categories*. In addition to his translations, Boethius wrote commentaries on four different logical works and composed five independent logical treatises. If we add to these, a few logical works composed prior to Boethius,[52] all these treatises taken together constituted the old logic (*logica vetus*). For many centuries – from the fifth to the tenth century – the old logic played a negligible role in Western Europe. But it became a subject of serious study in the eleventh and twelfth centuries, partly because there was a paucity of intellectual material and logic served to meet a growing need. Because logic is the embodiment of reason, the new emphasis on logic brought reason and reasoned argumentation to the fore. Gerbert of Aurillac (ca. 946–1003),

[47] Because I have discussed this in *God and Reason in the Middle Ages*, 46–48, I shall only present a brief summary here.

[48] John Marenbon, *From the Circle of Alcuin to the School of Auxerre: Logic, Theology and Philosophy in the Early Middle Ages* (Cambridge: Cambridge University Press, 1981), 139.

[49] From Eriugena's *On the Division of Nature* in *Periphyseon (De divisione naturae) liber primus*, ed. and trans. I. P. Sheldon-Williams, with the collaboration of Ludwig Bieler in *Scriptores Latini Hiberniae*, 7 (Dublin, 1968), 199. Cited from Peter Abelard, *A Dialogue of a Philosopher with a Jew, and a Christian*, translated by Pierre J. Payer (Toronto: Pontifical Institute of Mediaeval Studies, 1979), 82–83, n. 136.

[50] For a fuller treatment of Boethius and medieval logic, see Grant, *God and Reason in the Middle Ages*, 40–45.

[51] Boethius also translated Aristotle's *Sophistical Refutations, Prior Analytics*, and *Topics*, but these did not really circulate until the twelfth century.

[52] Among these works are Cicero's *Topics* and Marius Victorinus's *On Definitions*.

who became Pope Sylvester II (999–1003), taught the seven liberal arts at the cathedral school of Rheims, where he emphasized logic and may have been the first to teach the old Boethian logic. Although the old logic was eventually transformed into "modern logic" (*logica moderna*) in the fourteenth century, logic was regarded as an indispensable tool for the study of philosophy and natural philosophy. Two twelfth-century authors – Hugh of St. Victor (d. 1141) and John of Salisbury (ca. 1115–1180) – sang the praises of logic and helped establish it as the basic analytic instrument for students and scholars of the twelfth century. These two authors were not in any sense natural philosophers. Rather, they reflected on the requirements of a good and sound education and both saw the important role of logic and reason in achieving such goals. They were good indicators of the new outlook that had emerged in the twelfth century.

Hugh of St. Victor (d. 1141)
Hugh of St. Victor wrote his *Didascalicon* in the late 1120s and John of Salisbury completed his *Metalogicon* in 1159. What did they desire to convey to their contemporaries and to future generations? In his translation of Hugh's *Didascalicon*, Jerome Taylor inserted a subtitle that describes the *Didascalicon* as "A Medieval Guide to the Arts," a guide that "provided intellectual and practical orientation for students of varying ages and levels of attainment who came in numbers to the open school of the newly founded Abbey of Saint Victor" in Paris.[53] The *Didascalicon* also was serviceable for teachers who taught in the numerous cathedral schools scattered through Europe. A noteworthy aspect of Hugh's treatise is the fact that it is divided into two seemingly unrelated parts, the first concerned with secular writings, the second with divine writings. In the preface, Hugh explains that in the first part he:

enumerates the origin of all the arts, and then their description and division, that is, how each art either contains some other or is contained by some other, thus dividing up philosophy from the peak down to the lowest members. Then it enumerates the authors of the arts and afterwards makes clear which of these arts ought principally to be read; then, likewise, it reveals in what order and in what manner. Finally, it lays down for students their discipline of life, and thus the first part concludes.[54]

"In the second part," Hugh declares, now turning his attention to theology,

it determines what writings ought to be called divine, and next, the number and order of the Divine Books, and their authors, and the interpretations of the names of these Books. It then treats certain characteristics of Divine Scripture which are very important. Then it shows how Sacred Scripture ought to be read by the man who

---

[53] See *The "Didascalicon" of Hugh of St. Victor: A Medieval Guide to the Arts*, translated from the Latin with an Introduction and Notes by Jerome Taylor (New York: Columbia University Press, 1961), 3.
[54] *Didascalicon*, Preface, ibid., 44.

seeks in it the correction of his morals and a form of living. Finally, it instructs the man who reads in it for love of knowledge, and thus the second part too comes to a close.[55]

Hugh of St. Victor laid great emphasis on the seven liberal arts where "the foundation of all learning is to be found."[56] Of the seven arts, three – grammar, dialectic, and rhetoric – came to be called the trivium. Hugh regards logic as broader than the trivium. He distinguishes "rational logic" from "linguistic logic," explaining that:

Rational logic, which is called argumentative, contains dialectic and rhetoric. Linguistic logic stands as genus to grammar, dialectic, and rhetoric, thus containing argumentative logic as a subdivision.[57]

Logic, according to Hugh, "ought to be read first by those beginning the study of philosophy, for it teaches the nature of words and concepts, without both of which no treatise of philosophy can be explained rationally."[58] All knowledge is contained in the four subdivisions of philosophy, of which logic is one, the others being theoretical, practical, or ethical, and mechanical.[59] Hugh divides theoretical knowledge into theology, mathematics, and physics.[60] Physics, as Hugh understood it, is akin to natural philosophy. "The business of physics," Hugh explains,

is to analyze the compounded actualities of things into their elements. For the actualities of the world's physical objects are not pure but are compounded of pure actualities which, although they nowhere exist as such, physics nonetheless considers as pure and as such. Thus, physics considers the pure actuality of fire, or earth, or air, or water, and, from a consideration of the nature of each in itself, determines the constitution and operation of something compounded of them.[61]

Thus, one can determine the nature of a compound, not by examining the compound, but by knowing about the pure elements, or actualities, that form the compound.

Physics is the only subject area that is concerned solely with things, "while all the other disciplines are concerned with concepts of things,"[62] as are logic and mathematics. These two disciplines "are prior to physics in the order of learning and serve physics, so to say, as tools – so that every person ought to be acquainted with them before he turns his attention to physics."[63] And then in a forceful tribute to the power of reason and abstract thought, Hugh explains that "these two sciences" – logic and mathematics – "base their considerations not upon the physical actualities of things, of which we have deceptive experience, but upon reason alone, in which unshakeable truth stands fast, and that then with reason itself to lead them, they descend

---

[55] Ibid., 44–45.   [56] *Didascalicon*, bk. 3, ch. 4, 89.   [57] Ibid., bk. 1, ch. 11, 59.
[58] Ibid.   [59] Ibid., bk. 2, ch. 1, 62.   [60] Ibid.   [61] Ibid., bk. 2, ch. 17, 72.
[62] Ibid.   [63] Ibid., bk. 2, ch. 17, 73.

into the physical order."[64] This is a powerful testament to the superiority of reason and logic over experience and empirical procedures. To understand the nature and behavior of physical objects, one does not start with a direct examination of the physical object, but must use reason to analyze the properties of the object and then to "descend into the physical order."

John of Salisbury (ca. 1115–1180)

Far more than Hugh of St. Victor, his predecessor, John of Salisbury emphasized the indispensable role of logic in philosophy. In agreement with Hugh, John of Salisbury urged that logic be studied first so that it could then be used as a powerful instrument for the study of the whole of philosophy.[65] For John, "logic is 'rational' [philosophy]."[66] He regards as essential not only demonstrative logic but also probable dialectical arguments. Indeed, although "the sophist is satisfied with the mere appearance of probability," John is "loath to brand knowledge of sophistry as useless. For the latter provides considerable mental exercise, while it does most harm to ignoramuses who are unable to recognize it." From all this, John concludes that "one who will not embrace demonstrative and probable logic is no lover of the truth; nor is he even trying to know what is probable. Furthermore, since it is clear that virtue necessitates knowledge of the truth, one who despises such knowledge is reprobate."[67] John was deeply impressed with logic, which he believed gave great promise.

For it provides a mastery of invention and judgment, as well as supplies ability to divide, define, and prove with conviction. It is such an important part of philosophy that it serves the other parts in much the same way as the soul does the body. On the other hand, all philosophy that lacks the vital organizing principles of logic is lifeless and helpless. It is no more than just that this art should, as it does, attract such tremendous crowds from every quarter that more men are occupied in the study of logic alone than in all the other branches of that science which regulates human acts, words, and even thoughts, if they are to be as they should be. I refer to philosophy, without which everything is bereft of sense and savor, as well as false and immoral.[68]

In this passage, not only does John depict logic as vital to the well-being of philosophy, but he also presents a vividly graphic picture of its wide appeal as a subject of study in the twelfth century.

Despite the crucial role he assigned to logic, John did not view that discipline as an end in itself. "By itself," he insisted, "logic is practically useless. Only when it is associated with other studies does logic shine, and then by

---

[64] Ibid.

[65] See *The "Metalogicon" of John of Salisbury: A Twelfth-Century Defense of the Verbal and Logical Arts of the Trivium*, translated with an Introduction & Notes by Daniel D. McGarry (Gloucester, MA: Peter Smith, 1971), bk. 2, ch. 5, 82. For a more detailed discussion of John's defense of logic, see Grant, *God and Reason in the Middle Ages*, 48–51.

[66] Ibid. [67] Ibid., bk. 2, ch. 5, 83–84. [68] Ibid., bk. 2, ch. 6, 84.

a virtue that is communicated by them."[69] Logic is a tool for the analysis of the other aspects of philosophy, and not a subject to be studied for its own sake. John of Salisbury, however, parted company with many of his predecessors and contemporaries. For although he extolled the virtues of reason and reasoned argument, John was convinced that reason and logic ought not to be applied to the divine mysteries. In support of this position he cites Ecclesiasticus (iii, 22), who admonishes to "Seek not things that are beyond your reach, and do not fret over questions that exceed your comprehension."[70] John explains that "since not only man's senses, but even his reason frequently err, the law of God has made faith the primary and fundamental prerequisite for understanding of the truth."[71]

Thus, John appears to have opposed one of the most powerful intellectual movements of his time: the application of logic and reason to the faith. This had been going on since the eleventh century when Berengar of Tours (ca. 1000–1088) insisted on applying reason to faith, because reason is a gift of God.[72] In theology, Berengar regarded evidence as more important than authority. By applying logical arguments to the Eucharist, Berengar came to deny the act of transubstantiation, that is, he denied that the accidents of the bread could exist independently of their substance. In his dispute with Lanfanc of Bec, who attacked Berengar's interpretation of the Eucharist, Berengar wrote to Lanfranc that "it is incomparably superior to act by reason in the apprehension of truth; because this is so evident, no one will deny it except a person blinded by madness." Berengar urged all to use dialectic, or logic, in all things, "because to have recourse to dialectic is to have recourse to reason."[73]

Berengar did not oppose authoritative Christian texts but insisted that they be read with the aid of reason so they would be rendered intelligible. Berengar was, as Toivo Holopainen explains, "a representative of the Augustinian programme of faith in search of understanding: he applies reason to revealed doctrine, as it is conveyed by the sacred authorities, not in order to demolish it but in order to arrive at a coherent interpretation of it as a whole."[74]

Anselm of Canterbury (1033–1109)
Others in the eleventh century followed in Berengar's path and one, at least – Anselm of Canterbury (1033–1109) – went beyond. Anselm, who became archbishop of Canterbury in 1093, accepted St. Augustine's view that belief was essential for the understanding of faith. In order to understand, however, Anselm thought it essential to apply reason to theology. It was Anselm, in

---

[69] Ibid., bk. 4, ch. 28, 244.   [70] Ibid., bk. 4, ch. 41, 272.   [71] Ibid., bk. 4, ch. 41, 273.
[72] For a detailed discussion, see Grant, God and Reason in the Middle Ages, 51–52.
[73] Grant, ibid., 52. The translations are drawn from Toivo J. Holopainen, Dialectic and Theology in the Eleventh Century (Leiden: E. J. Brill, 1996), 109, 116.
[74] Holopainen, ibid., 118; also cited in Grant, God and Reason, 52.

a treatise titled *Proslogium*, who formulated the famous ontological proof for the existence of God.[75] Anselm's approach toward theology is, however, nicely illustrated in the opening lines of another treatise, the *Monologium*, in which Anselm informs his readers that he wrote this treatise at the request of his fellow monks, who asked him to produce a meditation for them on the Being of God that was based solely on reason and in which Scripture played no part.

It is in accordance with their wish, rather than with my ability, that they have pre-scribed such a form for the writing of this meditation; in order that nothing in Scripture should be urged on the authority of Scripture itself, but that whatever the conclusion of independent investigation should declare to be true, in an unadorned style, with common proofs and with a simple argument, be briefly enforced by the cogency of reason, and plainly expounded in the light of truth.[76]

Anselm was more of a rationalist in theology than was Berengar of Tours. Although reason was vital for both of these authors:

for Berengar the primary task of reason in theology is to function as a means of interpreting the authoritative writings of the Church. For Anselm, the primary task of reason in theology is to construct rational demonstrations for articles of faith. Because of his rational method, Anselm appears to be more of a rationalist than the schoolmaster of Tours.[77]

Anselm is generally assumed to have laid the foundations for the transfor-mation of theology into a science in the thirteenth century.

Peter Abelard (1079–1142)

The process begun in the eleventh century by the likes of Berengar of Tours and Anselm of Canterbury was brought to its greatest heights by Peter Abelard (1079–1142) in the first half of the twelfth century.[78] Peter Abelard wrote treatises on logic and was primarily a logician. But he was also a the-ologian, one who applied reason, and his knowledge of logic, to theology. Peter thought it important to consider plausible alternatives in dealing with any problem and especially in theological problems. In his famous theological treatise, *Yes and No (Sic et Non)*, Peter encouraged his students to think for themselves and to arrive at their own answers to theological problems. To facilitate this objective, Peter responded to 158 questions he posed in *Yes and*

---

[75] For a summary and discussion of the ontological proof, see Grant, *God and Reason in the Middle Ages*, 53–56.

[76] From St. Anselm, *Proslogium; Monologium; An Appendix in Behalf of the Fool Guanilon; and Cur Deus Homo*, translated from the Latin by Sidney Norton Deane, with an Introduc-tion, Bibliography, and Reprints of the Opinions of the Leading Philosophers and Writers on the Ontological Argument (La Salle, IL: Open Court Publishing Co., 1944), 35.

[77] Holopainen, *Dialectic and Theology*, 132.

[78] I rely here on my fuller treatment of Abelard in Grant, *God and Reason in the Middle Ages*, 57–62.

No by presenting the affirmative and negative opinions (yes and no) for each question and refusing to choose between them. He derived his answers for each side by drawing on the writings of the Church Fathers and by this means succeeded in showing that the Fathers were themselves in disagreement on many basic questions, often contradicting one another. At the conclusion of his Prologue to *Yes and No*, Peter explains his purpose:

I present here a collection of statements of the Holy Fathers in the order in which I have remembered them. The discrepancies which these texts seem to contain raise certain questions which should present a challenge to my young readers to summon up all their zeal to establish the truth and in doing so to gain increased perspicacity. For the prime source of wisdom has been defined as continuous and penetrating enquiry. The most brilliant of all philosophers, Aristotle, encouraged his students to undertake this task with every ounce of their curiosity. In the section on the category of relation he says: "It is foolish to make confident statements about these matters if one does not devote a lot of time to them. It is useful practice to question every detail."[79] By raising questions we begin to enquire, and by enquiring we attain the truth, and, as the Truth has in fact said: "Seek, and ye shall find; knock, and it shall be opened unto you." He demonstrated this to us by His own moral example when he was found at the age of twelve "sitting in the midst of the doctors both hearing them and asking them questions." He who is the Light itself, the full and perfect wisdom of God, desired by His questioning to give his disciples an example before He became a model for teachers in His preaching. When, therefore, I adduce passages from the scriptures it should spur and incite my readers to enquire into the truth and the greater the authority of these passages, the more earnest this enquiry should be.[80]

In his important Prologue, Abelard reveals a new sense of inquiry that seems to have captured Western Europe. Not only did he believe that "by raising questions we begin to enquire, and by enquiring we attain the truth," but in the same Prologue he explains how students and scholars should approach and deal with what different writers say; how to determine the meanings of words in different contexts; how to determine difficult texts and passages; and how to weigh the arguments of one authority against another. In order to arrive at truth, all texts are open to criticism and analysis.[81] Thus did Abelard emphasize reason as no others had done before.

---

[79] In the Oxford translation of J. L. Ackrill, Aristotle says (*Categories* ch. 7, 8b.23–24): "It is perhaps hard to make firm statements on such questions without having examined them many times. Still, to have gone through the various difficulties is not unprofitable." The Latin text Abelard used came to him from Boethius's translation.

[80] The translation is by Anders Piltz, *The World of Medieval Learning*, translated into English by David Jones (Totowa, NJ: Barnes & Noble Books, 1981), 82. The translation was made from the version in Migne, *Patrologia Latina*, vol. 178, col. 1349. In *Peter Abailard Sic Et Non: A Critical Edition* by Blanche Boyer and Richard McKeon, see Prologue, 103–104, lines 330–346.

[81] See Piltz, *The World of Medieval Learning*, 81, and L. Minio-Paluello, "Abailard, Pierre," in *Dictionary of Scientific Biography*, vol. 1, 2, where Minio-Paluello lists seven methodological principles proclaimed by Abelard.

By the twelfth century, the application of logic and reason to theology had ushered in the era of systematic, scholastic theology. Scholars tried to bring order out of the contradictory statements of the Church Fathers and they began to cite one another's opinions, thus adding a new set of respected authorities beyond the Bible and the Fathers. A characteristic feature of the new breed of theologians was their eagerness to analyze and speculate about ideas.

HOSTILE RECEPTION OF THE NEW THEOLOGY

As the new theologians became a significant intellectual force in the study of theology, they met resistance and much hostility from those theologians who had a very different idea of how theology should be studied and presented. Those who opposed the new approach to theology were often called "monastic theologians," largely because they were members of monastic orders. Monastic theologians emphasized contemplation rather than analysis. They were suspicious of the application of logic and the liberal arts to theology. Among their number were such eminent theologians as Rupert of Deutz (1070–ca. 1129), Walter of St. Victor (d. 1180), William of St. Thierry, and St. Bernard of Clairvaux (1090–1153). Easily the most significant member of this group was Bernard of Clairvaux, who is famous for his hostile reaction to Peter Abelard, whom he pursued relentlessly as a heretic until he succeeded in having Peter's works condemned at the Council of Sens in 1140.

Bernard was convinced that Peter gave voice to heretical opinions because he was excessively rationalistic. In a letter to a Cardinal of the Church, Bernard said of Peter Abelard that

He has defiled the Church; he has infected with his own blight the minds of simple people. He tries to explore with his reason what the devout mind grasps at once with a vigorous faith. Faith believes, it does not dispute. But this man, apparently holding God suspect, will not believe anything until he has first examined it with his reason.[82]

Although Bernard and Abelard were reconciled before the latter's death in 1142, Bernard had clearly triumphed over Abelard. In the long term, however, Abelard's approach to faith and learning was victorious. For it was Abelard's use of reason and logic in all things, but especially theology, that swept the day by the end of the twelfth century. The end of Bernard's way – and that of his colleagues and sympathizers – was virtually guaranteed by the appearance of the *Four Books of Sentences* by Peter Lombard (ca. 1095–1160) between 1155 and 1158, a few years after Bernard's death in 1153. The *Sentences* of Peter Lombard was a monumental theological treatise that became the basic textbook in schools of theology until the end of

---

[82] *The Life and Letters of St. Bernard of Clairvaux*, newly translated by Bruno Scott James (London: Burns Oates, 1953), letter 249, p. 328. The letter was to a Cardinal Haimeric.

the seventeenth century, a period of nearly five centuries. The major themes in its four books were concerned with God (bk. 1), the creation (bk. 2), the Incarnation (bk. 3), and the sacraments (bk. 4). "His method was to propose a doctrinal thesis or question, to bring forward authorities for and against this thesis from Scripture, the Councils, the Canons, and the Fathers, and then give judgment on the issue."[83] Each book was comprised of a series of questions to which Peter Lombard, in contrast to Peter Abelard, supplied answers, relying most heavily on the works of St. Augustine. Our knowledge of medieval theology is derived from the large number of extant commentaries on the *Sentences*.

The period during which medieval theologians wrote commentaries on the *Sentences* of Peter Lombard may be viewed as the second stage in the evolution of theology in the Middle Ages. The first stage was, as we saw, the period during the eleventh and twelfth centuries when theology was rationalized in a struggle between an emergent new breed of theologian and the old antianalytic brand of monastic theologian. That struggle was won by 1200 when the first universities at Paris and Oxford appeared. The second stage presupposes the advent of Peter Lombard's *Four Books of Sentences* as the theological textbook in schools of theology and the virtual completion of the process of translation into Latin of Greek and Arabic texts in science and natural philosophy of which the most important aspect was the translation of Aristotle's works in logic and natural philosophy (to be discussed later). All this made possible the second stage of theological evolution, which is most notable for the pervasive use of logic and natural philosophy in the analysis and resolution of theological questions.

Before the translations were completed and before the second stage of theological evolution occurred, the pre-Aristotelian natural philosophy reached its climax in the first sixty to seventy years of the twelfth century. It will be useful to have some idea of how the new attitudes toward learning and theology affected the way scholars in that period did natural philosophy while relying on the old Latin learning.

NATURAL PHILOSOPHY IN THE TWELFTH CENTURY

Among scholars in the twelfth century who devoted all or part of a treatise to what might reasonably be regarded as natural philosophy – although none used such a term or expression – are Thierry of Chartres (d. 1151), Clarenbaldus of Arras, Daniel of Morley, and the three authors whose major works will be discussed here, namely, Adelard of Bath (fl. 1116–1142), Bernard Silvester (fl. 1150), and William of Conches (ca. 1090–d. after 1154). The formats employed to convey natural philosophy varied: Thierry

---

[83] David Knowles, *The Evolution of Medieval Thought* (Baltimore: Helicon Press, 1962), 179–180.

of Chartres embedded his natural philosophy in a commentary on the intro-
ductory chapters of Genesis, in a work titled *Treatise on the Works of the Six
Days (Tractatus de sex dierum operibus)*;[84] Adelard of Bath and William of
Conches embedded their natural philosophy in a dialogue format, whereas
Bernard Silvester used poetry to describe the development of the cosmos.[85]
By coming to grips with the problem of the creation of the world and of man,
whether in poetry or prose, these three were doing, as William of Conches
put it, what "almost all moderns" were doing,[86] from which it appears there
was a widespread interest in explaining physical aspects of the world, especially its creation. Let us see what Adelard of Bath, Bernard Silvester, and
William of Conches regarded as important in their descriptions of the world.

Adelard of Bath (ca. 1080–d.ca. 1152)

Adelard traveled in the Arab world, learning Arabic sufficiently well to
serve as a translator of Euclid's *Elements*, the *Astronomical Tables* of al-
Khwarizmi, and a few astrological treatises.[87] Adelard's version of natural
philosophy is titled *Natural Questions* and consists of seventy-six chapters,
most of which inquire why, how, or whether something occurred. In the preface to his treatise, Adelard informs us that the genesis of his treatise was a
certain nephew of his "who, in investigating the causes of things, was tying
them in knots rather than unraveling them." The nephew, knowing that
Adelard had traveled among the Arabs, requested his uncle to present some
things he had learned from his studies among the Arabs. At the conclusion of
his preface, Adelard declares that he will start with the chapter headings and
then, he says, "I shall reply to my nephew's questions concerning the causes
of things."[88] After listing the titles of the seventy-six chapters, Adelard provides some further preliminary information before launching into his first
chapter. He emphasizes that he will present knowledge he obtained from the
Arabs, but, as Marshall Clagett has observed, "no Arabic author is mentioned by name or quoted directly."[89] Although a clepsydra that Adelard

---

[84] See Nikolaus M. Häring, "Thierry of Chartres," in *Dictionary of Scientific Biography*, vol.
13 (1976), 340.

[85] See Brian Stock, *Myth and Science in the Twelfth Century: A Study of Bernard Silvester*
(Princeton, NJ: Princeton University Press, 1972), 227. See also p. 237, where Stock mentions
the wide variety of formats that were used in twelfth-century natural philosophy.

[86] The Latin phrase is *fere omnes modernos*; see Stock, ibid., 228.

[87] See Marshall Clagett's brief article on "Adelard of Bath" in *Dictionary of Scientific Biography*, vol. 1, 62.

[88] From Adelard of Bath's *Natural Questions* in *Adelard of Bath, Conversations with His
Nephew On the Same and the Different, Questions on Natural Science, and On Birds*, edited
and translated by Charles Burnett, with the collaboration of Italo Ronca, Pedro Mantas
Espana and Baudouin van den Abeele (Cambridge: Cambridge University Press, 1998), 83.

[89] Clagett, "Adelard of Bath," ibid., 61.

mentions may have been derived from Arabic sources,[90] he fails completely to deliver anything of substance about Arabic science and natural philosophy in a manner his readers might have expected. And yet he presents himself as a modern who is breaking from tradition, as he explains in his preface, when he deprecates the "present generation," which "suffers from this ingrained fault, that it thinks that nothing should be accepted which is discovered by the 'moderns.'"[91] Adelard's remarks are part of a general pattern among those twelfth century authors who were rebelling against authority and tradition.

After presenting the seventy-six chapter headings, Adelard declares: "The chapter headings end. The book begins. This is how the causes of things work," thus reiterating his task as the explanation of the causes of things. Just before the first chapter commences, but after the dialogue between Adelard and his nephew begins, Adelard explains that they will "start from the lowest objects and end with the highest," which is, indeed, how he proceeds, as the following description of chapter topics reveals.[92]

Chapters 1–6: plants and how they grow.
Chapters 7–13: animals.
Chapter 14 is titled: "Whether opinion is founded in an animate body."
Chapters 15–47: humans and their behavior.
Chapters 48–50: earth and earthquakes.
Chapters 51–58: waters in their various manifestations as oceans, rivers, springs, etc.
Chapters 59–63: winds.
Chapters 64–68: thunder and lightning.
Chapters 69–76: the moon (69–70); planets (71–72); and stars (73–76).

Adelard chose to begin his ascent from "lowest to highest" with plants, rather than inanimate matter. The questions he posed often seem rather odd and all of his questions could easily have been replaced by other questions of a similar kind. Adelard's successive questions usually bear little relationship to one another, as we readily detect from the following examples. On plants, Adelard asks:

1. The reason why plants grow without a seed being sown beforehand.
2. In what way some plants are to be called hot, when they are all more earthy than fiery.
3. How plants of contrary natures grow in the same spot.[93]

---

[90] See Clagett, ibid.; see also Edward Grant, *Much Ado about Nothing: Theories of Space and Vacuum from the Middle Ages to the Scientific Revolution* (Cambridge: Cambridge University Press, 1981), 83.
[91] Adelard of Bath, *Natural Questions*, preface, in *Adelard of Bath: Conversations with His Nephew*, 83.
[92] For a description of the content of some of Adelard's chapters, see Louise Cochrane, *Adelard of Bath: The First English Scientist* (London: British Museum Press, 1994), 41–52.
[93] *Adelard of Bath: Conversations with His Nephew*, ibid., 85.

In his questions about animals, Adelard included the following successive, unrelated questions:

10. Why not all those which drink urinate.
11. Why some animals have a stomach, others not.
12. Why some of them see more clearly by night.
13. Whether brute animals have souls.[94]

The section on human behavior has the following sequence of unrelated questions:

31. By what nature we smell, taste and touch.
32. By what nature joy is the cause of weeping.
33. By what nature we breathe out from the same mouth now hot air, now cold.
34. Why fanning generates coldness, if movement generates heat.
35. Why the fingers are created uneven.
36. Why the palm is hollow.
37. Why men do not walk as soon as they are born, when brute animals do this.
38. Why men have such soft limbs.[95]

Each of the three questions about the earth is distinct from the others:

48. Why, or by what nature, the globe of the earth is held up in the middle of the air.
49. Where, if the globe of the earth were bored through, a rock thrown into the hole would end up.
50. From what cause an earthquake occurs.[96]

As the final example of unrelated questions, I cite the following from Adelard's questions about waters:

51. Why sea water is salty.
52. Why the flows and ebbs of the tide occur.
53. How the Ocean is not increased by the flowing in of rivers.
54. Why some rivers are not salty.
55. How the flow of rivers can be perpetual.
56. From what cause waters arise on the tops of mountains.
57. Whether there are any true springs.
58. Why water does not flow out from a full vessel which is open at the bottom, unless a higher opening is uncovered.[97]

---

[94] Ibid.   [95] Ibid., 87.   [96] Ibid., 87–89.

[97] Ibid., 89. Question 58 is about the clepsydra and may have been derived from Arabic sources. For the brief description, see page 195.

Most of the seventy-six questions are similarly unrelated. In a few instances, there is continuity in successive questions, as in the following sequence:

23. What opinion should be held about sight.
24. Whether the visual spirit is a substance or an accident.
25. How that spirit goes to a star and returns in such a brief space of time.
26. How, when the eye is closed, the visual spirit is not left outside.
27. How the same spirit does not get in the way of itself if it returns at the same time as it goes out.
28. By what means the soul receives forms from that spirit.

This sequence is all about sight and the visual rays that produce it. Hence the transitions from one question to another are smooth and plausible. But where the relationship between successive questions in other sequences is far more remote, Adelard nevertheless makes an effort to bridge the gap and link the questions, however arbitrary it might appear. Here is how Adelard managed the transition in questions 31, 32, and 33, quoted earlier.

Adelard arranges it so that transitions always occur at the beginning of each chapter, and that his nephew always ushers in the new chapter, or problem, in response to Adelard's concluding remarks in the question immediately preceding. At the beginning of chapter 31, the Nephew, after hearing Adelard discuss sight and vision in the preceding questions, declares that "I want to hear about everything else which pertains to the senses."[98] Adelard then launches into a discussion about sense, taste, and touch. After a brief exchange at the end of chapter 31, where the nephew says they might be ready for higher topics and Adelard agrees, the nephew replies that "at present we should deal with these more down-to-earth matters," and then recalls to Adelard that when the latter returned from the orient, he [the nephew] wept tears of joy," so that what "should have been the cause of laughter for me, cast me down into weeping. Since this is said to happen not just to me," the nephew continues, I believe that the causes of things went awry rather than that I did." With this, Adelard presents a brief analysis of this problem, which is the theme of chapter 32 ("By what nature joy is the cause of weeping").

At the beginning of chapter 33, the title of which is "By what nature we breathe out from the same mouth now hot air, now cold," the Nephew, after asking about tears of joy, which is only an occasional occurrence, says that he now "shall bring into the open what happens to us in practice every day: by what reason from the same mouth, when a person wishes, he can blow hot breath, but when it pleases him, the opposite." Adelard responds to his nephew's query and the nephew then provides the transition to the next chapter (34).

All of the chapters in Adelard's *Natural Questions* are linked successively in this way, sometimes in a seemingly plausible manner, but more often the linkage of two successive chapters is arbitrary. But whether natural or arbitrary, Adelard had the good sense to recognize the importance of connecting each chapter to its immediate predecessor. In this way, he conveyed the illusion of a cohesive treatise.

Perhaps the most noteworthy aspect of Adelard's treatise is its emphasis on reason and rationality. In a number of passages, Adelard emphasizes the importance of following reason over authority. "Those who trust only in the name of an ancient authority," he declares,

do not understand that reason has been given to each single individual in order to discern between true and false with reason as the prime judge. For unless it were the duty of reason to be everybody's judge, she would have been given to each person in vain. . . . Moreover, those who are called "authorities" did not obtain their initial trust among lesser men, except in that they followed reason, and whoever ignores or neglects reason, should worthily be thought blind. However, I do not state categorically that in my judgement authority should be spurned. Rather, I assert that first, reason should be sought, and when it is found, an authority, if one is at hand, should be added later. But authority alone cannot win credibility for a philosopher, nor should it be adduced for this purpose.[99]

In a subsequent chapter, Adelard's nephew points out man's absence of innate defensive weapons and lack of mobility to escape danger and considers it strange that such weapons are denied man, who is supposedly favored by God over brute animals. To which Adelard replies:

I agree that man is dearer to the Creator than the other animals. Nevertheless it is not appropriate either for arms to be innate in him, or for very swift flight to be attached to him. For he has that which is much better and more worthy than these – I mean reason, by which he excels the very brute animals so that they are tamed by it, and once tamed, bridles are put on them, and once bridled, they are put to various tasks. Thus you see how much the gift of reason is superior to bodily instruments.[100]

In a chapter titled "Why the nose is placed above the mouth" (chapter 19), the nephew finds it difficult to comprehend why the Creator placed the unclean and moist nose above the mouth. To this complaint, Adelard replies: "you do not understand that nothing natural is unclean or unfitting. Whatever is against the reason of nature, although to the sight it is decorated with surface beauty, is rightly said to be both unclean and flawed in itself."[101]

---

[99] *Natural Questions*, chapter 6, *Adelard of Bath*, ibid., 103–105.
[100] *Natural Questions*, chapter 15, ibid., 121.
[101] *Natural Questions*, chapter 19, ibid., 127–129.

Adelard emphasizes that natural explanations are always preferred over unnecessary appeals to God. "I am not slighting God's role," he explains,

For whatever exists is from him and through him. Nevertheless that dependence <on God> is not <to be taken> in blanket fashion, without distinction. One should attend to this distinction, as far as human knowledge can go; but in the case where human knowledge completely fails, the matter should be referred to God. Thus since we do not yet grow pale with lack of knowledge, let us return to reason.[102]

In explaining "By what nature joy is the cause of weeping" (chapter 32), Adelard offers his solutions and then declares that "the causes of things conform with reason."[103] Toward the end of the treatise, in chapter 69, Adelard's nephew is converted to the utility and power of reason when he tells Adelard that:

you have nursed my ignorance at the breasts of reason, and since halter-like opinion has been thrown away, let reason lead the way, wherever she wishes! I shall follow. For, in the encouraging words of Socrates, as virtue and friendship bring things which are almost impossible into the realms of possibility, so reason leads the one who has despaired of understanding things, into thinking about them clearly. Having put on the wings of reason, let us ascend to the stars.... [104]

Adelard of Bath drew his themes and material from a variety of sources, especially from Plato's *Timaeus*.[105] The themes he includes were probably popular and regarded as important by his contemporaries. Most of the chapters are relatively brief, and there is little serious presentation of alternative arguments. But Adelard's treatise is important because it clearly reveals him as one of those twelfth-century authors who sought to make reason, rather than secular and Church authorities, the decisive element in an argument. But in one important sense, Adelard differs radically from our next author, Bernard Silvester. Adelard describes a world already in existence and does not regard it as his task to include, at some point, an account of the creation of the world. But other authors, following in the tradition of Plato's *Timaeus*, thought it essential to do so. One of them was Bernard Silvester.

Bernard Silvester (fl. 1145–1153)

The *Cosmographia* is divided into two parts. Bernard calls the first *Megacosmos*, "the Greater Universe," and titles the second part *Microcosmos*, or "the Lesser Universe." After a brief dedicatory passage in honor of his friend, Thierry of Chartres, Bernard presents a concise but useful summary

---

[102] *Natural Questions*, chapter 4, ibid., 97–99.
[103] *Natural Questions*, chapter 32, ibid., 157.
[104] *Natural Questions*, chapter 69, ibid., 209.
[105] These are given in notes by Charles Burnett, the editor of Adelard's *Natural Questions*. See also Louise Cochrane, *Adelard of Bath: The First English Scientist*, 43–44.

account of the two parts of his treatise.[106] We learn that in the *Megacosmos*, Nature complains to Noys, or Divine Providence,

about the confused state of the primal matter, or Hyle, and pleads that the universe be more beautifully wrought. Noys, moved by her prayers, assents willingly to her appeal, and straightway separates the four elements from one another. She sets the nine hierarchies of angels in the heavens; fixes the stars in the firmament; arranges the signs of the Zodiac and sets the seven planetary orbs in motion beneath them; sets the four cardinal winds in mutual opposition. Then follow the creation of living creatures and an account of the position of earth at the center of things. Then famous mountains are described, followed by the characteristics of animal life. Next are the famous rivers, followed by the characteristics of trees. Then the varieties of scents and spices are described. Next the kinds of vegetables, the characteristics of grains, and then the powers of herbs. Then the kinds of swimming creatures, followed by the race of birds. Then the source of life in animate creatures is discussed. Thus in the first book is described the ordered disposition of the elements.[107]

Bernard describes Noys's formation of the four elements in the second chapter of the *Megacosmus*, where we read:

When the mother of all life thus gave scope to the fullness of her generative capacities, and opened forth the womb of her fecundity to the production of life, there straightway took place, from this source and within it, the origin of the created essences, the birth of the elements.

From the confused and turbulent depths the power of fire emerged first, and instantly dissipated the primeval darkness with darting flame. Earth appeared next, distinguished by no such lightness or radiance, but stable in tendency, and of a more concrete corporeity; for she was destined to reclaim, once their earthly round was completed, the returning stream of all those creatures which would be born of her. Forth came the gleaming substance of clear water, whose level and shimmering surface gave back rival images when darkened by the intrusion of shadows. Then the vast region of the air was interposed, volatile and subject to change; now giving itself to shadows, now gleaming at the infusion of light, now growing crisp with frost, now languid with heat.

When each of these bodies had taken up the abode to which it was most readily drawn by material affinity, the earth rested firm, fire darted far above, and air and water assumed intermediate positions.[108]

Following the separation of the four elements, Bernard elaborates in chapter 3 what he had outlined in his prefatory account. In chapter 4, he tells us that the firmament and movement of the stars cause the elements to produce change in the world.[109] And in that same chapter, Bernard proclaims

---

[106] For a concise but very helpful description of Bernard's *Cosmographia*, see Brian Stock, *Myth and Science in the Twelfth Century*, 14–17.

[107] The *"Cosmographia"* of Bernardus Silvestris. A translation with Introduction and Notes by Winthrop Wetherbee (New York: Columbia University Press, 1973), 65–66.

[108] Bernard Silvester, *Cosmographia*, bk. 1, ch. 2, 72.   [109] Ibid., bk. 1, ch. 4, 88.

the cosmos eternal when he says: "The totality of creatures, the universe, is never to be subjected to the infirmity of old age or sundered by ultimate destruction."[110] We are left to ponder how Bernard would have reconciled this with the Christian doctrine of creation.

Bernard firmly believes, however, that the universe he describes is astrologically determined. "For the firmament," he declares

is inscribed with stars, and prefigures all that may come to pass through decree of fate. It foretells through signs by what means and to what end the movement of the stars determines the course of the ages. For that sequence of events which ages to come and the measured course of time will wholly unfold has a prior existence in the stars.[111]

In this passage, Bernard reflects the increasing importance of astrology in the twelfth century, not yet in its horoscopic aspects, but "as a set of beliefs about physical influence within the cosmos,"[112] which derived from the translations of a few Arabic astrological treatises, especially Albumasar's *Introduction to Astrology*, translated in 1133 and again in 1140. That the celestial region exerted a powerful influence on the terrestrial region was an integral part of the natural philosophies of Plato and Aristotle, although neither discussed astrology as such. But their fundamental idea that the celestial motions caused the movements of the sublunar elements became ubiquitous, an idea that would receive powerful reinforcement when the *Tetrabiblos* of Claudius Ptolemy, or *Quadripartitum*, as it was known in the Middle Ages, was translated in the 1130s,[113] and became the most influential and authoritative astrological treatise in the Latin West during the late Middle Ages and Renaissance. The major impact of astrology, and works like the *Tetrabiblos* and *Introduction to Astrology*, did not occur until the thirteenth century.[114]

The *Cosmographia* is a mixture of poetry and prose and "is both a dramatic *myth* enacted by a group of allegorical personifications, and a resulting *model* of universal order, relating the macro- to the microcosm. In other words, there is both a story of the creation of the world and of man and a resulting design whose parts are analyzed in relation to each other."[115] In his prefatory summary account for book 2, the *Microcosmos*, Bernard uses four main allegorical characters: Noys, Nature, Urania, and Physis. Before creating man, Noys informs Nature to find Urania, "queen of the stars," and

[110] Ibid., bk. 1, ch. 4, 87.   [111] Ibid., bk. 1, ch. 3, 76.

[112] David C. Lindberg, *The Beginnings of Western Science*, 274.

[113] For a good, brief summary of the *Tetrabiblos*, see Olaf Pedersen, *A Survey of the Almagest* (Odense: Odense University Press, 1974), 400–403.

[114] See Laura Ackerman Smoller, *History, Prophecy, and the Stars: The Christian Astrology of Pierre d'Ailly, 1350–1420* (Princeton, NJ: Princeton University Press, 1994), 29–30.

[115] Brian Stock, *Myth and Science in the Twelfth Century*, 14.

Physis, "who is deeply versed in the nature of earthly life."[116] When found by Nature, Urania joins her and together they find Physis:

dwelling in the very bosom of the flourishing earth amid the odors of spices, attended by her two daughters Theory and Practice. They explain why they have come. Suddenly Noys is present there, and having made her will known to them she assigns to the three powers three kinds of speculative knowledge, and urges them to the creation of man. Physis then forms man out of the remainder of the four elements and, beginning with the head and working limb by limb, completes her work with the feet.

Although the *Cosmographia* embodies a considerable degree of rationalism and Bernard used much natural philosophy that he drew from available encyclopedic works, he "never lost sight of the fact that he was writing a work of literature," so that it is not surprising that "the mythical predominates over the scientific."[117] Because of its poetic form and its strong allegorical features, Bernard Silvester's *Cosmographia* represents an unusual kind of treatise in natural philosophy, one that was not destined to have many imitators.

William of Conches (ca. 1090–d. after 1154)

It was probably during the period 1144 to 1149 that William wrote the *Dragmaticon Philosophiae*, which was a revision and expansion of his much earlier treatise, *Philosophia*.[118] In his Prologue to book 1, William pays tribute to the Duke of Normandy and Count of Anjou, who was Geoffrey Plantagenet. At the end of his Prologue, and after announcing that his first topic will be about substances, William explains: *"because an uninterrupted exposition produces boredom*, and boredom annoyance, we shall divide up our discourse in the form of a dialogue. You, therefore, most serene Duke, should ask the questions; let a philosopher who shall remain unnamed reply to them."[119] The *Dragmaticon* is a dialogue between the Duke, who is the Duke of Normandy and Count of Anjou, and an unnamed Philosopher, who is, of course, William of Conches. As he informs us, William chose the dialogue form because it is less boring than a straightforward prose account.

[116] These quotations and those to follow appear in Bernard's summary account of book 2 in Bernard Silvester, *Cosmographia*, 66.

[117] Brian Stock, *Myth and Science in the Twelfth Century*, 30.

[118] For the translation of William's *Dragmaticon*, see *William of Conches: A Dialogue on Natural Philosophy ("Dragmaticon Philosophiae")*, Translation of the New Latin Critical Text with a Short Introduction and Explanatory Notes by Italo Ronca and Matthew Curr (Notre Dame, IN: University of Notre Dame Press, 1997). For the dating of the treatise, see p. xvii. Whether William called his treatise *Dragmaticon* is unclear, as is the meaning of the term. See ibid., xx–xxiii.

[119] William of Conches, *Dragmaticon*, Prologue, ibid.

With William of Conches, we return to the dialogue form used by Adelard of Bath.

A glance at the topics considered in the six books of the *Dragmaticon*, reveals that instead of proceeding from lowest to highest, as Adelard did, William moves from highest to lowest in books 2 to 6, using the first book to define certain terms (substance, infinite, element, nature) and to mention briefly demons, angels, and the Creator.[120] The Philosopher declares, "Everything is the work of either the Creator, or nature, or a craftsman." He explains that "The work of the Creator is the creation of the elements and the souls from nothing, bringing the dead back to life, causing a virgin to give birth, and such like." The Duke then requests that the Philosopher define nature before explaining how nature works. In response, the Philosopher observes that "As Cicero says, 'it is difficult to define nature'; however, as the term is understood here, nature is a certain force implanted in things, producing similar from similar. It is, therefore, the work of nature that men are born of men, asses from asses, and so on."

In William's Platonic world, God creates the four elements from a pre-existing Chaos, which he describes in book 2. At the end of the Prologue to book 3, the Philosopher asks the Duke to "please ask whatever you consider worthwhile discussing." The Duke now takes readers to the celestial region and asks about the waters above the heavens. "I remember," he says,

that you said earlier that the world is structured in the same way as an egg[121] and, as there is nothing of the egg beyond its shell, so there is nothing of the world beyond fire. Now by this statement you seem to contradict divine and human philosophy, according to which heaven covers all things, there is a firmament in which the stars are fixed, and there are waters above the heavens, as we find in the verse, "Bless the Lord, O waters above the heavens."[122] The firmament is in the middle of the waters, from which it necessarily follows that waters lie above and below that firmament, as in, "He divided the waters which were below the firmament from those which were above the firmament."[123]

In his reply to the Duke, the Philosopher explains away the biblical waters above the firmament by arguing that "he who said that 'the waters are above the heaven' or 'above the firmament' called the air heaven and firmament or,

[120] For a more detailed account of William of Conches's natural philosophy, see Dorothy Elford, "William of Conches," in Peter Dronke, ed., *A History of Twelfth-Century Western Philosophy* (Cambridge: Cambridge University Press, 1988), 308–327.

[121] In book 2, ch. 2, para. 8, p. 25, the Philosopher declares that: "As corroborated by [natural] philosophers, the configuration of our world resembles that of an egg. As in the middle of the egg is the yolk and on every side of it the white, around the white the skin, around which is the shell, outside of which there is nothing more of the egg; so in the middle of the world there is the earth, all around it from every part water flows, around the water there is air, around which is fire, outside which there is nothing."

[122] Daniel 3:60.

[123] Given by Ronca and Curr as Psalm 148.4 (Vulg.). See *Dragmaticon*, 186, n. 7.

what is truer, the lower part of the air, *above which the waters are suspended as vapor* in the clouds."[124] According to William of Conches, the Philosopher, the biblical waters above the firmament are really vaporized waters suspended above the lower part of the air, which has been called "heaven and firmament." Thus, the waters are above the airy "firmament," but below the sphere of the fixed stars. In response, the Duke invokes Venerable Bede who declared the waters above the firmament to be frozen waters.

In a reply that clearly shows that he belongs with the new rationalist, anti-authoritarian natural philosophers, the Philosopher replies, "In those matters that pertain to the Catholic faith or moral instruction, it is not allowed to contradict Bede or any other of the holy fathers. If, however, they err in those matters that pertain to physics, it is permitted to state an opposite view. For although greater than we, they were only human."[125] The Duke persists in his claim that the waters above the firmament are frozen "and changed into crystals of ice so that, although touched by fire, they cannot be dissolved."[126] The Philosopher explains why this cannot be, and the Duke, in apparent frustration, responds: "You attribute everything to the qualities of things and nothing to the Creator. Surely the Creator was able to place the waters there [above the ether], freeze them, and keep them suspended, contrary to nature?"[127]

In another startling, dramatic response, the Philosopher declares:

What is more foolish than to assume that something exists simply because the Creator is able to make it? Does He make whatever He can? Therefore, whoever says that God makes anything contrary to nature should either see that it is so with his own eyes, or show the reason for its being so, or demonstrate the advantage of its being so.[128]

Some twenty years earlier, William was attacked on theological grounds by William of St. Thierry, who wrote about him to St. Bernard of Clairvaux. What offended William of St. Thierry were William of Conches's views on the Trinity and his interpretation of the creation of woman from one of Adam's ribs.[129] We see the traces of this assault in William of Conches's *Dragmaticon*, where, in book 1, chapter 3, William includes an "Author's Confession of Faith" in which he carefully expounds his views on the Trinity and other matters of faith and then concludes with these words:

Thus we believe, approving some propositions with human reason, others, although possibly contrary to human reason, we yet believe and profess with absolute certainty

---

[124] William of Conches, *Dragmaticon* bk. 3, ch. 2, para. [2], 38.

[125] Ibid., bk. 3, ch. 2, [3], 38–39.   [126] Ibid., bk. 3, ch. 2, [7], 40.

[127] Ibid., bk. 3, ch. 2, [8], 40.   [128] Ibid.

[129] On this, see Lynn Thorndike, *A History of Magic and Experimental Science during the First Thirteen Centuries of Our Era*, 8 vols. (New York: Columbia University Press, 1923–1958), vol. 2 (1929), ch. 37 ("William of Conches"), 59–60.

because they were written by men to whom the Spirit had revealed them: men who professed neither to lie nor to affirm anything but certainty. But if any religious person should read this small work of ours, and something in it should appear to deviate from the faith, he should correct it either by spoken or by written word, and we will not object to altering it.[130]

Despite his Confession of Faith in book 1, William boldly denied the waters above the celestial firmament in book three, choosing to interpret the term "firmament" as subcelestial, lying just above the lower air. He did this probably knowing that St. Augustine, in his commentary on the Book of Genesis, had said of the waters above the firmament that "whatever the nature of that water and whatever the manner of its being there, we must not doubt that it does exist in that place. The authority of Scripture in this matter is greater than all human ingenuity."[131] If he were pressed, William would undoubtedly have argued that he was consistent with St. Augustine, who did not trouble to mention the location of the firmament. By locating the waters above the firmament, even if the latter was no longer celestial, William could be said to have complied with Augustine's requirements.

For the remainder of book 3, the Philosopher describes the creation of the stars from the four elements (he rejects Aristotle's fifth element from which Aristotle composed the celestial ether); from the motion of the newly created stars he describes the process that leads to the creation of animals and man; and concludes with a description of the motions of the stars and firmament, followed by a discussion of eleven celestial circles, including the Zodiac and Equinox.

At the conclusion of the Prologue to book 4, the Duke declares that "since you have now dealt with the firmament and the fixed stars as far as was relevant to our present task, please now deal with the planets." After devoting book 4 to the planets, William, at the close of the Prologue to book 5, has the Duke request the Philosopher "to discuss the phenomena that take place in the air,"[132] thus descending from the heavens toward the earth. In this chapter, William describes winds, rain, the rainbow, hail and snow, thunder, lightning, shooting stars, comets,[133] water, ocean tides, and a few other manifestations of water. The fifth, and final, book is in two parts: the first is about the earth; the second describes the characteristics and behavior of man.

William of Conches drew his ideas from a large variety of Latin sources, but, with perhaps one exception, did not have access to the large body of

---

130 William of Conches, *Dragmaticon*, bk. 1, ch. 3, [1–5], 7–8.

131 St. Augustine, *The Literal Meaning of Genesis: De Genesi ad litteram*, ed. and tr. John Hammond Taylor, in Johannes Quasten, Walter J. Burghardt, and Thomas Comerford Lawler, eds., *Ancient Christian Writers: The Works of the Fathers in Translation*, 2 vols. (vols. 41, 42) (New York: Newman, 1982), vol. 41, 52.

132 William of Conches, *Dragmaticon*, bk. 5, 1, [6], 92.

133 Shooting stars and comets appear in this chapter because they were regarded as sublunar phenomena.

Greco-Arabic science and natural philosophy that was in process of translation into Latin even as he wrote. The one exception is Abu Mashar's *Introduction to Astrology*, with which he was apparently familiar and that, as we shall see in the next chapter, included some of Aristotle's basic ideas about the world. But despite mentioning Aristotle a few times in the *Dragmaticon*, William, as we saw, actually rejects Abu Mashar's acceptance of Aristotle's celestial ether and confines his world to the four elements.

Throughout the *Dragmaticon*, William of Conches, in the guise of the Philosopher, responds to the Duke's questions in a completely naturalistic and rationalistic manner. We see this in the following exchange. After a lengthy series of questions concerned with sexual intercourse, the Duke asserts that he "will not venture to inquire any further about sexual intercourse because the subject is not quite decent, but ask instead that you proceed to other topics." To this, the Philosopher replies that "Nothing that is natural is indecent: for it is a gift of the Creator. However, because our hypocrites, more in abhorrence of the name than of the thing itself, avoid talking of such things, we should leave the subject for the time being and discuss conception."[134]

If we described additional treatises in natural philosophy written in the twelfth century we would find similar discussions based on the same Latin tradition, with Calcidius's Latin translation of Plato's *Timaeus* exerting a major influence. Although Adelard of Bath, Bernard of Silvester, and William of Conches, and all of their other twelfth-century colleagues, had to rely on a relatively meager body of natural philosophical literature, they had already developed a critical, rationalistic attitude that often prompted them to reject traditional authoritarian opinions and interpretations. The approach they developed and nourished was their legacy to the scholastic tradition that would be built on the new knowledge that entered Western Europe via translations in the twelfth and thirteenth centuries and that became the basis for the university curriculum of the late Middle Ages.

---

[134] William of Conches, *Dragmaticon* bk. 6, ch. 8, [14], 138.

6

Translations in the Twelfth and Thirteenth Centuries

The critical spirit of inquiry that developed and even flourished in Western Europe during the twelfth century was, as we saw, confined to interpreting and elaborating a by-then traditional body of Latin learning that was largely Platonic and Neoplatonic. But even as they exercised their intellects on the old learning, they had become aware that there was a body of learning of which they were ignorant. As Christians slowly wrested control of much of Spain and Sicily from Islamic rule, they came into contact with Islamic culture and the Arabic language. They not only learned of a large body of learning in the Arabic language but also that there were treatises in Greek. We may assume that the literature in the Greek and Arabic languages ranged over the whole spectrum of learning, extending from the humanities and literature to science and natural philosophy. Western interest in this body of literature focused almost exclusively on logic, science, and natural philosophy, largely ignoring the rest of it. The translations from Arabic and Greek to Latin occurred during the twelfth and thirteenth centuries. The lengthy process represents what is probably the greatest intellectual expropriation of knowledge by one culture and civilization from other cultures and civilizations.

THE WORLD OF THE TRANSLATORS

The translators came from all parts of Europe and worked alone or collaboratively. Although there was some royal and ecclesiastical patronage, the West had nothing like the great translation center in Baghdad, called the House of Wisdom (see Chapter 4), which supported and sponsored translations into Arabic. Without large-scale patronage, most translators had to find means of support. Some were ecclesiastics, or civil servants, or teachers, and even physicians. They often were drawn to intellectual centers because it was there that they would find treatises to translate and, as they often said in the prefaces to their translations, thereby relieve the "poverty of the Latins" in knowledge of science and natural philosophy.[1]

---

[1] See Grant, *The Foundations of Modern Science in the Middle Ages*, 23; also Scott Montgomery, *Science in Translation: Movements of Knowledge through Cultures and Time* (Chicago: University of Chicago Press, 2000), 143.

Most of the translators who went to Spain or Sicily to translate from Arabic to Latin did not know Arabic when they departed on their missions. But they eventually gained some degree of mastery of that language. Those who acquired sufficient knowledge of Arabic could translate without much dependence on others. But those who did not could choose from a few options. They could seek help from an expert in Arabic; or, if they knew Spanish, they could find someone who knew Arabic and Spanish and could translate the Arabic work into Spanish, from which it could be rendered into Latin by the initial translator. Spanish Jews and Mozarabs – that is, Christian Arabs whose native language was Arabic – played a significant role, as most knew Arabic and the Spanish vernacular, with the Jews also knowing Hebrew. A Jew with these linguistic skills, for example, might help a Latin translator by translating an Arabic work into Hebrew and then, if he also knew Latin, could also cooperate with the original translator in making a final translation from Hebrew into Latin. Or they might use the Spanish vernacular as an intermediary, as happened in the translation of Avicenna's *De anima* (*On the Heaven*) from Arabic to Latin. On this occasion, Dominicus Gundissalinus worked with the Jew Avendauth, who explained that "he put the text into the vulgar tongue one word at a time, while the archdeacon Domincus [Gundissalinus] converted the individual words into Latin."[2] Because Latin and Greek are cognate languages, translations from Greek into Latin did not ordinarily demand more complex tactics, although an intermediate vernacular language could have served to bridge the linguistic gap between Greek and Latin, where the translator's Greek might have been weak.

In the translations that were done from Greek and Arabic into Latin, the translators employed different methods. In a direct translation, rather than a paraphrase, translators sought to capture the substance of the work while also preserving the sense of the words. To achieve this the method used most frequently was a word-by-word (*verbum de verbo*) translation. The result of such a method was to produce literal translations where one tried to account for every word in the text that is translated. This worked far better for translations from Greek to Latin, cognate languages, than for translations from Arabic to Latin, two languages that were unrelated. By contrast, some translators sought to communicate the substance and sense (*ad sensum*) of a work and thus avoid a word-by-word rendition.

The essential objective for all translators, however, should have been to transfer faithfully the meaning of the text from the one language to another. How successful were medieval translators? Although there was great variation in the quality and intelligibility of translations, "viewed as a whole, . . . translations provided Western Christendom with an adequate

---

[2] Cited by David C, Lindberg, "The Transmission of Greek and Arabic Learning to the West," in David C. Lindberg, ed., *Science in the Middle Ages* (Chicago: University of Chicago Press, 1978), 70. Lindberg's article is an excellent summary account of medieval translations.

knowledge of the Greek and Arabic intellectual achievement – and thus with the basic materials out of which its own system of philosophy and natural science would be constructed."[3]

TRANSLATIONS FROM ARABIC AND GREEK IN THE TWELFTH AND THIRTEENTH CENTURIES

In what follows, I may mention a few translations of works from the exact sciences, especially astronomy and mathematics, but the focus will be overwhelmingly on natural philosophy, primarily on the works of Aristotle and his commentators.[4] Translations of works in natural philosophy from Arabic made the earliest impact in the West and will be considered first.

Translations from Arabic into Latin

Indeed, Aristotle's ideas reached the West before the works that contained those ideas. The vehicle for this transmission was a work on astrology known as Introduction to Astronomy (Introductorium in Astronomiam), written originally in Arabic by Abū Maʻshar al-Balkhī, Jaʻfar ibn Muḥammad (787–886), known in the West as Albumasar, as he will be cited here. Of the two Latin translations of Albumasar's Introduction, one was a literal translation made in 1133 by John of Seville, the other an abridged version made in 1140 by Hermann of Carinthia. In writing his astrological treatise, Albumasar sought to give it a scientific foundation by incorporating ideas from Aristotle's books on natural philosophy, although Aristotle never discussed astrology.[5] We saw earlier (Chapter 4) that Albumasar had initially been hostile to Aristotle's natural philosophy, but eventually came to accept Aristotle's ideas about the structure of the cosmos because they enabled him to provide what he considered a rigorously scientific foundation for astrology. But he did not derive his ideas about Aristotle's natural philosophy from Arabic translations of the latter's relevant works, but rather from literature that was compiled by a group known as the "Sabaeans" of Harran,

---

[3] Lindberg, ibid., 79.

[4] For a thorough summary of the translations of Aristotle's works from Arabic and Greek, see Bernard G. Dod, "Aristoteles latinus," in Norman Kretzmann, Anthony Kenny, and Jan Pinborg, ed., The Cambridge History of Later Medieval Philosophy from the Rediscovery of Aristotle to the Disintegration of Scholasticism 1100–1600 (Cambridge: Cambridge University Press, 1982), 45–79. The article concludes with a detailed catalogue of translations from Greek and Arabic that lists the work, translator, date, and number of surviving manuscripts for each translation (see pp. 74–79).

[5] See Richard Lemay, Abu Mashar and Latin Aristotelianism in the Twelfth Century: The Recovery of Aristotle's Natural Philosophy through Arabic Astrology, Publication of the Faculty of Arts and Sciences, Oriental Series No. 28 (Beirut: American University of Beirut, 1962), xxix. See also David Pingree, "Abū Maʻshar al-Balkhī, Jaʻfar ibn Muḥammad," in Charles C. Gillispie, ed., Dictionary of Scientific Biography, vol. 1 (1970), 33.

or Harranian prophets.[6] As David Pingree explains, "The religious view of the Harranians...assumes an Aristotelian physical universe in which the four Empedoclean elements are confined to the sublunar world, and the celestial spheres consist of a fifth element,"[7] namely, Aristotle's celestial ether.

Although Albumasar introduced some of Aristotle's ideas about physics and cosmology before Aristotle's natural philosophy had been translated into Latin, the impact of it was hardly significant. Indeed, we saw earlier that William of Conches (Chapter 5) rejected the Aristotelian concept of a celestial ether and four sublunar elements, which he presumably derived from Albumasar's *Introduction to Astrology*, in favor of four elements that comprised both the heavens and the sublunar region. It is correct to assert that "In his general cosmology William therefore remained a Platonist in opposition to the Aristotelian view of the heavens which was growing in importance in his times."[8]

Despite the dissemination of Ptolemy's *Tetrabiblos* (or *Quadripartitum*) and Albumasar's *Introduction to Astrology* in the latter half of the twelfth century, and perhaps a few other astrological treatises of Arabic origin, scholars in Western Europe were still basically dependent on the old Latin learning. Albumasar could have done little more than whet their appetites for the real thing. By the second half of the twelfth century, the effort to bring Greco-Arabic learning into Western Europe was well under way. The part of it relevant to natural philosophy was mostly centered on Aristotle's works on which we shall now largely focus, beginning with translations from Arabic to Latin.

Although many scientific works were translated from Arabic to Latin in the first half of the twelfth century by such translators as Plato of Tivoli, Adelard of Bath, Robert of Chester, Hermann of Carinthia, Dominicus Gundissalinus, Peter Alfonso, John of Seville, and others, the earliest translations of Aristotle's works on natural philosophy appear to have occurred in Spain in the latter half of the twelfth century. By far the most prominent translator of Aristotle's natural philosophy was Gerard of Cremona (ca. 1114–1187), the most prolific translator from Arabic to Latin of works on science, medicine, and natural philosophy. From a brief biobibliography that his students appended to Gerard's translation of one of Galen's medical treatises (*Tegni* or *Ars parva*), we learn that Gerard, unable to locate a copy of Ptolemy's *Almagest* among the Latins, came to Toledo in search of that great astronomical treatise. Once there, Gerard not only found the *Almagest*,

---

[6] This is Pingree's interpretation, which he opposes to Lemay's interpretation, the latter assuming that Albumasar derived these ideas directly from Aristotle's works.

[7] Pingree, "Abū Maʿshar al-Balkhī," *Dictionary of Scientific Biography*, vol. 1, 34.

[8] Lemay, *Abu Mashar and Latin Aristotelianism in the Twelfth Century*, 186. On pages 157–188, Lemay has an extended discussion of William of Conches as a natural philosopher.

which he translated, but, his students inform us, "seeing the abundance of books in Arabic on every subject, and regretting the poverty of the Latins in these things, he learned the Arabic language, in order to be able to translate."[9] And translate he did! In approximately thirty years, he converted seventy-one works from Arabic to Latin. What is remarkable about Gerard's monumental effort is the range of his achievements, which included works in mathematics, astronomy, medicine, alchemy, and, under the rubric of philosophy, he translated most of Aristotle's natural philosophy. Included among his translations are Aristotle's *Physics, On the Heavens (De caelo), On Generation and Corruption, Meteorology, Books I–III* (the fourth had already been translated).[10] Gerard also translated the *Posterior Analytics*, Aristotle's important work on the theory of science.

Did Gerard do all of his own translations? This is not easy to answer. Translations were sometimes, if not often, collaborative affairs. In his translation of Ptolemy's *Almagest*, Gerard was assisted by a Mozarab named Gallipus Mixtarabe. But we have no other names of any possible translating assistants, and it is noteworthy that the students who wrote the biobibliography, mentioned earlier, make no mention of any one who might have assisted Gerard.[11]

The next series of translations of Aristotle's natural philosophy occurs in the thirteenth century and involves not only Aristotle but also Averroes (Ibn Rushd), the great Islamic commentator on the works of Aristotle, who was born in Cordova and became not only the greatest commentator on Aristotle's works but also was a famous physician.[12] Modern scholars have thus far attributed thirty-eight commentaries to Averroes, a very large number indeed. As we saw in Chapter 4, the reason for this unusual activity derives from the fact that, Averroes wrote three different kinds of commentaries: an epitome, or short version, which was essentially a paraphrase or summary account of Aristotle's text and therefore not really a commentary. It is designed to explain the content of an Aristotelian treatise. Like the short version, the second, or middle commentary, is also not a direct translation, but it gives the substance of an Aristotelian text sequentially intermingled with Averroes's own comments, as will be seen in a few examples later. The Long Commentary includes the whole of Aristotle's text presented in successive segments, with each segment followed by a commentary.

---

[9] See "A List of Translations Made from Arabic into Latin in the Twelfth Century: Gerard of Cremona (ca. 114–1187)," translation, introduction, and annotation by Michael McVaugh in Edward Grant, *A Source Book in Medieval Science* (Cambridge, MA: Harvard University Press, 1974), 35–38.

[10] For all this, I rely on McVaugh's translation of the entire list of translations attributed to Gerard by his students (in Grant, *Source Book,* 36–38).

[11] See Richard Lemay, "Gerard of Cremona," in *Dictionary of Scientific Biography,* vol. 15, Supplement I (1978), 174.

[12] For further details about Averroes, see Chapter 4.

According to Sarton, "these three commentaries corresponded to different stages in education, the short one being studied first, then the intermediate, finally the long one.... The long commentary was an innovation, probably inspired by Qur'ānic exegesis, the original texts upon which it was based being fully quoted and carefully separated from the glosses."[13]

A closer description of the three types will be useful. In his translation of Averroes's *Epitome* and *Middle Commentary of Aristotle's De Generatione et Corruptione*, Samuel Kurland describes the manner in which Averroes used each of these two types. The Epitome of the *De Generatione et Corruptione*, Kurland explains,

is not a commentary in the strict sense of the term but rather an independent brief restatement of the contents of the original work, preceded by a few remarks about the scope of its subject matter and about its exact place within the framework of the other Aristotelian works.[14]

To clarify Aristotle's thought for his readers, Averroes would even rearrange the order of Aristotle's arguments. His goal was to elucidate Aristotle's thought as much as he could.

The format of most of Averroes's Middle, or Intermediate, Commentaries was that of a running commentary where Averroes summarized, or even translated, a brief portion of Aristotle's text, followed, without any break or interruption, by a brief or lengthy interpretive comment. "In these commentaries," Wolfson explains, "the text of Aristotle, sometimes translated and sometimes paraphrased, was interspersed with Averroes's own comments and discussion. To a reader unacquainted with the text of Aristotle's own works, it would often be difficult to distinguish within those Intermediate Commentaries between Aristotle's original statements and Averroes' elaboration."[15]

In his long commentaries, however, as we saw earlier, Averroes included Aristotle's text but separated it from his commentary. His method was to present a segment of Aristotle's text and then to explain it at some length. On completion of his comments, he presented the very next portion of text and then commented upon that. By following this procedure, he moved successively through the entire text, commenting upon it section by section. A good example of this technique can be seen in the modern edition of

---

[13] Sarton, *Introduction to the History of Science*, vol. 2, part 1, 356; see also Harry A. Wolfson, "Revised Plan for the Publication of a *Corpus Commentariorum Averrois in Aristotelem*," *Speculum* 38 (Jan. 1963), 90. In what follows on Averroes, I rely heavily on Wolfson's article.

[14] *Averroes on Aristotle's "De Generatione et Corruptione" Middle Commentary and Epitome*, translated from the original Arabic and the Hebrew and Latin versions, with Notes and Introduction by Samuel Kurland (Cambridge, MA: Mediaeval Academy of America, 1958), xvi.

[15] Harry A. Wolfson, *Crescas' Critique of Aristotle: Problems of Aristotle's "Physics" in Jewish and Arabic Philosophy* (Cambridge, MA: Harvard University Press, 1929), 8.

Michael Scot's Latin translation of Averroes's *Commentary on Aristotle's "De anima."* [16]

The linguistic history of Averroes's commentaries is of considerable interest. Of the thirty-eight titles thus far attributed to Averroes, twenty-eight exist in the original Arabic, whereas thirty-six exist in Hebrew translations. The Hebrew role in the preservation of Averroes's works is even more extensive. Of the twenty-eight commentaries extant in Arabic, fifteen survive in the original Arabic, "four are both in Arabic and in Hebrew characters, and nine are only in Hebrew characters." [17]

Over the period of the Middle Ages to the end of the sixteenth century, Latin translations were made of thirty-four of the thirty-eight commentaries. Of these, fifteen were medieval translations made in the thirteenth century directly from Arabic, whereas the other nineteen Latin translations were made from Hebrew in the sixteenth century. [18] Although Michael Scot was not the only scholar who translated Averroes's commentaries into Latin, he was apparently largely, if not solely, responsible for translating Averroes's commentaries on natural philosophy. Michael translated the long, or great, commentaries on *On the Heavens (De caelo)*, *On the Soul (De anima)*, *Physics*, and also the *Metaphysics*. [19] He also translated Averroes's Middle Commentaries on the *Meteorology* and *On Generation and Corruption*, and did Averroes's epitome of Aristotle's *Treatise on Natural Things (Parva naturalia)*. [20] Of all these translations, however, only the great commentary on *De caelo* is with reasonable certainty by Michael Scot. The others are with a good degree of plausibility also thought to be by Michael Scot. [21]

Latin scholars who had access to Averroes's long commentaries in Michael Scot's translations could read not only Aristotle's text of the *Physics*, say, or *On the Heavens*, but also Averroes's elaborate commentaries on each segment of Aristotle's text. But scholars, who had only an epitome or middle commentary before them, could not easily distinguish Aristotle's words and thoughts from those of Averroes. There was no clear and obvious demarcation between the two. Nevertheless, because Averroes's objective in the epitome and Middle Commentary was to present Aristotle's thoughts as accurately and clearly as possible, it is likely that readers would have absorbed Aristotle's arguments and ideas in a reasonable manner and in

---

[16] *Averrois Cordubensis Commentarium Magnum in Aristotelis De anima libros*, recensuit F. Stuart Crawford (Cambridge, MA: The Mediaeval Academy of America, 1953).

[17] Wolfson, "Revised Plan," ibid., 94.   [18] Wolfson, ibid.

[19] Minio-Paluello suggests that Theodore of Antioch, not Michael Scot, may have been the translator of Averroes's long commentary on the *Physics* and argues further that there is no evidence to support the claim that Michael Scot is the translator of the long commentary on the *Metaphysics*. See Lorenzo Minio-Paluello, "Michael Scot," *Dictionary of Scientific Biography*, vol. 9, 362–363.

[20] Wolfson, "Revised Plan," 94.

[21] See Dod, "Aristoteles latinus," in Kretzmann, et al., eds., *The Cambridge History of Later Medieval Philosophy*, 59.

the sequence in which Aristotle presented them, although Averroes did take some liberties with the text.

Between the translations from Arabic to Latin of Gerard of Cremona and Michael Scot, Aristotle's natural philosophy became wholly available by the mid-thirteenth century. With Gerard's translations of Aristotle's *Meteorology* and *On Generation and Corruption* and Michael Scot's translation of Averroes's long commentaries on *On the Heavens* and *On the Soul*, and perhaps of the *Physics* (Gerard also had made a complete translation of the *Physics*), the Latin West had virtually the whole of Aristotle's natural philosophy. And, moreover, they had the very helpful lengthy commentaries that accompanied the full translations in the long commentaries. But even the short and middle commentaries provided usable summaries and paraphrases of the content of Aristotle's *Meteorology* and *On Generation and Corruption*. Indeed, although the short and middle commentaries were not direct translations, they were faithful summaries and paraphrases with additional helpful insights from Averroes to guide Latin readers.

But it was well known that, all things being equal, a Latin translation made from the Greek was likely to be more accurate than one made from Arabic. To know this is to know nothing more than that Greek and Latin are cognate languages. As the Latin translations from Arabic were made during the twelfth and thirteenth centuries, so also were parallel efforts made to translate the corpus of Greek natural philosophy and science into Latin.

Translations from Greek into Latin

The first fruits of translations from the Greek were probably an outgrowth of the revival of interest in Aristotle in Constantinople during the eleventh century. It was in that famous city that James of Venice[22] (d. after 1147) had the best opportunities to gain access to Aristotle's Greek texts. James was the first to translate Aristotle's Greek works into Latin, doing the *Physics*, *On the Soul*, the *Parva naturalia*, and at least part of the *Metaphysics*. He also translated the *Posterior Analytics* and other logical treatises.[23] James's translations endured. His translations of the logical treatises were regarded as "the 'authentic' texts for the next three centuries."[24]

Although James of Venice marks the real beginning of the translations from Greek to Latin in the first half of the twelfth century, William of Moerbeke (ca. 1215–ca. 1286) brought the process to a grand climax in the last half of the thirteenth century. Moerbeke translated at least forty-eight treatises.

---

[22] Also known as Iacobus Veneticus Grecus. Minio-Paluello explains that "'Grecus' could mean either that he spent much of his life in some Greek-speaking part of the Byzantine Empire or that he was of Greek descent." See Lorenzo Minio-Paluello, "James of Venice," in *Dictionary of Scientific Biography*, vol. 7 (1973), 65.

[23] See Minio-Paluello, "James of Venice," ibid., 66; see also Dod, "Aristoteles latinus," in *The Cambridge History of Later Medieval Philosophy*, 54–55.

[24] Ibid.

Included in this number were seven treatises on mathematics and mechanics by Archimedes, translated for the first time into Latin.[25] He also translated for the first time from Greek into Latin, Aristotle's biological works, including *The History of Animals, On the Parts of Animals*, and *On the Generation of Animals*, and translated *On the Motion of Animals* and *On the Progress of Animals* into Latin for the first time from any language. He also translated the four books of the *Meteorology*.

In translating the rest of Aristotle's natural philosophy, Moerbeke found it useful to revise or expand earlier translations. Thus, he revised James of Venice's translation of Aristotle's *Physics* and *On the Soul* (the latter in 1268 at Viterbo, the Papal Court).[26] He also revised, added to, and completed earlier translations of the *Parva naturalia, On the Heavens,*[27] *On Generation and Corruption*, and the *Metaphysics*. In addition, Moerbeke translated commentaries on Aristotle's works by Greek commentators from late antiquity. Thus he translated John Philoponus's *Commentary On the Soul*, and Simplicius's *Commentary On the Heavens*. One of the earliest beneficiaries of Moerbeke's translations was Thomas Aquinas who used Morebeke's translations of Aristotle's *On the Soul* and *Physics*; Simplicius's *Commentary on On the Heavens*; Themistius's *Commentary on On the Soul*, and others.[28] With Moerbeke's monumental contributions, all of Aristotle's natural philosophy was available by the last quarter of the thirteenth century in translations from Greek and Arabic.

HOW TRUSTWORTHY ARE ARISTOTLE'S TRANSLATED TEXTS?

In the second chapter, we confronted the problem of Aristotle's authorship – and the nature of that authorship – of the many treatises attributed to him. Now we must cope with the problem of reliability.

An important question arises. Because translations from both Arabic and Greek circulated and there were in some instances two and even three translations of the same work from Greek or Arabic, how reliable could their

---

[25] For a brief biography and list of Moerbeke's translations, see Lorenzo Minio-Paluello, "Moerbeke, William of," in *Dictionary of Scientific Biography*, vol. 9 (1974), 434–440; and Marshall Clagett, *Archimedes in the Middle Ages*, Vol. 2: *The Translations from the Greek by William of Moerbeke* (Philadelphia: American Philosophical Society, 1976), 3–13 and 28–31, n. 1. A list of Moerbeke's translations also appears in Edward Grant, *A Source Book in Medieval Science* (Cambridge, MA: Harvard University Press, 1974), 39–41.

[26] He also revised James of Venice's earlier translation of the *Posterior Analytics*. See Minio-Paluello, "James of Venice," ibid., 437 and Grant, *Source Book*, 40.

[27] Moerbeke revised the first two books of Robert Grosseteste's earlier translation and translated books 3 and 4 anew. See Grant, *Source Book*, 40.

[28] See James A. Weisheipl, O.P., *Friar Thomas d'Aquino: His Life, Thought, and Work* (Garden City, NY: Doubleday & Co., 1974), 152.

knowledge of Aristotle's treatises have been? Indeed, as Scott Montgomery has recently shown, the problem with translations is not merely one of converting a text from one language to another, but of transmitting that text from one culture to another. If we confine our remarks to Aristotle's works, Montgomery would argue that the Aristotelian texts translated in the Middle Ages from Arabic and Greek were so far removed from what might qualify as Aristotle's original Greek texts that we are in fact dealing with essentially different treatises. The numerous translations of Aristotle's individual works – from Greek into Syriac and Arabic and then from Arabic into Hebrew or Spanish and finally into Latin; or from Greek into Latin – must inevitably have altered the meanings of the texts as they passed from language to language and culture to culture. When one adds to this, the innumerable copyist errors that were inevitably and inexorably intruded into Aristotle's texts as these were multiplied and disseminated through Western Europe before the invention of printing, we seem driven to conclude that a different Aristotle lurked in every version of what purported to be one and the same treatise.

Under these difficult circumstances, one naturally inquires how scholars in the Middle Ages could have arrived at any consensus about Aristotle's ideas and interpretations. And yet somehow they did. Harry Wolfson, who was master of the Greek, Latin, Arabic, and Hebrew languages, and knew much about Aristotle, medieval philosophy, and translations, wrote this illuminating passage in his book about the sixteenth-century Jewish Aristotelian scholar, Hasdai Crescas:

Aristotle was unknown to Crescas in the original Greek. He was also unknown to him in the Arabic translations. He was known to him only through the Hebrew translations which were made from the Arabic. It would be, however, rash to conclude on the basis of this fact that his knowledge of Aristotle was hazy and vague and inaccurate, for, contrary to the prevalent opinion among students of the history of philosophy, the translations of Aristotle both in Arabic and in Hebrew have preserved to a remarkable degree not only clear cut analyses of the text of Aristotle's works but also the exact meaning of his terminology and forms of expression. The literalness and faithfulness with which the successive translators from one language into another performed their task, coupled with a living tradition of Aristotelian scholarship, which can be shown to have continued uninterruptedly from the days of the Lyceum through the Syriac, Arabic and Hebrew schools of philosophy, enabled Crescas to obtain a pretty accurate knowledge of Aristotle's writings.[29]

Wolfson's remarks about translations into Arabic and Hebrew apply equally, and probably even more so, to translations of Aristotle's texts into Latin. Once Aristotle's treatises on logic and natural philosophy were adopted as the basic curriculum in the arts faculties of medieval universities in the course of the thirteenth century, they became the common property of students and teachers over the whole of Europe for almost four centuries.

[29] Harry A. Wolfson, *Crescas' Critique of Aristotle*, 7.

Students read very similar Aristotelian texts at the Universities of Oxford and Paris and the scholars who wrote commentaries or questions on those same texts confronted much the same issues and understood one another's interpretations and analyses, because they shared a common heritage of Aristotle's works.

PSEUDO-ARISTOTLE: WORKS FALSELY ATTRIBUTED TO ARISTOTLE

During the late Middle Ages, a large number of treatises, perhaps as many as one hundred, appeared in the West that were falsely attributed to Aristotle.[30] A few of these were in fact derived from Greek originals, as, for example, *On the Universe (De mundo)*, *Mechanics*, *Problems*, *On Plants*, *On Colors*, and others. These were treatises that derived from Greek sources, perhaps even from Aristotle's school. But they have been judged by scholars to be spurious. Nevertheless, they seem sufficiently akin to the genuine works so that editors of Aristotle's collected works include them regularly as part of the Aristotelian corpus. Indeed, sixteen of the thirty works that appear in the *Complete Works of Aristotle* (Revised Oxford translation) are judged spurious, including those mentioned earlier in this paragraph.[31] But few Latin treatises falsely ascribed to Aristotle in the Middle Ages were derived from Greek texts.[32] Many spurious treatises – perhaps thirty to forty – were translated from Arabic to Latin.[33] The most famous of these was the *Secret of Secrets (Secretum secretorum)*, of which Schmitt says:

This best known and most widely distributed of all spurious Aristotelian works had an enormous *fortuna* stretching from the tenth to the seventeenth century or even eighteenth century, from Iceland, Scotland and Portugal to North Africa, the Middle East, and Muscovy. Read by philosophers, kings, alchemists and charlatans, humanists and courtiers. It can claim an influence far greater than many of the most influential of genuine Aristotelian works. In addition to being extant in more Latin manuscripts than any work – spurious or genuine – attributed to Aristotle during the Middle Ages it was also translated into far more vernacular languages than any other Aristotelian work.[34]

[30] On this topic, I follow C. B. Schmitt, "Pseudo-Aristotle in the Latin Middle Ages," in Jill Kraye, W. F. Ryan and C. B. Schmitt, eds., *Pseudo-Aristotle in the Middle Ages* (London: the Warburg Institute, University of London, 1986), 3–14.

[31] The other treatises falsely attributed to Aristotle but included in *The Complete Works of Aristotle* are: *On Breath*, *On Things Heard*, *Physiognomics*, *On Marvellous Things Heard*, *On Indivisible Lines*, *The Situations and Names of Winds*, *On Melissus, Xenophanes, and Gorgias*, *Magna Moralia*, *On Virtues and Vices*, *Economics*, and *Rhetoric to Alexander*. All told, sixteen spurious treatises are included in *The Complete Works of Aristotle, The Revised Oxford Translation*.

[32] Schmitt, ibid., 4.   [33] Ibid., 5.

[34] Ibid., 4–5. The *Secret of Secrets* is extant in approximately six hundred manuscripts, an astonishing number. See Schmitt, ibid., 8.

The greater number of spurious Aristotelian treatises had a minimal inter-relationship with the much smaller corpus of Aristotle's genuine works. Of the pseudo-Aristotle treatises translated from Arabic to Latin only one – *On the Causes of the Properties of the Elements* (*De causis proprietatum elementorum*) – was incorporated into the standard body of works on Aristotle's natural philosophy that was used in the medieval universities.[35] Rarely did spurious works appear in codices of genuine Aristotelian works. They tended to appear in codices that included astrological, astronomical, medical, or alchemical treatises.[36] In light of the fact that the spurious treatises were thought to be as authentic as the genuine corpus of Aristotle's works, the sharp separation between the two types of treatise is truly striking. Why did this occur?

The answer is likely to lie in the fact that "with relatively few exceptions the Latin based *spuria* circulated among and were read by different social and intellectual groups than were the genuine works. . . . Relatively few of the *spuria* (i.e. works such as the *Liber de causis* and *De causis proprietatum elementorum*) appealed to the university culture, while many in the category of pseudo-science circulated among a much different group."[37] It is highly likely that the spurious works were simply not in the same category of excellence as were Aristotle's genuine works in natural philosophy. There were numerous anomalies among the spurious works. Occasionally, a part of the work is attributed to Aristotle but not the whole work.[38] In some instances, the spurious nature becomes obvious when the anonymous author mentions Aristotle by name. Averroes also probably played a role in excluding the spurious treatises from the university milieu. Scholars and students in the medieval universities would not have found the spurious treatises mentioned in Averroes's commentaries, whereas Averroes had occasion to mention the genuine works frequently.

A striking feature of the numerous pseudo-Aristotle treatises is that only rarely were commentaries written on them. Despite the great popularity of the *Secret of Secrets*, there is apparently only one commentary on it, written by Roger Bacon.[39] One of the few pseudo-Aristotle treatises on which some commentaries were written is the *Book on Causes* (*Liber de causis*), a treatise on which both Albertus Magnus and Thomas Aquinas commented. But, for the most part, there is an almost complete absence of commentaries on the spurious treatises that were not derived from Greek originals. The absence of commentaries is probably explicable by the fact that the spurious treatises were not part of the university curriculum. Teaching masters were not required, or expected, to teach or write about them; consequently, students were not expected or required to read the spurious texts and may have known little about them.

---

[35] Ibid., 9.   [36] Ibid., 8.   [37] Ibid.   [38] Ibid., 7.   [39] Ibid., 9.

By the end of the thirteenth century, the medieval age of translation was at an end.[40] The Middle Ages had received all of Aristotle's works and many works of Aristotle's commentators in both Greek and Arabic, including commentaries by John Philoponus, Simplicius, and Themistius representing the former language, and Avicenna and Averroes representing the latter. We must now see how the new natural philosophy was absorbed, used, and elaborated in its new linguistic format: the Latin language.

[40] A new wave of translations from Greek to Latin occurred in the fifteenth and sixteenth centuries – the period usually regarded as the Renaissance. These translations included the works of numerous ancient Greek authors whose works were virtually unknown in the Middle Ages, among whom Plato was the most prominent.

7

Natural Philosophy after the Translations
Its Role and Place in the Late Middle Ages

By the beginning of the thirteenth century, the major treatises of Greek and Arabic origin in science and natural philosophy were available in Latin translation.[1] It is almost certain that lectures were being given on at least some of Aristotle's books on natural philosophy at the new universities of Oxford and Paris in the first decade of the thirteenth century.[2] This is evident from the fact that in 1210, the Parisian Synod decreed, among other things, that "no lectures are to be held in Paris either publicly or privately using Aristotle's books on natural philosophy or the commentaries, and we forbid all this under pain of excommunication."[3] The prohibition of 1210 is a good indication that Aristotle's works on natural philosophy were readily available, for otherwise there would have been no need to ban public and private lectures on them. Although the University of Paris is not specifically mentioned in the decree, it is virtually certain that mention of public lectures is a reference to lectures at the university. To place medieval natural philosophy in its proper context, it is essential to describe the structure and character of the universities, which were indeed the "proper context" for Aristotle's natural philosophy and the commentaries on, and elaborations of, it. Indeed, it is almost as if the universities of Oxford and Paris, and their numerous successors, came into being to serve this function.

THE MEDIEVAL UNIVERSITY

In the course of the eleventh and twelfth centuries, Europe had rapidly become urbanized. Activities usual and normal to urban environments developed in these new or expanded cities. But it was not the usual and normal

---

[1] See D. A. Callus, O. P., "Introduction of Aristotelian Learning to Oxford," *Proceedings of the British Academy* 29 (1943), 229.

[2] Bernard G. Dod, "Aristoteles latinus," in Norman Kretzmann, Anthony Kenny, and Jan Pinborg, eds., *The Cambridge History of Later Medieval Philosophy from the Rediscovery of Aristotle to the Disintegration of Scholasticism 1100–1600* (Cambridge: Cambridge University Press, 1982), 70.

[3] Cited in Enzo Maccagnolo, "David of Dinant and the Beginnings of Aristotelianism in Paris," translated by Jonathan Hunt in Peter Dronke, ed., *A History of Twelfth-Century Western Philosophy* (Cambridge: Cambridge University Press, 1988), 429–430.

activities of urban life that fashioned the crucial events that made possible
the new intellectual life that forms the basis of our study. The key event that
made the new intellectual life of Western Europe different than anything that
had gone before is the emergence of the university as a unique and vital institution.[4] Not only was it unique in the history of Western Europe, but it also
was unique in the history of the world. The universities that already existed
around 1200 have a continuous history with their sister institutions of today.[5]
Of course, they did not appear full-blown in 1200. Their institutional development must be sought in the twelfth century, and even in earlier events –
for example, in the cathedral schools of the eleventh and twelfth centuries,
to identify one vital element. I shall now describe some of the salient features
that made the medieval university such an extraordinary institution.[6]

No university could be founded without a congregation of teachers and
students in relatively large numbers. Why this occurred in cities such as Paris,
Oxford, and Bologna, where the first three universities were established,
rather than London, Rome, and other possibilities, is simply unknown.[7]
The schools that would be transformed into universities attracted teachers
and student from all over Western Europe. Indeed, this was a characteristic
feature of the medieval university. Thus, students and teachers were usually
foreigners in the cities in which they taught and studied and therefore without the rights and privileges granted to the citizens of that community. To
overcome this serious obstacle to a stable existence, the masters and students
realized that their prospects would be greatly enhanced if they formed themselves into a corporation, or *universitas*, as had been done by numerous craft
and merchant guilds.

The corporation was an important medieval legal concept that was built
on Roman antecedents.[8] Church and state in the Middle Ages permitted

---

4 I shall draw on my earlier, lengthier account of the medieval university in Edward Grant,
   *The Foundations of Modern Science in the Middle Ages: Their Religious, Institutional, and
   Intellectual Contexts* (Cambridge: Cambridge University Press, 1996), ch. 3, 33–53.
5 This is made evident in a four-volume work on the universities titled *A History of the University in Europe* under the general editorship of Walter Rüegg. The four volumes in the
   series are: I: *Universities in the Middle Ages*; II: *Universities in Early Modern Europe (1500–
   1800)*; III: *Universities in the Nineteenth and Early Twentieth Centuries (1800–1945)*; IV:
   Universities from 1945 to the Present.
6 For detailed accounts of the history of the medieval university, see Hilde de Ridder-Symoens,
   ed., *A History of the University in Europe*: Vol. I: *Universities in the Middle Ages* (Cambridge:
   Cambridge University Press, 1992); Alan B. Cobban, *The Medieval English Universities:
   Oxford and Cambridge to c. 1500* (Berkeley: University of California Press, 1988); and Gordon Leff, *Paris and Oxford Universities in the Thirteenth and Fourteenth Centuries* (New
   York: John Wiley & Sons, 1968). For a study focused on the beginnings of the university, see
   Stephen C. Ferruolo, *The Origins of the University: The Schools of Paris and Their Critics,
   1100–1215* (Stanford, CA: Stanford University Press, 1985).
7 See Walter Rüegg, "Themes," in Hilde de Ridder-Symoens, ed. *A History of the University
   in Europe*: Vol. I: *Universities in the Middle Ages*, 14.
8 I have drawn on my brief account of the corporation in Edward Grant, *God and Reason in
   the Middle Ages* (Cambridge: Cambridge University Press, 2001), 98–100. My brief summary

individuals in a given profession or trade to form a corporation that was a fictional entity with various legal rights. For example, the members of the corporation had the right to elect their own officers. The candidate with the majority of votes was regarded as elected. The corporation was thus a form of representative government. Elected officials had the right to represent the corporation in the law courts or before official state and church authorities. Corporations could own property, sue or be sued in the courts, draw up contracts, and perform other functions in behalf of its members. To look out for the interests of their members, corporations formulated laws and statutes that were binding on their members.

Corporate status was a boon to universities. The privileges mentioned in the preceding paragraph effectively conferred autonomy on all entities legally regarded as corporations. As educational institutions, the universities enjoyed autonomy and were thus able to control their own affairs. As members of the university community, or corporation, students and teachers enjoyed various privileges and were also expected to adhere to certain restrictions. Despite their legal autonomy, universities in the thirteenth and fourteenth centuries were subject to some ecclesiastical control for which they received ecclesiastical benefits.

The students (often themselves called "clerks" and distinguished by their tonsure), like clerics already in minor orders or monks or priests, were subject to certain restrictions, being forbidden to bear arms or wear "secular" clothes, and usually to marry. But they also claimed to be answerable only to ecclesiastical courts and to enjoy the exemptions from taxation and military service allowed to the clergy. In comparison with ordinary clerics, indeed, members of universities enjoyed exceptional advantages. Their person and property were theoretically exempt from seizure or arbitrary summons (*ius non trahi*). To a great extent they were not subject to the jurisdiction of the local ecclesiastical authorities and could appeal directly to the pope or his representative. They were entitled to enjoy the income of ecclesiastical benefices without being resident in them.[9]

The legal fiction of the corporation enabled universities to become virtually permanent institutions. Many of them have continued operations to the present day, including Bologna, Paris, and Oxford, the first three universities that have now endured for more than eight hundred years.

The corporate character of the medieval university was so pervasive that it extended to its various subdivisions, such as faculties, colleges, and nations. Thus each of the four faculties of a complete university – the faculties of arts, medicine, law, and theology – was a corporate entity with various rights and privileges within the overall university corporation. Any teacher or student

was drawn from Harold J. Berman, *Law and Revolution: The Formation of the Western Legal Tradition* (Cambridge, MA: Harvard University Press, 1983), 215–221.

9 Aleksander Gieysztor, ch. 4: "Managment and Resources," in Hilde de Ridder-Symoens, ed., *A History of the University in Europe: Vol. I: Universities in the Middle Ages*, 109.

might legitimately belong to two or more distinct corporations of the university.

A university that drew its students from various parts of Europe was usually called a *studium generale* and its graduates usually had the right to teach at any university in Europe, a right known as "the license to teach anywhere" (*licentia ubique docendi*).[10] A degree from a university that drew its students primarily from the city in which it was located did not confer that right. Teaching was the primary activity of every university. Universities were never regarded as research institutions, although many teachers who commented on the works of Aristotle, or wrote separate tracts on this or that theme in natural philosophy, drew on the writings of many other authors and in that sense were doing research. But that was not their primary function.

The corporate structure of the medieval university provided substantial stability and thus allowed the teaching of natural philosophy to develop as the basis of all university learning in the four faculties that comprised a major university, namely, arts, theology, medicine, and law. The last three faculties constituted graduate level instruction for professional degrees in theology, medicine, and law. To matriculate for a degree in one of these higher faculties, the prospective student was ordinarily expected to have acquired a Bachelor of Arts degree as well as a Master of Arts degree, the latter customarily requiring two additional years of lectures and study beyond the four years required for the bachelor's degree. It was while pursuing their arts degrees that future theologians, physicians, and lawyers acquired an unusual degree of familiarity with logic and natural philosophy. It was their shared experience in the arts faculty that gave to all university students for approximately four centuries a common intellectual experience, and a common intellectual heritage. Whatever their future careers might be – whether as professors in arts, or practitioners or professors of theology, law or medicine, or as clerks in royal courts or municipalities – all students acquired a basic knowledge of logic and natural philosophy, subjects that were valued for their own sakes and also regarded as useful, if not indispensable, in the higher disciplines of theology, law, and medicine.

THE IMPACT OF ARISTOTELIAN NATURAL PHILOSOPHY IN THE EARLY THIRTEENTH CENTURY TO 1240

The natural philosophy that was taught and studied in medieval universities was primarily based on the works of Aristotle that, as we saw in the preceding chapter, had been translated from Arabic and Greek into Latin. Although William of Moerbeke translated many of Aristotle's treatises from Greek to

[10] See Ferruolo, *The Origins of the University*, 4, and Gieysztor, "Management and Resources," in Ridder-Symoens, ed., *Universities in the Middle Ages*, 108.

Latin in the 1260s, Aristotle's natural books – the *Physics*, *On the Heavens*, *On the Soul*, *Meteorology*, *On Generation and Corruption*, and the *Small Books on Natural Things* (*Parva naturalia*) – were available much earlier in the thirteenth century in translations from Greek and Arabic, as is obvious from the ban on Aristotle's books in Paris in 1210.[11]

There is little direct evidence of substantive use of Aristotle's books in the first decade of the thirteenth century. Around 1200, for example, Daniel of Morley, at Oxford, wrote a treatise titled *Book on the Natures of Inferior and Superior Things* (*Liber de naturis inferiorum et superiorum*) in which he seems to cite from Aristotle's natural books (*Physics*, *De caelo*, and the *De sensu et sensato*). It has been shown, however, that the quotations are not from Aristotle directly but derive rather from a translation of Avicenna's commentary on Aristotle's *De caelo et mundo*.[12]

Evidence that Aristotle was read at Oxford before the 1240s and at Paris before 1255 is meager.[13] At Paris, the bans of 1210 and 1215 had an impact in discouraging the reading of Aristotle's natural philosophy and metaphysics, although his logical works and his treatise on *Ethics* were read and discussed at the university. But even if Aristotle's books on natural philosophy were excluded from the curriculum at the University of Paris, those same works were being read by some theologians who found occasion to cite them in their theological treatises. Among these theologians were William of Auxerre, who, in his *Summa aurea* (composed ca. 1215–1220), cited from Aristotle's *On the Soul* and *Physics;* Philip the Chancellor who, in his *Summa de bono* (composed ca. 1230–1236), quoted from Aristotle's *On the Soul*, *Physics*, *De animalibus*, *Metaphysics*, *On the Heavens*, and *On Generation and Corruption;* and in William of Auvergne's *On the Universe* (*De Universo*) and *On the Soul* (*De anima*) (both composed between 1231 and 1236), in which a similar range of quotations appears.[14]

In Oxford, Alexander Neckham (or Nequam) and Alfred of Sareshel cited Aristotle's works in the early decades of the thirteenth century. The former mentioned a few of Aristotle's works but did not utilize them much.[15] D. A. Callus regards Alfred of Sareshel as "the pioneer in the long list of medieval commentators on the *libri naturales*," namely, Aristotle's books on natural philosophy. Alfred wrote glosses on Aristotle's *On the Heavens* (De caelo et mundo), *On Generation and Corruption* (De generatione et corruptione),

---

[11] See Callus, "Introduction of Aristotelian Learning to Oxford," *Proceedings of the British Academy* 29 (1943), 232.

[12] Callus, ibid., 234.   [13] I rely here on Dod, "Aristoteles Latinus," 71–73.

[14] See Dod, ibid., 72. Callus, "Introduction of Aristotelian Learning to Oxford," 231, n. 4, lists the number of times some of these authors quoted from each of Aristotle's works. See also Gilson, *A History of Christian Philosophy in the Middle Ages*, 260.

[15] See Charles Homer Haskins, *Studies in the History of Mediaeval Science* (Cambridge, MA: Harvard University Press, 1924), 128.

On the Soul (De anima), and *Meteorology*. Not only was Alfred of Sareshel a pioneer among medieval commentators on Aristotle, but he also "links up the tradition of the twelfth with that of the thirteenth century."[16]

The extant record indicates that the incorporation of Aristotle's thoughts and ideas into contemporary literature, and the composition of commentaries in a variety of forms and contexts was a gradual, even slow process. In this process, theologians were as apt to use quotations and ideas from Aristotle's natural books – that is, his books on natural philosophy – as were secular authors. William of Auvergne, as we saw, was one of those theologians who used Aristotle's books, doing so in his *On the Universe (De universo)*, composed between 1231 and 1236. William used natural philosophy to serve the ends of theology and in this sense was in the Augustinian tradition that made philosophy the handmaiden of theology.

William of Auvergne (ca. 1180–1249)

As is all too often the case with medieval authors, little is known of William's life. He was probably born between 1180 and 1190 and became a canon of the cathedral of Notre Dame by 1223. Two years later, in 1225, he was a professor theology at the University of Paris and in 1228 was named bishop of Paris, a post he held until his death in 1249.[17] William was a prolific author, devoting most of his works to theology and theological problems.[18] His most important treatise was *Magesterium divinale et sapientale (Teaching on God in the Mode of Wisdom)*, an incredibly lengthy work in seven parts of which the second part, *On the Universe of Creatures (De universo creaturarum)*, was devoted to the world of nature.

William makes it clear that he was philosophizing and not relying on appeals to divine law or Scripture. In the fourth part of the *Magesterium*, called *Cur Deus homo (Why God Became Man)*, he informs his readers that he will proceed "through the paths of proof and investigation . . . because only in this way can one take care of the errors of the educated."[19] Not even Aristotle was to be taken as authoritative. William declares: "Let it not enter your mind that I want to use the words of Aristotle as authoritative for the proof of those things which I am about to say, for I know that a proof from an authority is only dialectical and can only produce belief, though it is my aim, both in this treatise and wherever I can, to produce demonstrative certitude,

---

[16] Callus, "Introduction of Aristotelian Learning to Oxford," 238; see also Haskins, *Studies in the History of Mediaeval Science*, 128–129.

[17] See Roland J. Teske, S. J., *William of Auvergne: The Universe of Creatures: Selections Translated from the Latin with an Introduction and Notes* (Milwaukee: Marquette University Press, 1998), 13. For a few additional details of William's life, see ibid.

[18] See the list of his works in Teske, tr., *William of Auvergne*, 14.

[19] Translated by Teske, *William of Auvergne*, 16.

after which you are left without any trace of doubt."[20] Steven Marrone explains that, in his *Magesterium divinale*, William generally "proposed to defend all of truth against the errors of the impious philosophers, and he said he would do so without making any appeal to Christian faith or the authority of divine revelation, relying only on the way of natural reason and logical proof."[21] By his strong emphasis on reason in problems involving natural philosophy and theology, William reveals how firmly he was linked to those theologians and natural philosophers of the twelfth and early thirteenth century who, as we saw, regularly emphasized reason in the problems they considered.

The *Magesterium divinale* was largely devoted to theological problems. *On the Universe* (*De universo*), however – the second part of that very lengthy treatise – includes theological problems that utilize natural philosophy. It is here that William frequently cites Aristotle and Avicenna. William's knowledge of Aristotle's works was rather superficial, however, and he occasionally confuses Aristotle's ideas with those of Avicenna. Indeed, he seems to have been more familiar with the latter than the former. The *De universo* has a complex format. It is divided into two principal parts, of which the first is concerned with the corporeal, physical universe, and the second is devoted to spiritual aspects of the world. The first principal part is divided into two parts, the first of which is divided into three distinct treatises.

The first part of the first principal part of the *De universo* is primarily concerned with the creation of the world. The first treatise of the first principal part (chapters 2 to 10) is directed against the heretical Cathars whom William calls Manichees. The Manichees were dualists, who believed that there are two gods, one good, the other evil, and that there are two universes.[22] In the second treatise of the first principal part (chapters 11 to 16), William seeks to demonstrate that there is only one world, not two. In his defense of one world, we might have expected William to rely heavily on Aristotle's *On the Heavens* (*De caelo*), where, in book 1, chapter 9, Aristotle argues that it is impossible for more than one world to exist. In the course of his argument, Aristotle declares that "there neither is, nor can come into being, any body outside the heaven."[23] Without any mention of Aristotle, William may have had him in mind, when he declares: "those who maintain only one world claim that there is no body outside it." Although William may have derived this claim from Aristotle, parts of the rest of his argument appear to draw on

[20] Translated by Teske, ibid.
[21] See Steven P. Marrone, *William of Auvergne and Robert Grosseteste: New Ideas of Truth in the Early Thirteenth Century* (Princeton, NJ: Princeton University Press, 1983), 27.
[22] *De Universo*, First Principal Part, First Treatise, chapter 2, in Teske, *William of Auvergne*, 34.
[23] *On the Heavens*, 1.9.278b.22–24; translated by J. L. Stocks, in *The Complete Works of Aristotle*.

Michael Scot's *Commentary on the Sphere of Sacrobosco*, which was written about the time William composed his *On the Universe*.[24] William opposes those who argue that there are other worlds besides our own and that void spaces lie between them. This cannot be, argues William, because we know that void spaces are impossible, a claim he probably derived from Aristotle. He then presents a series of arguments to show why void is impossible. "I say, then," he argues,

that a void can either be cut or divided or it cannot. If it cannot, it is impossible that motion come about in it or through it, just as if air and water were utterly indivisible and uncutable, it would be impossible that there be motion through one of them, since every body that moves through another cuts a path for itself by such a division in it or through it. But those who maintain a vacuum or a void say that all motion takes place in it or through it.

Moreover, on this view, the void or vacuum will be the most solid and strong of all bodies, since it will be neither penetrable nor divisible or cutable in any way, and it could not yield to any other body. How then will it be a void or vacuum?[25]

William rejects the existence of other worlds, in part because other worlds would necessarily have void spaces lying between them. He also argues for the finitude of the world and against the possibility of an infinite world, citing Aristotle's *De caelo et mundo*.[26]

In the third treatise of the first principal part (chapters 17 to 27), William deals with the manner in which God created the creatures of the world, emphasizing that God creates through his word, which is His thought. William attempts to refute Aristotle's ideas about creation, but the opinions he attacks are really those of Avicenna. He explains the creation of the first intelligence and how the second intelligence proceeds from the first intelligence, and the third from the second, and so on to the tenth intelligence. Thus did William of Auvergne falsely attribute Avicenna's doctrine of the emanation of intelligences to Aristotle, who held no such doctrine.

In the second part of the first principal part, William considers the eternity of the world (chapters 1 to 11). Here William cites, and refutes, Aristotle's arguments in favor of an eternal world.

The second principal part of *De Universo* is exclusively concerned with the spiritual, as opposed to the corporeal, world, the focus of the first

---

[24] For the Latin text of Michael Scot's commentary, see Lynn Thorndike, *The "Sphere" of Sacrobosco and Its Commentators* (Chicago: University of Chicago Press, 1949), 247–342.

[25] *De Universo*, First Principal Part, the second treatise, chapter 14, in Teske, *William of Auvergne*, 63.

[26] See *Guilielmi Alverni Episcopi Parisiensis, mathematici perfectissimi, eximii philosophi, ac theology praesantissimi, Opera Omnia* Tomus Primus (Parisiis: Apud Ludovicum Billaine, 1674), *De Universo*...In duas partes principales divisum, primae partis De universo, pars I, ch. XIV, pp. 607, col. 2–610, col. 1.

principal part. In a preface to the second principal part, William explains that:

> it is necessary that the science of the spiritual universe is more noble than the science of the corporeal or bodily universe to the extent that the spiritual nature is recognized as more excellent than any bodily nature.[27]

Because the spiritual world is nobler than the corporeal world, William urges everyone to study it, but censures the philosophers who have thus far ignored it, largely because of "its profundity, the shortsightedness of the human intellect, and the remoteness of those substances from our ordinary life."

> And for these three reasons this universe was neglected and left aside by those philosophers who preceded me. For it seems incredible that men who were completely dedicated to the pursuits of philosophy and who were most eager researchers of the sciences and who were most fervent with the love of philosophy held so noble a science in contempt and passed it over out of laziness or negligence.[28]

In his analysis of the spiritual universe, William explains, in the first chapter of that section, that the spiritual world consists of three parts: (1) the intelligences that are separated from matter, and in which Aristotle believed; (2) the good angels, which the Greeks call "good demons" (*kalodaemones*); and (3) the bad angels, or devils, which the Greeks call "evil demons" (*kakodaemones*).[29] As he describes and analyzes these aspects of the spiritual world, William finds occasion to use natural philosophy to support his arguments, as when he speaks of the movements and velocities of the heavenly bodies and also mentions that light does not move instantaneously.[30]

How does William of Auvergne's *On the Universe* relate to natural philosophy? Is it primarily a treatise on natural philosophy, or is it essentially a theological work? I believe that it is really a theological treatise in which natural philosophy is made to play a role in the elucidation of theological problems. Some of the problems that William raised were central to both theology and natural philosophy, as is obvious in his discussion of the eternity of the world and whether more than one world can exist. In the course of his analysis of these problems, he found occasion to introduce the concept of void space and to utilize that concept in his arguments against the existence of other worlds.

In all this, William is not interested in doing natural philosophy for its own sake, but in utilizing it for theological purposes, as, for example, to destroy the heretical arguments of the Cathars, or Manichees, as he calls them, in favor of the existence of more than one world; or he argues against

---

[27] *De Universo*, The first part of the second principal part, Preface, 139 (Teske tr.)
[28] Ibid.
[29] *De Universo*, The first part of the second principal part, chapter 1, 140 (Teske tr.)
[30] See William A. Wallace, O. P., "William of Auvergne (Guilelmus Arvernus or Alvernus)," in *Dictionary of Scientific Biography*, vol. 14 (1976), 388.

Aristotle's advocacy of the eternity of the world. But it is obvious that his consideration of the corporeal world is but a prelude to the more important task of describing the spiritual world.

William of Auvergne marks a transition to the new world of scholastic philosophy that was already emerging in his day. By the 1230s, when he wrote his lengthy treatise, Aristotle was becoming more and more widely known, despite obstacles to the study of his works that had been posed at the University of Paris. William found occasions to cite from most of Aristotle's works on natural philosophy. But his familiarity with them sometimes lacked depth and understanding, as is evident from the fact that some of the citations he attributes to Aristotle were really drawn from Avicenna.

Within a relatively short period after William of Auvergne wrote his great theological treatise, wherein natural philosophy – albeit a natural philosophy that was now becoming increasingly Aristotelian – retained its traditional role as the handmaiden of theology, natural philosophy would be wholly transformed, as it became the central focus of the curriculum in the medieval university. The natural philosophy that came to dominate the universities was Aristotle's natural philosophy as embedded in his natural books, namely, his *Physics*, *On the Heavens*, *On the Soul*, *On Generation and Corruption*, *Meteorology*, and the *The Short Physical Treatises*.[31]

UNIVERSITY LECTURES ON NATURAL PHILOSOPHY

The ban on teaching Aristotle's natural philosophy at the University of Paris was clearly over by 1255, when the faculty of arts made lectures on the whole of Aristotle's natural philosophy mandatory.[32] Despite the ban, lectures on Aristotle's natural philosophy had already been given at the University of Paris during the 1240s, when Roger Bacon lectured on Aristotle's *Metaphysics*, *Physics*, *On Generation and Corruption*, *On the Soul*, and perhaps *On the Heavens*.[33] The extant texts attributed to Roger Bacon – for example, his *Questions on the Physics*, *Questions on On the Soul*, and *Questions on Generation and Corruption*, to name only a few – are probably based directly on his lectures. Perhaps some ten years or so earlier, around 1230, at Oxford University, Robert Grosseteste wrote a commentary on Aristotle's

---

[31] Also see Chapter 2.

[32] For the document listing the treatises in logic and natural philosophy and when they were to be read and discussed, see Lynn Thorndike, *University Records and Life in the Middle Ages* (New York: Columbia University Press, 1944), Selection 28, 64–66. The Latin text is cited by Maurice de Wulf, "The Teaching of Philosophy and the Classification of the Sciences in the Thirteenth Century," *The Philosophical Review* 27 (1918), 361. See also Lindberg, *The Beginnings of Western Science*, 218.

[33] David C. Lindberg, ed. and tr., *Roger Bacon's Philosophy of Nature: A Critical Edition, with English Translation, Introduction, and Notes, of "De multiplicatione specierum" and "De speculis comburentibus"* (Oxford: Clarendon Press, 1983), xvii. Bacon also lectured on Aristotle's *On Animals* (*De animalibus*) and on at least two pseudo-Aristotelian works.

Physics, and played a major role in introducing the new natural philosophy and logic into Oxford.[34] Other teaching masters who lectured on the works of Aristotle from the 1240s, and in the years and centuries to come, usually did much the same thing: they converted their classroom lectures into written commentaries of one form or another. It will prove useful to describe the relationship between the university classroom lectures on Aristotle's treatises on natural philosophy and the written texts on those same treatises.

By making Aristotle's natural books the basis of the study of natural philosophy, university teachers of that discipline had to devise methods for explicating those treatises to the many students who were required to study them.[35] A basic form of classroom instruction was to comment on a text. One form of this method was for the teacher to read a portion of text and to gloss, or explain, difficult terms or passages. If the master presented his glosses in written form, this kind of commentary would be designated a gloss (*glossa*). Or a master might explain a text section by section but provide a literal explanation for the meaning of each section. This method does not seem to have had a written counterpart.

More widely used than either of these two techniques was a third approach in which the lecturers separated the text from the commentary. After reading the text, the teacher would comment on that section and perhaps even add the opinions of other commentators. The primary aim of this kind of commentary was to explain the text section by section. The written form of this approach was the straightforward commentary, the kind that scholastic natural philosophers found in Averroes's numerous commentaries on the works of Aristotle. It was the model for commentaries by St. Thomas Aquinas, Walter Burley, and Nicole Oresme.

In another form of commentary, masters paraphrased Aristotle's treatises, often intermingling bits and pieces of text. They also added explanatory opinions and interpretations. Scholastic natural philosophers found such techniques employed in Avicenna's works. Albertus Magnus used this technique in his numerous paraphrases of Aristotle's treatises (for an example, see Chapter 8).

The most popular type of commentary literature in the Latin Middle Ages was the *questiones*, or questions, format, which is almost synonymous with the scholastic method of medieval philosophy (for examples, see Chapter 8). The method of posing questions evolved from the university's lecture system. Because teaching masters had a reasonable degree of freedom in covering the text of an Aristotelian treatise, some of them began to pose questions at the

---

[34] See A. C. Crombie, "Grosseteste, Robert," in *Dictionary of Scientific Biography*, vol. 5 (1972), 549.

[35] For a list of the different literary forms of the commentaries on Aristotle's philosophical works, see Charles H. Lohr, S. J., "Medieval Latin Aristotle Commentaries: Authors A–F," in *Traditio* 23 (1967), 313. I have discussed the different types of lectures and scholastic literary forms in *The Foundations of Modern Science in the Middle Ages* (Cambridge: Cambridge University Press, 1996), 40–42.

end of lectures. These questions concerned special problems and topics in the text itself. Gradually, the masters increased the number of thematic questions while reducing the sequential commentary until they had transformed the commentary into a series of questions with the commentary wholly eliminated. Thus, a course on Aristotle's *Physics* might consist of a series of questions on each of the eight books of that treatise. The lectures of masters were often transcribed and "published" in the sense that scribes at the university bookstore made a master copy of the questions from which other copies were made that were either sold or rented to students and teachers. Additional copies were often made from rental copies. We obtain our knowledge of the vast body of Aristotelian questions literature from copies that somehow survived and are presently preserved in the libraries of Europe and America. We shall say more about this later.

Another kind of question relevant to natural philosophy is the *quodlibetal* question, which is a direct outgrowth of the medieval system of university disputations of which there were two kinds. The "ordinary disputation" (*disputatio ordinaria*), which was presided over by a teaching master, was usually held once a week. The presiding master posed a question and the audience of students and masters split into two parts, one defending the affirmative, the other upholding the negative side. Another form of disputation was held once or twice a year and was a special occasion. In this context, the questions were posed by the audience to the master who was expected to answer them in the standard way, namely, to give the pros and cons and then to defend one side or the other. Within this format, any member of the audience could pose any question he pleased, for which reason this kind of disputation was called a *disputatio de quolibet*, namely, a disputation about anything whatever. Over the two-day period of the disputation, the master was expected to answer numerous questions. Many eminent medieval masters – for example, Godfrey of Fontaines (d. 1306), Duns Scotus, and William of Ockham – presided over quodlibetal disputes and published most, if not all, of the questions to which they had responded. The questions posed in a quodlibetal dispute could be on any topic whatever. Hence the aggregate of questions in a quodlibetal treatise ranged over many subjects and areas of natural philosophy and might even go into areas that lay outside of natural philosophy.

But there is one more type of treatise on Aristotle's natural philosophy that is worthy of mention. This is the summary account of a wide range of themes in Aristotle's natural philosophy. Such treatises were made to provide students, or perhaps anyone interested, with an introduction to the subject. One of the most unusual and important works in this category is a very lengthy unpublished, anonymous treatise of 236 manuscript folios. It was probably written around the middle of the fourteenth century. Not only did the anonymous author wish to make the subject intelligible to young readers, who, as he explains, usually found Aristotle difficult to understand, but he also wished to inform his readers about topics that Aristotle had not treated but that had been introduced into natural philosophy by the anonymous author's contemporaries and immediate predecessors.[36] Because this important, anonymous treatise reveals much about the status of natural philosophy at the high point of its development in the late Middle Ages, more will be said about it later in this chapter.

Natural philosophy was not, however, confined to summary accounts, commentaries, and questions on Aristotle's treatises. A large body of medieval natural philosophy was written in the form of tractates (*tractatus*), or treatises, on special topics or themes. The themes were often drawn from Aristotle's works. For example, a number of treatises on problems of motion have been preserved, among which we might mention Thomas Bradwardine's *Treatise on Proportions or on the Proportions of the Speeds of Motions* (*Tractatus proportionum seu de proportionibus velocitatum in motibus*)[37] and Nicole Oresme's *On Ratios of Ratios* (*De proportionibus proportionum*).[38] In a work titled *Treatise on the Configuration of Qualities and Motions*,[39] Oresme considered how qualities varied, representing the variations geometrically. Variation of qualities was a problem derived ultimately from Aristotle's *Categories*.

Once we leave Aristotle's works and the various commentaries on them, we are confronted with a problem. Did natural philosophy extend beyond Aristotle's natural philosophy? And if so, what did it embrace? Were alchemy and astrology regarded as part of natural philosophy? Did alchemists and astrologers believe they were doing natural philosophy? Or were they engaged in some other activity?

THE CLASSIFICATION OF THE SCIENCES AND THE SUBJECT OF NATURAL PHILOSOPHY

Along with the Greek and Arabic treatises that poured into Western Europe in the twelfth and thirteenth century was a translation of an Arabic treatise

---

[36] The manuscript that contains this treatise is Bibliothèque Nationale 6752, fols. 1r–236r. Lynn Thorndike, *A History of Magic and Experimental Science*, 8 vols. (New York: Columbia University Press, 1923–1958), vol. 3, 570. Thorndike devotes chapter 33 (pp. 568–584) of vol. 3, to the anonymous treatise. The chapter is titled "An Anonymous Treatise in Six books on Metaphysics and Natural Philosophy."

[37] For the edition and translation of Bradwardine's treatise, see H. Lamar Crosby, Jr., ed. and tr., *Thomas of Bradwardine His "Tractatus de Proportionibus: Its Significance for the Development of Mathematical Physics* (Madison: University of Wisconsin Press, 1955).

[38] The Latin text and English translation appear in Edward Grant, ed. and tr., *Nicole Oresme "De proportionibus proportionum" and "Ad pauca respicientes"* (Madison: University of Wisconsin Press, 1966).

[39] See *Nicole Oresme and the Medieval Geometry of Qualities, A Treatise on the Uniformity and Difformity of Intensities Known as "Tractatus de configurationibus qualitatum et motuum,"* edited with an Introduction, English Translations, and Commentary by Marshall Clagett (Madison: University of Wisconsin Press, 1968).

by al-Fārābī, titled in Latin *De scientiis* (*On the Sciences*). In this treatise, al-Fārābī describes how the different sciences were organized and related to one another, including natural philosophy, which the translators called *natural science* (*scientia naturalis*). A measure of its importance may be gleaned from the fact that three separate translations of the Arabic text into Latin were made by three different translators – John of Seville, Dominicus Gundissalinus (or Domingo Gundisalvo), and Gerard of Cremona. Indeed, Dominicus Gundissalinus not only translated al-Fārābī's *On the Sciences*, but around 1150 he also wrote a treatise on the classification of the sciences, titled *On the Division of Philosophy* (*De divisione philosophiae*), which draws on al-Fārābī's text, but also incorporates numerous other sources into his treatise.[40] With the translation of al-Fārābī's text, and Gundissalinus's additions to it under another title, Western Europe had before it, for the first time, a sophisticated classification of the sciences based on the works of Greek and Arabic authors, especially those by Aristotle.

Dominicus Gundissalinus (fl. 1140)

Like al-Fārābī, Gundissalinus distinguishes mathematics, and the sciences that are mathematical, from natural science. Taking the latter first, Gundissalinus defines natural science, or natural philosophy, as "the science considering only things unabstracted and with motion."[41] He elaborates further by quoting from Avicenna's *Metaphysics* that "the matter of natural science is body ... according to what is subjected to motion and rest and change."[42] And then, following al-Fārābī rather closely, Gundissalinus declares that natural science is divided into eight parts, which are, as we shall see, rooted in Aristotle's natural books, including a few treatises falsely attributed to Aristotle. Here is my summary account of Gundissalinus's description of the eight parts:

1. The first part, according to Gundissalinus, is based on what is taught in Aristotle's *Physics* treating all natural bodies, both simple and compound.
2. The second part of natural science is wholly concerned with whether simple bodies exist, and if they do exist which bodies are they, and what is the number of them. This is a study about what the world is, what the parts of it are and how many there are, and whether there

---

[40] See Claudia Kren, "Gundissalinus, Dominicus," in *Dictionary of Scientific Biography*, vol. 5 (1972), 592.

[41] I draw on the translation of parts of Gundissalinus's *De divisione philosophiae* by Marshall Clagett and Edward Grant in Edward Grant, ed., *A Source Book in Medieval Science* (Cambridge, MA: Harvard University Press, 1974), 63. (In our translation, he is called by his Spanish name, Domingo Gundisalvo.)

[42] Ibid.

are three or five parts. And this is also a study of the heaven and how it is distinguished from the rest of the world and that the matter of the heaven is one. Gundissalinus informs us that this and other aspects of the world are the subject matter of the four books of Aristotle's *On the Heavens and World.*

3. The third part of natural science is based on Aristotle's *On Generation and Corruption* and is therefore about the generation and corruption of bodies.

4. The fourth part is based on the first three (of four) books of Aristotle's *Meteorology*[43] and, according to Gundissalinus, is "about the principles of actions and passions that are proper to the elements only without the bodies composed of them."[44]

5. The fifth part is based on the fourth book of the *Meteorology* and "is a consideration of bodies compounded of the elements and of those things [constituted] of similar or dissimilar parts."[45]

6. "The sixth part is a consideration of what is shared by all compound bodies of similar parts, which are not parts of a compound body of different parts."[46] Such bodies are minerals that are the subject matter of the book titled *On Minerals,* a treatise falsely attributed to Aristotle.

7. The seventh part of natural science is about plants and is taught in the book *On Plants,* also falsely ascribed to Aristotle.

8. The eighth, and final, part is about animals and their properties and is about "compound bodies of different parts." As the basis for this part, Gundissalinus cites "the book entitled *On Animals (De animalibus),*[47] in the book *On the Soul (De anima),* and in those books which continue to the end of the natural books."[48]

Thus did Gundissalinus include by name and subject matter, all of the basic books that comprised Aristotle's natural philosophy.[49] He concludes his section on natural philosophy by explaining that "The 'end' of natural

---

43 Gundissalinus refers to the *Meteorology* as *On Phenomena of the Upper Regions (De impressionibus superioribus).*

44 Ibid., 64.   45 Ibid.   46 Ibid.

47 The title *On Animals* is a generic title for the following treatises by Aristotle: *History of Animals, On the Parts of Animals,* and *On the Generation of Animals.* In the thirteenth century (no later than 1220), Michael Scot translated these three treatises on animals with the title *On Animals.* See Grant, *Source Book in Medieval Science,* 681, n. 1.

48 Marshall Clagett and Edward Grant in Grant, ed., *A Source Book in Medieval Science,* 64. Just what Gundissalinus means by "the end of the natural books" is unclear. For a translation directly from the Arabic of al-Fārābī's division of natural philosophy into eight parts, see Marshall Clagett, "Some General Aspects of Physics in the Middle Ages," *Isis* 39 (May 1948), 33–34.

49 Gundissalinus did not include Aristotle's collection of brief treatises that bear the title *Parva naturalia (The Short Physical Treatises).* These are usually regarded as part of Aristotle's natural philosophy.

science is the cognition of natural bodies" and "The 'instrument' of this science is the dialectical syllogism, which consists of truths and probables." And, finally, Gundissalinus emphasizes that "This science . . . is called 'physical,' that is, 'natural,' because it intends to treat only of natural things which are subject to the motion of nature. Moreover, it is to be read and learned after logic."[50]

For Aristotle, as we saw in Chapter 2, mathematics is distinct from natural philosophy, as are the sciences that use mathematics. Indeed, the exact sciences, fall between natural philosophy and mathematics, for which reason they were usually called "middle sciences." Al-Fārābī called them "doctrinal sciences" in which category he includes seven sciences, two of which are purely mathematical, namely, arithmetic and geometry; and five are mathematical sciences, namely, the mathematical sciences of optics (*scientia aspectuum*), the science of the stars (*scientia stellarum doctrinalis*), which al-Fārābī says is really two sciences, describing what we would call astronomy and astrology; music (*scientia musice*), weights (*scientia ponderum*) and the science of devices (*scientia ingeniorum*).[51] It is noteworthy that for al-Fārābī, neither astronomy nor astrology is part of natural philosophy.

Gundissalinus makes a few significant changes in al-Fārābī's classification. The first is the exclusion of music, which is not mentioned. And, more radically, Gundissalinus includes medicine as part of natural philosophy. Indeed, he describes it as "the first species of natural science."[52] Medicine, Gundissalinus declares, is "one of the species of theoretical natural science, for it is unabstracted and with motion,"[53] which precisely fits the definition of natural philosophy he gave earlier, namely, "the science considering only things unabstracted and with motion" (see earlier). Thus Gundissalinus departed not only from al-Fārābī but also from Aristotle, who makes no mention of medicine in connection with natural philosophy. Over the medieval centuries, physicians usually regarded their science as part of natural philosophy.

Robert Kilwardby (d. 1279)

But not all scholastic natural philosophers agreed with Gundissalinus. Around 1250, approximately a century after Dominicus Gundissalinus wrote *On the Division of Philosophy*, Robert Kilwardby (d. 1279) of the Dominican Order wrote *On the Order of the Sciences* (*De ortu scientiarum*),

---

[50] Clagett and Grant in Grant, *Source Book*, 65.
[51] See *Al-Farabi Catalogo de las Ciencias*, edición y traducción castellana por Ángel Gonzalez Palencia, second edition (Madrid: Instituto Miguel Asín, 1953), 119. By the "science of devices" (*scientia ingeniorum*), al-Fārābī means the making of machines by the use of arithmetic and geometry. Among the devices he mentions are mirrors, musical instruments, and weapons. See ibid., 154–156.
[52] Ibid., 68.　　　[53] Ibid.

a lengthy treatise on the classification of the sciences.[54] Kilwardby excludes medicine from natural philosophy and instead makes it one of seven mechanical sciences, alongside fabric-making, armament, commerce, agriculture, hunting, and theatrics.[55] Kilwardby explicitly acknowledges that he drew the seven mechanical arts directly from the *Didascalicon* of Hugh of St. Victor, who wrote his treatise in the late 1120s.[56]

For Kilwardby, as for almost all who described and classified natural philosophy, or natural science, the subject of that discipline was mobile bodies, which could be treated in the most general way, or could be treated in terms of special parts.[57] Natural philosophy is treated in the first way in Aristotle's *Physics*, where Aristotle treats mobile body universally, ranging over such topics as matter, form and privation, nature, cause, motion, place, time. These and other topics are applicable to all mobiles, and not just to some.

To understand Aristotle's treatment of mobile body in specialized and restricted ways, however, we must look at his other books. Every mobile body must be interpreted in one of two ways: it is either ungenerable and incorruptible, or generable and corruptible. Aristotle treats bodies in the category of ungenerable and incorruptible in the book *On the Heaven and the World* (*De caelo et mundo*). Kilwardby mentions Aristotle's five simple bodies, the first of which is the incorruptible celestial body that extends from the concave surface of the lunar orb to the convex surface of the eighth sphere (presumably the sphere of the fixed stars). It has a natural circular motion. The other four bodies are Aristotle's four sublunar elements, earth, water, air, and fire, all of which have natural up and down rectilinear motions. Earth is absolutely heavy and water is relatively heavy, whereas fire is absolutely light and air is relatively light.

Corruptible bodies are either simple, that is, elemental, or compound, that is, composed of simple elemental bodies. Elemental bodies are treated in *On Generation and Corruption*. Compound bodies are either inanimate or animate. Inanimate bodies are considered in Aristotle's *Meteors*, where phenomena of the upper atmosphere are described, including rain, hail, winds, thunder, and so on. Animate compound bodies are considered in

---

54 For biographical details of Kilwardby's life, see *Robert Kilwardby O. P. De ortu scientiarum*, ed. Albert P. Judy O. P. (Published jointly by the British Academy and The Pontifical Institute of Mediaeval Studies, 1976), x–xvii.

55 Kilwardby, *De ortu scientiarum*, 129, para. 363 and 130, para. 369. In para. 369, Kilwardby explains the different aspects of medicine, the description of which he took directly from Hugh of St. Victor, as we see in the title Kilwardby assigned to chapter 39: "On the division and species of mechanics according to Hugh of Saint Victor." Ibid., 129 (see next note).

56 For Hugh's seven mechanical arts, see *Didascalicon*, bk. 2, chs. 20 to 27 in *The "Didascalicon" of Hugh of St. Victor: A Medieval Guide to the Arts*, translated from the Latin with an Introduction and Notes by Jerome Taylor (New York: Columbia University Press, 1961), 74–79.

57 "Igitur corpus mobile potest considerari simpliciter et in generali, vel secundum partem sive in speciali." Kilwardby, *De ortu scientiarum* ch. 10, 24.

Aristotle's De *vegetabilibus* (or De *plantis*) and in On Animals (De *animalibus*). Because animated bodies have souls, other treatises by Aristotle explain their actions. Here Kilwardby mentions On the Soul (De anima) and the following four brief works from Aristotle's Short Physical Treatises (that is, the Parva naturalia): On sense and the sensed (De sensu et sensato), On Sleep and Wakefulness (De somno et vigilia), On Death and Life (De morte et vita), and On Memory and Remembrance (De memoria et reminiscentia).

Dominicus Gundissalinus and Robert Kilwardby included in their concept of natural philosophy much the same basic list of Aristotle's treatises, differing only in minor ways.[58] Both accepted Aristotle's threefold division of theoretical knowledge: natural philosophy or physics at the bottom; mathematics and the mathematical sciences next; and at the highest level metaphysics, known also as "first philosophy" or "theology." Kilwardby explains how they differ. Of the three divisions, "natural [science or philosophy] is the least abstract and the divine science [that is, metaphysics] has the greatest abstractness, and mathematics is of middling abstraction. Therefore, the subject of mathematical science will necessarily be midway between the subjects of natural science and divine [science]. But this can only be because of quantity, which I show as follows,"[59] after which (in the next paragraph) Kilwardby explains why it falls to mathematics to deal with quantity.

With mathematics (arithmetic and geometry) lying between metaphysics and natural philosophy, mathematical sciences such as optics (or perspective), astronomy, and the science of weights, also were regarded as distinct from metaphysics and natural philosophy.

Although Gundissalinus and Kilwardby are in general agreement in their understanding of natural philosophy, there are a few significant differences. Where Gundissalinus includes medicine in natural philosophy, Kilwardby excludes it. Moreover, Kilwardby includes perspective (or optics) within natural philosophy, choosing not to regard it as a mathematical science,[60] a judgment he undoubtedly based on Aristotle's remarks in the Physics that "while geometry investigates natural lines but not *qua* natural, optics investigates mathematical lines, but *qua* natural, not *qua* mathematical."[61]

[58] Gundissalinus includes the pseudo-Aristotelian On Minerals, whereas Kilwardby mentions four brief treatises from Aristotle's Parva naturalia.

[59] I shall give the entire Latin text for ch. 24, para. 165: "Ad hoc dicendum et primo et primum quod cum tres sint essentiales modi scientiae speculativae, ut dicit Aristoteles in VI Metaphysicae, scilicet naturalis, mathematicus et divinus, quorum naturalis et minimae abstractionis et divinus maximae, mathematicus est mediae abstractionis, quare subiectum scientiae mathematicae aliquid medium erit necessario inter subiectum scientiae naturalis et divinae. Sed hoc non potest esse nisi quantitas, quod si ostendo." Kilwardby, De ortu scientiarum, 64.

[60] In ch. 17, para. 117, Kilwardby says: "Et dicendum quod perspectiva non est ponenda quinta mathematica quia ipsa est vere naturalis scientia et multo verius quam mathematica." Kilwardby, De ortu scientiarum, 48.

[61] Aristotle, Physics 2.2.194a.10–12; tr. R. P. Hardie and R. K. Gaye in the Revised Oxford Translation.

The relationship of natural philosophy to metaphysics was a major concern for natural philosophers. Kilwardby made a genuine effort to explain their relationship. Both metaphysics and natural philosophy, he observes, consider composite substances and their principles, as well as matter, form, privation, properties, and accidents.[62] Moreover, they also treat created spirits, such as rational souls and the celestial mover. But each discipline considers these substances from a different standpoint. Kilwardby explains that a composed substance, with its matter and form, along with its magnitude, is the province of metaphysics, but its mutability and privation, which are accidents of substance, do not fall within the domain of metaphysics. Insofar as a substance is mutable, it – along with its matter, form, and principles – belongs to the domain of physics, or natural philosophy.

They also differ in their treatment of the celestial mover. Physics considers it insofar as it is a principle of motion, whereas metaphysics treats its properties divorced from motion. Metaphysics begins with a consideration of God and moves on to spiritual substances separated from physical bodies. Physics treats the five simple bodies (the four elements plus the fifth incorruptible celestial element from which the celestial bodies are composed). Thus, it treats bodies that move with natural circular and rectilinear motions and that are light or heavy or weightless. Physics also treats all bodies that are compounded of the four elements, because all such bodies are mobile. In brief, "insofar as substances are beings per se, they belong to metaphysics; insofar as they are mobile or mutable, they belong to physics." Despite their different domains, metaphysics and natural philosophy are related. Kilwardby explains that:

metaphysics begins its consideration long before natural [philosophy]; but where natural [philosophy] begins, metaphysics does not cease, but runs with natural philosophy to the end, although [it operates] in different ways, because metaphysics considers what there is of being in things, whether substances or accidents, or such things; physics [or natural philosophy], however, considers the special nature of a mobile or the mutability that inheres in it.[63]

Roger Bacon (ca. 1219–1292)

Roger Bacon also contributed to the discussion about the nature of natural philosophy, although somewhat unclearly. Bacon considered the matter in

---

[62] Kilwardby compares metaphysics and natural philosophy in ch. 28 on pp. 87–89 of *De ortu scientiarum.*

[63] "Sic igitur incipit consideratio metaphysica procul ante naturalem, et ubi incipit naturalis non desinit metaphysica sed currit cum ea usque ad finem, sed diversimode, quia metaphysicus considerat quod entitatis est in rebus sive substantiae sive accidentis secundum quod huiusmodi; physicus autem quod specialis naturae mobilis aut mutabilis eius inest." *Kilwardby, De ortu scientiarum,* ch. 28, p. 89.

his *Liber primus communium naturalium, or Communia naturalium.*[64] He believed that one should first study grammar and logic, then mathematics, followed by natural philosophy, and finally metaphysics.[65] Natural science, or natural philosophy, is concerned with

a principle of motion and rest, as in the parts of the elements fire, air, earth and water, and in all inanimate things made from them, as metals, stones, salts (*sales*), sulphur, *attramento*, and colors such as red lead and white lead and lapis lazuli, which is bright blue (*azurium*), and green (*viride*), and such things as are generated in the bowels of the earth. And similarly plants, such as herbs, trees, cabbages, reeds (*canne*), and bushes; and there are brute animals and man. In [all] these there is naturally a principle of motion and rest and thus there is in them a nature which is called a principle of both motion and rest (*et motus et status*). Indeed, all these things rest and are moved naturally, as is obvious with respect to local motion, and other motions, and with respect to generation and corruption, alteration, augmentation, and diminution.[66]

After briefly describing generation, corruption, alteration, augmentation, and diminution, Bacon explains that:

Similarly the celestial bodies have a principle of motion within themselves, but not a principle of rest as do the other [terrestrial bodies]. However, they [the celestial bodies] are said to rest with respect to the whole. Although parts are moved – because parts are moved from place to place – the whole [heaven] always remains in the same place. However, celestial bodies are not so properly natural things according to the description of nature given [earlier for terrestrial bodies], especially because they lack rest. But they are called natural things more because they cause motion and rest in elements and compounds (*in elementis et elementatis*). For the heavens and stars are not moved absolutely by a natural motion, but [are moved] voluntarily by intelligences. Nevertheless, they [that is, the celestial bodies] are the cause of natural motions and rest in [all] these inferior natural things.[67]

Bacon's description of natural philosophy would have been acceptable to almost all scholastic natural philosophers. He goes on to mention some of Aristotle's treatises that are devoted to an analysis of natural things. Thus in the *Physics*, Aristotle treats of "principles, motion, the infinite, place, vacuum, time, and other similar things";[68] and in his *De caelo et mundo* (*On the Heaven and the World*), he considers "the parts of the heavens and world." Bacon then lists the topics Aristotle considers in the *De caelo* and follows this with a lengthy list of topics Aristotle omits from that treatise. Thus, Aristotle "teaches nothing in particular about the substantial nature

---

[64] I draw on Robert Steele, ed., *Liber Primus Communium Naturalium Fratris Rogeri* in *Opera hactenus inedita Rogeri Baconi*, Fasc. II (Oxford: Clarendon Press, 1909). In his earlier *Questions on the Physics* probably composed in the 1240s, Bacon asked only about the subject matter of Aristotle's *Physics*, and not about natural philosophy in general. The subject of the *Physics*, he concluded, was "mobile body" (*corpus mobile*).

[65] Steele, ed., *Liber primus communium*, 1.   [66] My translation from ibid., 2.

[67] My translation from ibid., 2–3.   [68] My translation from ibid., 3.

of the heavens and stars, nor about the powers by which they act on inferior things, nor about the nature of light in them, and about the obscurity of eclipses, or about the motions of the planets, or how many there are; nor does he speak about how the orbs touch, or about their number,"[69] and so on. Aristotle follows a similar procedure in his other natural books, among which Bacon mentions Aristotle's *On Animals* (*De animalibus*). But Bacon then expands natural philosophy beyond anything Aristotle and most of his medieval followers envisioned, as will be seen, in Chapter 10.

Albertus Magnus (Albert the Great) (ca. 1200–1280) and Thomas Aquinas (1225–1274)

As many scholastic natural philosophers found it convenient to do, Albertus Magnus expressed his judgments about natural philosophy in the very first chapters of his commentary on Aristotle's *Physics*. He accepted Aristotle's threefold division of philosophy into natural science (that is, natural philosophy), or physics, metaphysics, and mathematics. Albertus explains that, in both being and definition, physics is always concerned with matter and motion. For "if anyone defines heaven or element or something composed of elements, as [for example] flesh or bone, he can never define it without matter, which is subject to motion."[70] Turning more specifically to natural science a few chapters later, Albertus informs the reader that "in every natural science, in so far as it is subject to motion, there is mobile body (*corpus mobile*)."[71] Becoming more specific, Albertus explains that it is not body that is the subject of natural science, but it is body "in so far as it is mobile body (*corpus mobile*) that is subjected to natural science. Thus we say that it is not body alone, but mobile body that is the subject of natural science."[72] But the same mobile body that is the subject of natural science is not the subject of that science if it is at rest, because rest is a privation of motion.[73] As if to validate this point, Albertus explains that although "every physical body is mobile, not every physical body is capable of rest, because what is of perpetual motion is not naturally capable of rest."[74]

---

[69] My translation from ibid., 4.

[70] My translation from *Alberti Magni Ordinis Fratrum Praedicatorum Episcopi Opera Omnia*, Tomus IV: Pars I: *Physica*, Pars I, libri 1–4, ed. Paul Hossfeld (Aschendorff; Monasterii Westfalorum, 1987), bk. 1, tract. 1, ch. 1, p. 2.

[71] "Hoc autem in omni scientia naturali absque dubio est corpus mobile, prout motui subicitur," Ibid., bk. 1, tract. 1, ch. 3, p. 5.

[72] "Quia ergo non inquantum corpus, sed inquantum corpus mobile subicitur scientiae naturali, ideo dicimus, quod non corpus tantum, sed corpus mobile est subiectum scientiae naturalis. Ibid.

[73] "Non tamen dicimus, quod corpus mobile et quiescibile vel corpus quiescibile sit subiectum physicae quia quies privatio motus est...." Ibid.

[74] "Possumus etiam adhuc aliter dicere, scilicet quia omne corpus physicum mobile est, sed non omne corpus physicum est quiescibile, quia id quod est perpetui motus, non est quiescibile secundum naturam." Presumably, Albertus is thinking here of the celestial orbs, which are in perpetual motion and never rest.

But if not every physical body is capable of rest, is it not possible "that not every physical body is mobile"? This situation arises "because the earth and its extreme parts, which are around the center of the sphere of the world, are not mobile, because in every sphere that is moved it is necessary that the center and each pole be immobile." But although the center and each pole may be without local motion, "they can be moved with a motion of alteration, and alteration is a physical motion just as local motion is."[75]

Albertus devotes a chapter to the numerous divisions of natural science, or natural philosophy, that are reflected in Aristotle's natural books. He explains that "since mobile body (*corpus mobile*) is the subject, it has to be considered in natural science with respect to all its differences and divisions."[76] The divisions and subdivisions Albertus presents are a testament to the scholastic penchant for subtle distinctions. Generally, Albertus's conception of natural philosophy is based on the division of matter into simple and compound mobile bodies each of which is further subdivided, after which compound bodies are divided into inanimate and animate bodies. At each stage and subdivision, Albertus cites the relevant Aristotelian treatises that deal with that level and complexity of mobile matter.

Thomas Aquinas presented a simpler interpretation of the structure of natural philosophy than did Albertus Magnus. He also describes the subject matter of natural philosophy in a way that was influential in the fourteenth century. At the outset, he declares "it is necessary in the beginning to decide what is the matter and subject of natural science,"[77] an objective that became the customary way of commencing commentaries on Aristotle's *Physics*. "Natural science, which is called physics," Thomas explains,

deals with those things which depend upon matter not only for their existence, but also for their definition.

And because everything which has matter is mobile, it follows that mobile being (*ens mobile*) is the subject of natural philosophy (*naturalis philosophiae*).[78] For natural philosophy is about natural things, and natural things are those whose principle

---

[75] I cite the Latin that is relevant to this paragraph: "Si quis autem fortasse dicat, quod nec omne corpus physicum mobile est, quia terra et maxime partes, quae sunt circa centrum sphaerae mundi, mobiles non sunt, quia in omni sphaera mota necesse est centrum et utrumque polum immobilia esse, dico, quod licet inquantum circa centrum sint immobiles sicut partes terrae vel etiam si detur tota terra est immobilis, non erunt immobilia, quae dicta sunt, nisi motu locali. Possunt autem moveri alterationis motu, et alteratio est motus physicus sicut et motus secundum locum." Ibid., 6.

[76] "Dicamus igitur quod cum corpus mobile sit subiectum, ipsum habet considerari in scientia naturali secundum omnes differentias et divisiones eius." Ibid.

[77] *Commentary on Aristotle's "Physics" by St. Thomas Aquinas*, tr. Richard J. Blackwell, Richard J. Spath, and W. Edmund Thirlkel (New Haven, CT: Yale University Press, 1963), 3.

[78] I cite the Latin expressions in this sentence from the edition of Thomas's Latin text. See *S. Thomae Aquinatis In octo libros De physico auditu sive Physicorum Aristotelis Commentaria*, ed. P. Fr. Angel – M. Pirotta O. P. (Neapoli (Italia): M. d'Auria Pontificius Editor, 1953), bk. 1, lectio 1, para. 4, 14, col. 1.

is nature. But nature is a principle of motion and rest in that in which it is. Therefore natural science deals with those things which have in them a principle of motion.[79]

Thomas's identification of the subject matter of natural philosophy as "mobile being" (*ens mobile*) became common in the fourteenth century. Thomas explains that Aristotle's *Physics* is where "mobile being" is considered "simply," that is, presumably in its most general terms. Moreover, it is in the *Physics* that Aristotle shows that "every mobile being is a body."[80]

Although Kilwardby and Bacon mentioned the major topics of discussion in Aristotle's other treatises, Thomas views these treatises from the standpoint of mobile being, that is, from the standpoint of motion. He explains that:

after *The Physics* there are other books of natural science in which the species of motion are treated. Thus in the *De caelo* we treat the mobile according to local motion, which is the first species of motion. In the *De Generatione*, we treat of motion to form and of the first mobile things, i.e., the elements, with respect to the common aspects of their changes. Their special changes are considered in the book *Meteororum*. In the book *De mineralibus*, we consider the mobile mixed bodies which are non-living. Living bodies are considered in the book, *De anima*, and the books which follow it.[81]

Thus did Thomas subdivide natural philosophy in a manner quite similar to Albertus Magnus. But where Albertus used the expression *mobile body* (*corpus mobile*), Thomas used *mobile being* (*ens mobile*), which, on the face of it, is of broader scope than *mobile body*, since the former expression embraces the motion of both bodies and spirits. Thomas's *ens mobile* was widely used in the fourteenth century. Its meaning, however, may not always have been what Thomas had in mind, as we see in an anonymous fourteenth century treatise on natural philosophy.

ANONYMOUS FOURTEENTH-CENTURY TREATISE ON NATURAL PHILOSOPHY

In the fourteenth century, an anonymous author of a lengthy treatise on natural philosophy had occasion to describe his understanding of natural philosophy. The content of this treatise is of great importance because it was probably composed around the middle of the fourteenth century and thus represents conceptions and interpretations that came at the high point of the development of scholastic natural philosophy. For this reason, I shall describe the attitudes about natural philosophy that the anonymous author sought to

---

[79] *Commentary on Aristotle's "Physics" by St. Thomas Aquinas*, 3–4.   [80] Ibid., 4.

[81] Ibid. By the "books which follow" *De anima*, Thomas probably means the biological treatises, namely, *On Animals* (*De animalibus*), and the psuedo-Aristotelian *De plantis*, or *De vegetabilibus*.

convey, even though, as a unique copy, few, if any, scholars may have read it.[82] In the opening words, the anonymous author – who shall hereafter be cited simply as "Anonymous" – explains that his rationale for writing this lengthy treatise is to make the difficult works of Aristotle comprehensible to young students:

> Because the texts of Aristotle are quite prolix and verbose and often filled with difficult words, making their study for young people difficult and time-consuming, it thus seems appropriate to summarily collect the opinions of Aristotle and other philosophers and bring them under brief compass so they are more easily understood.[83]

Aristotle's philosophy is threefold: "natural philosophy (*philosophia naturalis*), which is also called by another name, physics (*phisica*), which considers things conjoined to motion, as is obvious in the eighth [book] of the *Metaphysics*,"[84] moral philosophy, and metaphysics. Of these three parts of philosophy, natural philosophy and metaphysics are speculative sciences, whereas moral philosophy is a practical science. Anonymous decides to exclude moral philosophy and to consider only the speculative sciences: natural philosophy and metaphysics.

Because they are speculative sciences, natural philosophy and metaphysics are intimately related, as Robert Kilwardby also had believed. "Indeed," Anonymous argues, they can be mutually separated only with difficulty." For example, "when natural philosophy treats whether local motion is successive, it is also appropriate to investigate what such a motion is and whether it is distinguished from the mobile thing. The first consideration belongs to natural philosophy, and the second to metaphysics."[85] Thus, whether a motion is successive is the province of natural philosophy, but the more abstract concern, as to what that motion is and whether it is distinct from the body itself, belongs to metaphysics. "In many passages in natural philosophy," Anonymous explains, "Aristotle searches out many metaphysical conclusions and, conversely, in metaphysics he often searches for physical

---

[82] The treatise appears in Bibliothèque Nationale 6752 and consists of 236 folios written in a clear hand. The treatise has never been edited or translated, although it has been briefly discussed by Lynn Thorndike, "An Anonymous Treatise in Six books on Metaphysics and Natural Philosophy," in *A History of Magic and Experimental Science*, vol. 3, ch. 33, 568–584. On pages 761–766, Thorndike gives the Latin text of all the chapters in BN 6752.

[83] "Quia textus Aristotelis nimia prolixitate verborumque difficultate sepius in utili iuvenum (?) proficientium studia retardant temporaque detinent nimium, ideo congruum apparet ipsius Aristotelis aliorumque philosophorum sentencias summatim colligere ut sub brevi compendio que prius extensa erant facilius comprehendantur." Bibliothèque Nationale 6752, fol. 4r. All translations from the anonymous treatise are my own.

[84] "Philosophia naturalis que alio nomine phisica dicitur que considerat res coniunctas motui ut patet octavo methaphysice." Ibid.

[85] "Nam cum philosophus naturalis considerat de motu locali utrum sit succesivus bene etiam congrueret investigare quid sit talis motus et utrum distinguatur a re mobili. Prima tamen consideratio pertinet philosophie naturali et secunda metaphysice." Ibid., fols. 4r–4v.

truths. Therefore, although physics and metaphysics are distinguished, their conclusions are nevertheless frequently intermingled...."[86]

Anonymous has lengthy discussions about what metaphysics and natural philosophy are. Because it is the most universal science, anonymous considers metaphysics first. In typical scholastic fashion, he offers numerous possible descriptions of the subject matter of metaphysics, but prefers the interpretation of Aristotle and Averroes, which he regards as "more true."[87] From their interpretation, "it is obvious that being, insofar as it is being, ought to be called the subject of metaphysics."[88]

Among alternative opinions Anonymous rejects:

is that of those who believe that God is the subject of metaphysics, an opinion that can be strengthened by this, [namely,] that metaphysics is especially divine because it is about God, as is said in the first [book] of the *Metaphysics*. Nevertheless, this opinion is false, because metaphysics is a human science. Moreover, God cannot be investigated by human modes of thinking. Therefore, God ought not to be assigned as the subject of metaphysics.[89]

In turning to natural philosophy, or physics, Anonymous, as usual, offers a few alternatives. He cites the opinion of those who argue that natural philosophy takes its name from *nature* from which it follows that the subject of natural philosophy is nature, by which is meant all things. But natural philosophy does not deal with all things, as is evident from the fact that "it is not about God insofar as God exists, but insofar as He is a mover; nor is it about the triangle insofar as it is a triangle, but [only] insofar as it is a certain quantity that is locally mobile."[90] Natural philosophy is, therefore, not about the whole of nature, namely, the totality of things.

Another interpretation views motion as the subject matter of natural philosophy. But motion is a term that is narrower than nature. Moreover one may not properly call nature a principle of motion, "for it could equally be said that God is a principle of motion and consequently is held to be the subject of physics," which is unacceptable. Another interpretation holds

---

[86] "Unde Aristotelis in suis pluribus passibus philosophie naturalis multas metaphysicales venatur conclusiones et econverso in metaphysica phisicas sepius veritates scrutatur. Licet ergo phisica et metaphysica distinguantur verumtamen conclusiones ipsarum frequenter in processu mixtim notabuntur..." Ibid., 4v.

[87] "Rursus alia opinio fuit Aristotelis quam estimo veriorem." Ibid., 5v.

[88] "Ex quo patet quod ens, inquantum ens, debet dici subiectum metaphisice." Ibid., 6r.

[89] "Alia opinio fuit quorundam credentium quod deus esset subiectum methaphisice quorum opinio roborari potest ex eo quod methaphisica est maxima divina quia de deo est ut dicitur primo Metaphisice. Hoc tamen opinio est falsa quia methaphisica est humana scientia. Deus autem humano ingenio investigari non potest. Non debet ergo Deus pro subiecto methaphisice assignari." Ibid., 5v.

[90] "Sic enim intelligendo phisica non est de omnibus rebus quia non est de Deo in quantum Deus est, sed inquantum motor est; nec est de triangulo inquantum triangulus est, sed inquantum est quedam quantitas localiter mobilis." Ibid., 7r.

that mobile body is the subject matter of natural philosophy. "This opinion, however, seems false, because many non-bodies are considered by natural philosophy, just as God insofar as He is a mover, is also a rational soul."[91]

Anonymous seems to accept the approach of the "moderns," which holds that *mobile being (ens mobile)* – the term Thomas Aquinas used – is the subject of natural philosophy. "It seems," Anonymous explains, "that this opinion sufficiently signifies the one Aristotle held in the second [book] of the *Physics*, when he says that beings that are not in motion are not the consideration of physics, which is as if he says that beings that are in motion are [indeed] the consideration of physics."[92] One "modern" Anonymous may have had in mind is John Buridan, who, in his *Questions on Aristotle's Physics*, asked: "Whether mobile being *(ens mobile)* is the subject of the whole of natural science, or something other"[93] and, in his second conclusion, replied that "this term 'mobile being' *(ens mobile)* is the proper subject that should be assigned in natural science, because it is the most common term among the things considered in natural science and does not transcend the limits of natural science."[94] That is, "mobile being" is the broadest term applied to natural science because "mobile being" is coextensive with all the things that natural philosophy properly considers.

Why did Anonymous take *mobile being* as the subject of natural philosophy rather than mobile body, as did Robert Kilwardby and Roger Bacon? Perhaps because mobile being is a more inclusive category than mobile body. Mobile being includes not only bodies but also the motion of immaterial substances, such as angels. All things in motion, not just bodies, are the province of natural philosophy. But unlike those of his predecessors – Gundissalinus, Kilwardby, and Roger Bacon – who subsumed subordinate sciences under natural philosophy, Anonymous does not specify any other sciences as lying within the boundaries of natural philosophy, or subordinate to it. Perhaps some felt free to include sciences such as perspective and medicine, as part of natural philosophy, because, in some sense, they were all concerned with mobile bodies. But Anonymous, as we saw, would probably

---

91 "Hoc tamen opinio ideo falsa apparet quia multa non corpora sunt de consideratione philosophie naturalis sicut [fol. 7v] Deus inquantum motor est, etiam anima rationalis." Ibid., 6v–7r.

92 "Hanc opinionem satis videtur innuere Aristotelis secundo *Phisicorum* cum dicit quod entia non mota amplius non sunt phisice considerationis, quasi diceret quod entia mota sunt phisice considerationis." Ibid., 7v.

93 "Consequenter queritur tertio utrum ens mobile sit subiectum totius scientie naturalis vel quid aliud." John Buridan, *Acutissimi philosophi reverendi Magistri Johannis Buridani subtilissime questiones super octo Phisicorum libros Aristotelis diligenter recognite et revise a Magistro Johanne Dullaert de Gandavo antea nusquam impresse* (Paris, 1509) (Facsimile, entitled *Johannes Buridanus, Kommentar zur Aristotelischen Physik*. Frankfurt: Minerva, 1964), bk. 1, qu. 3, fol. 3v.

94 "Secunda conclusio est quod iste terminus ens mobile est subiectum proprium in scientia naturali assignandum quia est terminus communissimus inter considerate in scientia naturali et non transcendens limites scientie naturalis." Ibid., fol. 4r.

have disapproved, because mobile being includes both immaterial and material substances, whereas perspective, medicine, and other separate sciences, are only concerned with material being.

Another "modern," Marsilius of Inghen (ca. 1340–1396), argued as Buridan did, insisting that natural science, or natural philosophy, is about mobile being (*de ente mobili*). Marsilius elaborates further by observing that natural science has eight principal parts, each associated with one or more of Aristotle's treatises. The first part is found in Aristotle's *Physics*, which is concerned with mobile being insofar as it is mobile; the second part appears in Aristotle's *Book on the Heaven and World* (*Liber de caelo et mundo*), in which the place (*ad ubi*) of mobile being is the primary concern; the third part is found in Aristotle's *On Generation and Corruption*, in which the major concern is the form of mobile beings in general (*ad formam generaliter*); the fourth part appears in Aristotle's *Meteorology* (or *Meteors*), in which the concern is the form of imperfect mixed mobile being (*de ente mobili ad formam mixti imperfecti*); the fifth part is in *On Minerals* (*De mineralibus*), falsely ascribed to Aristotle but dealing with the forms of perfect inanimate mixed mobile beings (*ad formam mixti perfecti inanimati*); the sixth part is in Aristotle's *On the Soul* (*De anima*), which is concerned in general with the perfect mixed form of animated mobile being (*ad formam mixti perfecti animati in generali*), with the third book treating the intellective soul of animated mobile being; the seventh part appears in Aristotle's book *On Vegetables and Plants* (*Liber de vegetabilibus et plantis*), in which the subject of discussion is the vegetative soul in the form of a mixed perfect body (*ad formam mixti perfecti anima vegetativa*); and, finally, the eighth part is found in Aristotle's *On Animals* (*De animalibus*), which deals with the form of mixed perfect animated bodies with sensitive souls (*ad formam mixti perfecti animati anima sensitiva*).[95] Marsilius, like all who sought to

---

95 "Quantum ad quartum quod tota scientia naturalis determinat de ente mobili cuius sunt octo partes principales. Prima pars est *Liber Physicorum* cuius expositio ad presens intenditur in quo determinatur de ente mobili inquantum mobile et ly inquantum determinat sive denotat ibi rationem generalem considerandi. Secunda pars est *Liber Celi et Mundi* in quo determinatur de ente mobili quantum magis specialiter, scilicet de ente mobili ad ubi. Tertia pars est *Liber de Generatione* determinat de ente mobili ad formam generaliter. Quarta pars est *Liber Metheororum* qui determinat de ente mobili ad formam mixti inperfecti. Quinta pars est *Liber de Mineralibus* qui determinant de ente mobili ad formam mixti perfecti inanimati. Sexta pars et *Liber de Anima* in quo determinat de ente mobili ad formam mixti perfecti animati in generali. Et quantum ad tertium librum eius specialiter de animato anima intellectiva. Septima est *Liber de Vegetabilibus et Plantis* in quo determinatur de ente mobili ad formam mixti perfecti anima vegetativa. Octava et ultima est *Liber de Animalibus* in quo determinatur de ente mobili ad formam mixti perfecti animati anima sensitiva." Marsilius of Inghen, *Questiones subtilissime Johannis Marcilii Inguen super octo libros Physicorum secundum nominalium viam. Cum tabula in fine libri posita suum in lucem primum sortiuntur effectum* (Lyon, 1518. Facsimile, Frankfurt: Minerva, 1964), bk. 1, qu. 1, fol. 2v, col. 1. The work just cited is actually attributed to Johannes Marsilius of Inghen rather than to Marsilius of Inghen. Whether they are one and the same person is uncertain. My discussion is not affected by this consideration.

identify the subject matter and scope of natural philosophy, reveals what is obvious: scholastic natural philosophers viewed Aristotle's works as the basis of all natural philosophy. Although they frequently disagreed about particular details, medieval natural philosophers were generally agreed that natural philosophy was about bodies involved in motion and change, and for some, perhaps, it also was concerned with the motion of immaterial substances, such as angels, or God as an agent of motion and change.

Although he insisted that natural philosophy was about mobile being, John Buridan also recognized that it was more formally a collection of demonstrations in the form of conclusions about natural phenomena, as we discover in his *Questions on the Physics*, where he informs us that he "calls the totality of natural science a habit comprised of all conclusions demonstrated in the natural books by natural demonstrations, so that I intend to exclude those that are found in these natural books to be demonstrated by a superior habit, namely metaphysics. I do not care about those [demonstrations] at present because they are not integral to natural science."[96] In this passage, Buridan also parts company with Anonymous, and others, by excluding from natural philosophy the demonstrations of the superior discipline of metaphysics.

THE OCCULT SCIENCES AND NATURAL PHILOSOPHY

If we take Thomas Aquinas's popular definition of natural philosophy as "mobile being," magic, and occult sciences, such as astrology and alchemy, would form an integral part of natural philosophy, because they are all, in some sense, concerned with matter in motion. However, to understand the relationship between natural philosophy and magic, it is essential to realize that during the Middle Ages magic was condemned by the Church, because it was thought that those who engaged in it invoked demons to produce their effects. As Robert Kilwardby expressed it in his classification of the sciences (*De ortu scientiarum*):

Magic is not accepted as part of philosophy since it teaches every iniquity and malice; lying about the truth and truly causing injury, it seduces men's minds from divine religion, it prompts them to the cult of demons, it fosters corruption of morals, and it impels the minds of its devotees to very wickedness.[97]

---

[96] John Buridan, *Questions on the Physics*, bk. 1, qu. 2, in *Acutissimi philosophi reverendi Magistri Johannis Buridani subtilissime questiones super octo Phisicorum libros Aristotelis* (Paris, 1509. Facsimile, entitled *Johannes Buridanus, Kommentar zur Aristotelischen Physik.* Frankfurt: Minerva, 1964), fol. 3r, col. 1.

[97] Translated by Bert Hansen in "Magic, Bookish (Western European)," in Joseph R. Strayer, ed., *Dictionary of the Middle Ages*, vol. 8 (New York: Charles Scribner's Sons, 1987), 37. The passage occurs in Robert Kilwardby's *De ortu scientiarum* (London: British Academy, 1976), ch. 67, para. 662, 225. At the beginning of the section on magic, Kilwardby explains that he is following the words of Hugh of St. Victor, whose views were expressed in his *Didascalicon*, a well-known twelfth-century work.

Demonic magic played no role in natural philosophy as it was taught and written about in the universities. But many, if not most, of those who believed in the efficacy of magic denied the charge that they appealed to demons, and insisted that magic was part of nature. It is this aspect of magic that plays a role in medieval natural philosophy.

Natural Magic

In 1558, Giambattista della Porta (1535–1615) published a work titled *Four Books on Natural Magic (Magiae naturalis libri iiii)*. His vision of natural magic was a kind of practical natural philosophy that also involved the application of experiments and mathematics. Della Porta regarded the operations of the world as well ordered and rational. As the name implies, natural magic was meant to exclude the role of demons. The magical powers that della Porta describes are hidden in nature and are therefore occult. They are not made operative by ceremonial rituals and the state of mind of the natural magician. They operate as natural powers. Astrology was a vital aspect of natural magic. Della Porta's conception of natural magic drew on Greek Neoplatonic and Hermetic treatises that were translated in the fifteenth century.[98]

Although the term "natural magic" does not seem to have appeared in medieval Latin, a version of natural magic did emerge as a consequence of the influx of Greco-Arabic science and philosophy in the twelfth and thirteenth centuries, especially the works of Aristotle. In his description of the cosmos, Aristotle divided the world into celestial and terrestrial regions. The former was composed of a special substance, ether, which was eternal and incorruptible – indeed, Aristotle regarded it as divine – and moved with a natural circular motion. By contrast, the terrestrial region was filled with bodies comprised of one or more of four elements: earth, water, air, and fire. When ordered by their natural places in the scheme of things, the four elements would be arranged concentrically as follows: fire just below the moon; air just below fire; water just below air; and earth at the center of the world. But in the ordinary course of nature, these elemental bodies are intermingled into compounds that are constantly changing. As Aristotle explains it, they are always coming-to-be and passing-away. What causes and maintains this incessant process of change?[99] For Aristotle, celestial motion produced this change, and, more specifically, it was the sun, "for coming-to-be occurs as

---

[98] On Della Porta, see Wayne Shumaker, *The Occult Sciences in the Renaissance: A Study in Intellectual Patterns* (Berkeley: University of California Press, 1972), 108–120; and M. Howard Rienstra, "Porta, Giambattista Della," in *Dictionary of Scientific Biography*, vol. 11 (1975), 95–98.

[99] I draw on my discussion of this topic in Grant, *Planets, Stars, and Orbs*, 571–617.

the sun approaches and decay as it retreats."[100] It is the circular motion of the heavens that produces coming-to-be and passing away in an eternal, cyclical manner.[101]

Aristotle also believed that the incorruptible ether of the celestial region was nobler than all terrestrial things composed of various proportions and mixtures of the four elements that were always undergoing change. It was appropriate, therefore, that the nobler celestial region should cause changes in the less noble terrestrial region, although Aristotle did not explain how this was accomplished.

In the second century AD, Claudius Ptolemy, the author of the *Almagest*, reinforced Aristotle's interpretation when he declared, in the *Tetrabiblos*, that:

a certain power emanating from the ethereal substance is dispersed through and permeates the whole region about the earth, which throughout is subject to change, since, of the primary sublunar elements, fire and air are encompassed and changed by the motions in the ether, and in turn encompass and change all else, earth and water and the plants and animals therein.[102]

Thus did Ptolemy link Aristotle's ideas about the relationship of the celestial and terrestrial regions to astrology. Aristotle's conception that the incorruptible, material ether of the celestial region somehow caused a never-ending sequence of effects in the incessantly changing terrestrial region below the concave surface of the lunar sphere came to be universally accepted during the Middle Ages. Saint Bonaventure spoke for all his medieval colleagues when he asserted:

The reason why superior things act on inferior things...is because they are nobler bodies and excel in power, just as they excel with respect to location. And since the order of the universe is that the more powerful and superior should influence the less powerful and inferior, it is appropriate for the order of the universe that the celestial luminaries should influence the elements and elemental bodies.[103]

We can learn much about the medieval version of natural magic and how it was viewed by scholastic natural philosophers in the Middle Ages by examination of a treatise by Thomas Aquinas that bore the lengthy, but revealing, title "A Letter of Thomas Aquinas to a Certain Knight Beyond the

[100] Aristotle, *On Generation and Corruption*, 2.10.336b.17–18.
[101] Ibid., 2.11.338a.19–338b.5.
[102] Claudius Ptolemy, *Tetrabiblos*, ed. and tr. F. E. Robbins, Loeb Classical Library (Cambridge, MA: Harvard University Press, 1948), bk. 1, ch. 2, 5–7. See Grant, *Planets, Stars, and Orbs*, 572.
[103] Translated from the Latin text in Saint Bonaventure, *Opera Omnia* (Ad Claras Aquas [Quaracchi]: Collegium S. Bonaventurae, 1882–1901), vol. 2: *Commentaria in quattuor libros Sententiarum Magistri Petri Lombardi: In Secundum librum Sententiarum* (1885), 360, col. 2.

Mountains on the Occult Workings of Nature or Concerning the Causality of Heavenly Bodies."[104] Thomas distinguishes the actions of an elemental body, as, for example, a stone, when it moves naturally down toward the center of the earth. This occurs because earth is the dominant element of a stone, which, if unimpeded, will always falls toward the earth's center. Thomas explains "all actions and movement whatsoever of bodies composed of elements take place according to the property and power of the elements of which such bodies are made."[105] But, Thomas continues, "there are some workings of these bodies which cannot be caused by the powers of the elements: for example, the magnet attracts iron, and certain medicines purge particular humors, in definite parts of the body. Actions of this sort, therefore, must be traced to higher principles."[106]

By "higher principles," Thomas means superior agents, which can act on inferior bodies in two ways. In the first way, an inferior object receives a form from the superior agent that enables it to act, as when the sun illuminates the moon by sending light to it. In this instance, the moon receives the light, or form, from the sun and appropriates it to become an illuminated body. In a second way, the inferior agent "acts only through the power of the superior agent, without receiving a form for acting. It is moved only through the motion of the superior agent, as a carpenter uses a saw for cutting."[107]

For Thomas, and most medieval scholastics, occult phenomena were understood to be effects in bodies and objects that one could not explain on the basis of the ordinary behavior of the elements composing them. When such phenomena were identified, their cause must be sought in celestial bodies that are composed of an incorruptible ether, or fifth element; or their cause must be sought in immaterial, separated substances, such as intelligences (or angels), demons, or God. But these superior agents, whether celestial bodies or immaterial, separated substances, produce two kinds of occult phenomena: constant or inconstant. If constant, this signifies that the superior agent impressed on the elemental body a permanent form or principle. The most popular example of this kind of action is the attractive force of magnets. Magnets always attract iron and it was inferred that the superior agent implanted a permanent form in all magnets.

However, not all bodies of the same kind manifest the same constant behavior. The moon causes tides, but not all bodies of water are tidal; some relics of saints cure disease, but not all. The superior agents causing these phenomena chose to act inconstantly by affecting some bodies of the same

[104] *The Letter of Saint Thomas Aquinas "De Occultis Operibus Naturae Ad Quemdam Militem Ultramontanum*, translated by Joseph Bernard McAllister, M.A., S.T.B. Ph.D. diss. (Washington, DC: Catholic University of America Press, 1939).
[105] Ibid., 20. [106] Ibid., 21. [107] Ibid.

kind, but not others. In these instances, the superior agent does not impart a permanent form or principle to the bodies it affects. Rather, it uses these bodies "as a carpenter uses a saw for cutting." That is, certain bodies, chosen from among a given species of body, are used by the superior causative agent to perform as instrumentalities to produce a given effect. They cannot cause this effect from their own natures.[108]

Thomas thus identified the various sources of magical effects recognized by most medieval theologians and natural philosophers. These were celestial bodies (orbs, planets, and stars) and separated substances (God, intelligences or angels, and demons). For Thomas, and many other natural philosophers, numerous terrestrial effects attributed to the causative power of celestial bodies were regarded as magical because the effect did not derive from the natural powers of the elements comprising the body in question. But these magical effects were regarded as natural, because they were caused by the natural powers of celestial bodies. In the sixteenth and seventeenth centuries, such magical effects came to be called "natural magic." By contrast, magical effects caused by separated substances, such as God, angels, and demons were not natural, but supernatural. The magical effects produced by celestial bodies and by God and angels served natural or beneficial purposes, whereas the magical effects caused by demons were regarded as evil. Natural magic was sharply contrasted with demonic magic, which the Church regularly condemned.

In arguing that the incorruptible, nobler celestial bodies influenced the behavior of the corruptible and incessantly changeable terrestrial bodies, Aristotle and Ptolemy partially explained how the celestial region actually affected terrestrial bodies. Medieval natural philosophers identified three instrumentalities of celestial causative power: motion (*motus*); light (*lumen*); and influence (*influentia*),[109] the first two of which find counterparts in Aristotle. Following Averroes, many regarded motion as the most important. In his widely read treatise *On the Substance of an Orb* (*De substantia orbis*), Averroes explains that "if motion were destroyed, so would the heaven itself [be destroyed]. Indeed, the heaven exists because of its motion; and if celestial motion were destroyed, the motion of all inferior beings would be destroyed and so also would the world."[110] Celestial motion was the most important of the three instrumentalities, because it caused the other two: light and influence. Directly or indirectly, the celestial motions were thought to produce the two fundamental pairs of qualitative opposites: hotness and coldness; and wetness and dryness.[111] It was widely believed that these four qualities cause the changes that occur in the terrestrial region.

---

[108] Based on ibid., 79–80.
[109] I draw here on my account of "The instrumentalities of celestial action" in my *Planets, Stars, and Orbs*, 586–617. [110] See *Planets, Stars, and Orbs*, 588.
[111] For a detailed discussion, see Grant, ibid., 591–595.

Averroes's idea that the world would be destroyed if the celestial motions ceased was held by many natural philosophers. Opposition to this opinion developed in the thirteenth century, as can be seen in an article condemned in 1277 by the bishop of Paris. Article 156 (of 219 condemned articles) declared: "if the heaven should stand [still], fire would not act on tow [or flax], because nature would cease to operate."[112] This idea was deemed offensive because it made it appear that terrestrial actions were totally dependent on celestial motion and, therefore, perhaps independent of God's action. It attributed too much to the celestial motions. Major fourteenth-century natural philosophers such as John Buridan, Albert of Saxony, and Nicole Oresme insisted that generation and corruption would continue even if all celestial motion ceased. Nevertheless, they, and all other scholastics, regarded the celestial motions as the major factor in the continuous, normal operation of the world.

As they moved around, celestial bodies produced the other two instrumentalities, light and influence. The production of light from the sun and its essential role in life was obvious to all. Most believed that all other planets also received light from the sun, those receiving the most produced heat in the terrestrial region, whereas those that received less caused coldness.[113] Through the third instrumentality, influence, the planets played an even more varied role in terrestrial behavior. Indeed, terrestrial effects that were not directly attributable to celestial motion or light were assigned to celestial influence. More than any other celestial power or force, influence may be categorized as natural magic. Influence produced metals in the bowels of the earth, where light could not penetrate. Influences from the moon produced the tides. Magnetism also was an effect of celestial influences, an action that seemed obvious to many in the Middle Ages, because magnetism operated even in dense fogs and in the dark where light was absent.[114] Influence played a useful role in medieval natural philosophy: it offered a plausible explanation for a host of otherwise inexplicable, occult phenomena.

As occult phenomena, celestial influences were important in astrology. The alleged powers and properties of celestial bodies had a long history from ancient Mesopotamia to the Middle Ages. Their effects on terrestrial activity were viewed within the context of the concept of influence. Influences radiated down from the substantial forms of celestial bodies, just as light did, except that light is visible, and influences are not. Astrological effects could be viewed in the same manner as those influences that produced magnetic effects in terrestrial matter. It was a form of natural magic, because the power to produce the influences was inherent in the forms of celestial matter. Although there were a few dissenters who believed that celestial bodies

[112] See Grant, ibid., 595–596.   [113] Ibid., 603–605.   [114] Ibid., 612.

acted only by motion and light, and not by influences, they were unusual exceptions.[115]

The Role of Magic in Medieval Natural Philosophy

Did either of the major kinds of magic – demonic and natural – play a role in medieval natural philosophy? Did authors of treatises on natural philosophy include questions or topics on magic? I am unaware of any form of demonic magic playing a role in any questions treatise on the natural books of Aristotle. Questions about the celestial influences on terrestrial phenomena, most of which may be regarded as falling within natural magic, occurred occasionally in questions on Aristotle's *On the Heavens* and *Meteorology*, but not in questions on his *Physics*, *On the Soul*, and *On Generation and Corruption*. For example, in his *Commentary on the Sphere of Sacrobosco*, Michael Scot asks "whether, by their motions, celestial bodies act on inferior bodies."[116] Themon Judaeus considered a few questions relevant to natural magic in his *Questions on the Meteorology*, when he asked "whether the motion of the heavens causes hotness in inferior things"[117] and "whether the sun causes the winds to cease and [also] stimulates them."[118] Themon also posed an astrological question when he asked "whether a comet, or bearded star, signifies the death of princes, droughts, winds, and other bad things."[119] From his response to the last question – expressed in nine conclusions – Themon believes that comets produce all the bad effects mentioned in the question. For example, in the ninth and final conclusion, Themon declares that "a comet signifies the death of princes is proved by Aristotle in the text. Haly proves the same thing in [his commentary] on the *Centiloquium* of Ptolemy. The same thing is proved by numerous experiences, [namely] that always after the appearance of a comet numerous princes died. Again, this can

---

[115] Ibid., 613–614.

[116] "Utrum corpora supercelestia per suum motum agant in inferiora." In Lynn Thorndike, ed. and tr., *The "Sphere" of Sacrobosco and Its Commentators* (Chicago: University of Chicago Press, 1949), 314.

[117] Themon Judaeus, *Questions on the Meteorology*, bk. 1, qu. 6 in *Questiones et decisiones physicales insignium virorum: Alberti de Saxonia in octo libros Physicorum; tres libros De celo et mundo; duos libros De generatione et corruptione; Thimonis in quatuor libros Meteororum; Buridani in tres libros De anima; librum De sensu et sensato; librum De memoria et reminiscentia; librum De somno et vigilia; librum De longitudine et brevitate vite; librum De iuventute et senectute Aristotelis. Recognitae rursus et emendatae summa accuratione et iudicio Magistri Georgii Lokert Scotia quo sunt tractatus proportionum additi.* (Paris: Vaenundantur in aedibus Iodici Badii Ascensii et Conradi Resch, 1518), fols. 160v, col. 1–161r, col. 1. In Aristotle's *Meteorology*, see 1.3.341a.19–31.

[118] Ibid., bk. 2, qu. 6, fols. 174r, col. 1–174v, col. 1. In Aristotle's *Meteorology*, see 2.5.361b.14–24.

[119] "Utrum cometa vel stella comata significet mortem principum, siccitates, et ventos, et alia mala." Ibid., bk. 1, qu. 13, fols. 165v, col. 1–166r, col. 1.

be proved by natural reason: because princes live lives that are addicted to pleasure and drunkenness from which mode of living they produced colic."[120]

There were a few other questions about the influence of the celestial bodies on the terrestrial region, and one or two questions on astrological influences similar to Themon's question. The overall impact of magic and astrology on natural philosophy in the commentary literature on Aristotle's works was minimal. Nevertheless, we can see that in the single, explicit question on astrology that Themon Judaeus included in his *Questions on the Meteorology*, he reveals himself as a firm believer in the astrological impact of comets on terrestrial events and lives. Many natural philosophers may also have been believers in natural magic and astrology, and perhaps even in some forms of demonic magic. But whether or not they were believers, they usually revealed such opinions in thematic treatises, not in their Aristotelian commentaries and questions on natural philosophy. Nicole Oresme, who was convinced that all phenomena were explicable by natural causation, illustrates this tendency in his *Treatise on the Configuration of Qualities and Motions*. There, Oresme includes a lengthy section on the magical arts, which he divides into two distinct parts: necromancy, or demonic magic, and the magical art itself. Oresme regards demonic magic as real and irreducible to natural causes. He explains that:

there are other marvelous things so dissimilar to, and removed from, any natural way, that they cannot be reduced to a natural cause by any rational way. Such is the appearance of demons and their operations.... Accordingly, certain people err with excessive foolishness when they simply deny that spirits of this sort exist and when they say that such things can be produced naturally.[121]

Oresme calls the second part of magic "the magical art," an art he regards as false. Magicians who utilize this art support themselves "by false persuasion, by the application of things, or by the power of words, and sometimes by several or all of these things on which that art, which Pliny calls 'most fraudulent,' is founded."[122]

It was, however, because discussions of magic and its various forms did not find an appropriate place in Aristotle's natural books, that those who

---

[120] "Nona conclusio: quod cometa significant mortem principum. Probatur per Aristotelem in litera; idem probat Haly super *Centiloquium* Ptolomei. Idem probatur per experientias plures quod semper post apparitionem comete plures moriebantur principes. Item probatur ratione naturali: quia tunc principes vivunt delicate et crapulose ex quo victu efficiuntur colerici." Ibid., bk. 1, qu. 13, fol. 166r, col. 1. There is no corresponding reference to Aristotle's text, because Aristotle did not consider astrology in his *Meteorology*, or anywhere else.

[121] *Treatise on the Configuration of Qualities and Motions*, Part II, ch. 35 in Marshall Clagett, ed. and tr., *Nicole Oresme and the Medieval Geometry of Qualities*, 373–375.

[122] Ibid., Part II, ch. 26, 337–339.

wished to discuss this popular subject had to do so outside the university's intellectual orbit. Even then the subject of magic was fraught with difficulties. If expressed the wrong way, it could arouse Church authorities and cause an author major difficulties and problems. But magic was so easily intertwined with natural philosophy that treatises on magic may be properly regarded as lying within the domain of natural philosophy.

8

The Form and Content of Late Medieval
Natural Philosophy

To understand the substantive character of natural philosophy, it is essential to describe the kinds of literature in which medieval natural philosophers expressed their thoughts. In Chapter 7, I mentioned the two most basic forms of scholastic literature: (1) the commentary on a text and (2) the questions format in which the author, or commentator, formulates a series of questions on Aristotle's text, posing the questions sequentially in the order of the text. Examples from both of these types will illustrate the most fundamental methods of conveying natural philosophy to a broad audience.

Textual Commentaries on the Works of Aristotle

In the preceding chapter, I distinguished four varieties of textual commentaries. Because the first two methods were used primarily for teaching and conveyed little of the commentator's opinions, I shall illustrate only the third and fourth types. The third type was that in which the commentator separated his commentary from the text on which he was commenting. This could take two forms, the first of which involved a section-by-section sequential commentary on the text, while the second was a paraphrase of Aristotle's text.

The section-by-section sequential commentary was probably the most popular and probably the easiest to follow. In this method, the commentator cited a section of Aristotle's text followed by his commentary on that passage, a technique that derived from Averroes, the great twelfth-century Islamic commentator, who quoted a section of Aristotle's text followed by comments, in which he explained Aristotle's meaning and intent. He then presented another segment of text, explaining it in the same manner. Averroes used this method to explicate many of Aristotle's treatises. Although it involved the repetition of Aristotle's texts, it was very helpful for readers to have the text and the commentator's remarks presented sequentially through an entire work.

A variant version of the textual commentary just described dispensed with the relevant section of Aristotle's text, replacing it with the first few words of that passage – or "cue-words," as they are called by modern scholars. Thomas Aquinas used this format in his *Commentary on Aristotle's Physics*.

To illustrate his method, I shall cite passages 514 and 515 from his commentary on the fourth book of Aristotle's *Physics*, the section concerned with void space.

514. Next where he says, "For the fact of motion..." (214a 22), he explains why they posit a void.

He says that they accept the existence of a void for the same reason that they accept the existence of place, that is, because of motion, as was said above. This is done so that motion in respect to place may be saved, both by those who say that place is something beyond the bodies which are in place and by those who hold that a void exists. But for those who deny place and void, motion in respect to place does not occur. Thus they think that a void is a cause of motion in the way in which place is; namely, as that in which there is motion.

515. Next where he says, "But there is no necessity..." (214 a 26), he refutes the arguments of those who hold that a void exists. He does not intend here to give the true solution of the arguments given above, but to give for the present a solution from which it appears that these arguments do not conclude of necessity.

First, therefore, he refutes the arguments of those who hold a separated void, and secondly where he says, "And things can also be..." (214 a 33), the arguments of those who hold that there is a void in bodies.[1]

In the next paragraph, 516, Thomas summarizes Aristotle's refutation of the two arguments mentioned in paragraph 515.

As mentioned earlier, the paraphrase technique was a second basic kind of commentary. In this method, the commentator customarily intermingled Aristotle's phrases, and even individual words, with his own in an effort to convey Aristotle's meaning. As an illustration of the paraphrase technique with inclusions of bits and pieces of Aristotle's words, I cite a section from Albert the Great's (Albertus Magnus) *Commentary on Aristotle's Physics*. Albert drew his fragmentary textual inclusions from two Latin translations, one from Greek, the other from Arabic. The text I have chosen concerns Aristotle's discussion of the vacuum in the fourth book of the *Physics*. Here is the Aristotelian text as it appears in the Oxford English translation:[2]

Let us explain again that there is no void existing separately, as some maintain. If each of the simple bodies has a natural locomotion, e.g. fire upward and earth downward and towards the middle of the universe, it is clear that the void cannot be a cause of locomotion. What, then, *will* the void be a cause of? It is thought to be a cause of movement in respect of place, and it is not a cause of this.

Again, if a void is a sort of place deprived of body, when there is a void where will a body placed in it move to? It certainly cannot move into the whole of the void. The same argument applies as against those who think that place is something separate,

---

[1] *Commentary on Aristotle's "Physics" by St. Thomas Aquinas*, translated by Richard J. Blackwell, Richard J. Spath, and W. Edmund Thirlkel (New Haven, CT: Yale University Press, 1963), 229–230.

[2] The translation is by R. P. Hardie and R. K. Gaye. The passage occurs in bk. 4, ch. 8 (214b.12–27).

into which things are carried; viz. how will what is placed in it move, or rest? The same argument will apply to the void as to the up and down in place, as is natural enough since those who maintain the existence of the void make it a place.

And in what way will things be present either in place or in the void? For the result does not take place when a body is placed as a whole in a place conceived of as separate and permanent; for a part of it, unless it be placed apart, will not be in a place but in the whole. Further, if separate place does not exist, neither will void.

In translating Albert's Latin commentary on this passage, I shall follow the editor and indicate the places where Albert includes phrases from Aristotle's text, even where, as is often the case, Albert may only include one or two of Aristotle's words[3] (Aristotle's words are italicized). Albert selected words and phrases from the Latin translations he had before him and thus fashioned his paraphrase into an integrated account formed from Aristotle's words and phrases and his own.

Ch. 4: On the arguments that prove a vacuum is not the cause of local motion.

Because we have already shown that the arguments of the ancients do not prove the existence of vacuum, we show here plainly and absolutely that a vacuum cannot be the cause of local motion in any way. Those who assumed that it was the cause of motion were of two minds. Some said that it is a cause of motion just as something in which, and to which, there is motion; others, however, said that it is the cause of local motion as a mover and with regard to this second opinion, we shall say more below. *However, since there is no vacuum* that is *divided* and *separated from bodies*, as to something to which there is a motion, *as some say, we would say this, repeating* the discussion otherwise than *before. If indeed,* we should say, just as the physicists truly say, that there *is a change of place* of all simple bodies having successive rectilinear motion, *just as the natural* change of place of *fire, indeed,* is to be moved *upward,* [and] *of earth downward and toward the middle [of the world],* then *it is obvious that a vacuum* cannot be *the cause of a change of place.* Indeed, we have shown above in our *Treatise on Place* that a place has a natural affinity for the located thing and thus it has the power to draw to itself what is located in it when that thing is moved toward it. However, such a natural affinity and power cannot be shown to be in a vacuum, since a vacuum is nothing but dimensions which are everywhere of one kind and there is never in vacua a higher nature, or the nature of a [material] medium. *Therefore of what* change is *vacuum the cause? It was seen indeed* from the arguments of the ancients that *it is the cause* of change *of place;* but we have now shown *that it is not* the cause of such a change.

Furthermore, if something is also assumed to be a vacuum as a place, which has the power to draw the located thing in the manner stated [earlier] about place, and it is said that a vacuum is a place *deprived of body, since the vacuum is* separated from the body, then it is necessary that *the body posited in it is moved everywhere* in it, that is, into some part of its space which is said to be void; *for indeed it cannot* be moved *into every* part of this void space, because, as we said, the motion of simple

3 The translation is made from *Alberti Magni Ordinis Fratrum Praedicatorum Episcopi Opera Omnia*: Tomus IV, Pars I: *Physica*, libri 1–4, ed. Paul Hossfeld (Monasterii Westfalorum: Aschendorff, 1987), bk. 4, tract. 2, ch. 4, 237–238.

bodies is only to one part of space. *Moreover,* there will be this argument which we present here, commonly against those who assume that a vacuum is a separate space and against those who assume that *a place is* the dimensions of a separate space *in which* that which is moved *is received.* For although that space which is said to be a vacuum or place is assumed to have the attractive power of the contained body, nevertheless, since that space is nothing but a quantity which is of one kind in every part of it, it is necessary that the attractive power be of one mode and one kind in all the parts of that space. And by assuming this, *it is sought how* that which *is posited* in it *would be moved or* how *it will remain* [in one place?]; for it will be moved either to every part or to no part *and,* similarly, it will either remain in every part or in none. Similarly, *it is sought* how *up and down* are found in it; for [up and down] do not seem to be in it except with respect to us, just as dimensions in mathematics. *Concerning the vacuum,* however, *the same argument rightly applies,* since *a vacuum is also assumed to be a* separate *place* that is deprived of body.

Albert's commentary on this segment of Aristotle's text continues on for another paragraph, but enough has been presented to reveal the manner in which Albert paraphrased Aristotle's text.

The Questions ("Questiones") Form of Commentary on Aristotle's Works

The genre of scholastic treatises on natural philosophy that take the form of a sequence of questions on a text of Aristotle's developed from the "ordinary disputation" (*disputatio ordinario*) that occurred regularly in the medieval universities during the late Middle Ages.[4] In these disputations, the teaching master proposed a question for his class. Students were chosen to defend the affirmative and negative sides. After the presentation of both sides, the master was expected to resolve, or "determine," the question by proposing a solution. This became the skeletal frame of all questions that were included in questions treatises, or *questiones,* as they were called. This is obvious from the following six-step outline of a typical question:

1. Statement of the question.
2. Arguments opposed to the author's position, usually referred to as the "principal arguments" (*rationes principales*).
3. Assertion of one or more opinions opposed to the "principal arguments," often accompanied by an appeal to a major authority, usually Aristotle.
4. Clarification about the meaning of the question or any of its terms; an optional step.
5. Author's main arguments, which were presented in a variety of ways. Sometimes, the arguments were given as ordinally numbered conclusions (*conclusiones*); or they were not identified as conclusions,

---

4 I am here following my discussion in Edward Grant, *God and Reason in the Middle Ages* (Cambridge: Cambridge University Press, 2001), 105–107.

but were numbered ordinally; or they were left unnumbered, but presented one after the other (as in Buridan's question, later).

6. Brief refutation of each of the principal arguments presented in the second step.

To illustrate the medieval approach to typical questions, I present the response to a single question by John Buridan (ca. 1300–1358), who, in his *Questions on On the Heavens*, considered the possible diurnal rotation of the earth.

JOHN BURIDAN: *ON THE POSSIBILITY OF OTHER WORLDS*

Translated by Edward Grant

Next, I ask, whether it is possible that other worlds exist.[5]

[1] It is argued yes, [that more worlds can exist], because "world" (*mundus*) and "this world" are as universal and singular. For this term "world" is a common term according to both the grammarians and logicians. But, nevertheless, the common term and the discrete [or singular] term differ only in the sense that the common term is more suitable for predication of several things. Therefore, this term "world" is suitable for predication of several [worlds], which would not be true unless there could be several worlds; therefore, etc. [several possible worlds can exist].

[2] Again, several gods can exist, therefore several worlds can exist. The consequence is known because by the method [*ratione*] by which one god can make one world, another god can make another world by the same method. The antecedent is proved because in the second [book] of *De anima*[6] it is said that a perfected and undamaged [living] thing is able, by its nature, to generate a thing like itself. Therefore since god is most perfect and in no way damaged, it follows that he could generate a likeness of himself.

[3] Again, if God could make this world, He could, by a parity of reasoning, make another, since He is not now of less power than He was then. And so there could be several worlds.

[4] Again, if the world is conceded to be at least good, the possibility for making it better ought not to be denied. But it would be better that many worlds exist, or even many gods, than one only because, other things being equal, more good things are better than one. Therefore the possibility for several worlds or gods ought not to be denied.

[5] Again, if the world does not remain the same this year and the next year, it is obvious that several different worlds will exist. But the world does not remain the same this year and next year, because many parts of it are corrupted in these lower

---

[5] My translation from the Latin text in Ernest A. Moody, ed., *Ioannis Buridani Quaestiones super libris quattuor De caelo et mundo* (Cambridge, MA: Mediaeval Academy of America, 1942), bk. 1, qu. 19, pp. 87–90. For a fine discussion of the traditional problem concerning the existence of other worlds, see Steven J. Dick, *Plurality of Worlds: The Extraterrestrial Debate from Democritus to Kant* (Cambridge: Cambridge University Press, 1982). Medieval views are covered in chapter 2, pp. 23–43; for Buridan, see pp. 29–30.

[6] Aristotle, *De anima* 2.415a.25–415b.3.

things and many others are generated. And so there will be another world, because the whole is not the same if the parts are not the same.[7]

Aristotle [however] determines the opposite.

It must briefly be noted that "world" (*mundus*) can be taken in many ways. In one way as the totality (*universitate*) of all beings; thus the world is called "universe" (*universum*). "World" is taken in another way for generable and corruptible things and in another way for perpetual things; and so it is that we distinguish world into this inferior world and into a superior world. And yet "world" is taken in many other ways that are not relevant to our present discussion. But "world" is taken in another way that is pertinent to our present discussion, [namely] as the totality *(congregato)* of heavy and light [bodies] which appear to us *and* [also] the celestial spheres that contain these heavy and light [bodies]. And it is about such a world that the question – whether it is possible that several worlds exist – inquires.

And with regard to this, it must be noted that a plurality of such worlds can be imagined in two ways: in one way existing simultaneously, as if outside this world one other such world existed now;[8] in another way, they exist successively, namely one after the other.

CONCLUSIONS

[1] With regard to the first way of imagining a world, Aristotle holds that a plurality of worlds is not possible because he believes that this implies a contradiction, namely that the earth of one world would be moved naturally to the middle of another world. But this was previously discussed in another question.[9]

[2] But he [also] argues in another manner because if several worlds did exist simultaneously, it would follow that several first principles would exist, namely

---

[7] These five arguments represent the "principal arguments" (*rationes principales*) in favor of the proposition that other worlds could exist. At the conclusion of the question, Buridan will reject them point-by-point. With occasional exceptions, it was customary to present at the outset the major arguments for the position that would ultimately be rejected.

[8] In his discussion of the same question, Albert of Saxony (*Questions on De celo*, bk. 1, question 11 in *Questiones et decisiones physicales insignium virorum: Alberti de Saxonia in octo libros Physicorum; tres Libros De celo et mundo... Aristotelis. Recognitae rursus et emendatae summa accuratione et iudicio Magistri Georgii Lokert Scotia quo sunt tractatus proportionum additi* [Paris, 1518]), distinguished simultaneously existing concentric worlds, eccentric worlds, and worlds that are distinct and separated from each other "as several globes placed in a sack" (fol. 95r, col. D).

[9] In bk. 1, qu. 18, Buridan considered this very question, namely, "whether, if there were a plurality of worlds, the earth of one world would be moved naturally to the middle [or center] of another world" (Moody ed., pp. 83–87). Aristotle had assumed (*De caelo*, 1.8.276a.18–277b.26) that if other worlds existed, particles of earth from one world would indeed move to the center of another world, a consequence that involved the element earth rising upward contrary to its natural tendency to move downward. So absurd did this seem, that it was in and of itself a sufficient basis for Aristotle's rejection of other worlds. But Buridan, and other medieval scholastics (for example, Albert of Saxony and Nicole Oresme), disagreed with Aristotle and argued that the earth of each world would remain at the center of its own world (see, Moody ed., p. 87). On the assumption that all the hypothetical worlds are identical, which also was Aristotle's assumption, Buridan concluded that the earth of a given world would be dominated and determined by the laws of that world and would have no tendency to move toward another world beyond its own.

several gods. But this is impossible, as is obvious from the twelfth [book] of the Metaphysics.[10] But a consequence is proved from this because God is most simple and Aristotle believed that from one such most simple [God] several things could not arise except by mediating one thing by another; but these several worlds, which would be alike, could not arise by mediating one from another.[11] Thus they could not derive from a simple unique God. But you know that this argument is not valid because we believe on faith that God could make a world, indeed a plurality of worlds, and He could also destroy them again.[12]

But then let us inquire whether there can be a plurality of successive worlds. It must be noted that this can be understood in many ways: in one way where different worlds succeed each other with respect to their total diversity; in another way, with respect to their partial diversity. And again, partial diversity can be taken in two ways: in one way with respect to the most important parts; in another way with respect to the least important parts.

[3] I say, briefly, that concerning total diversity, different worlds can be made successively by the divine power, but not by natural power because celestial bodies are not generable or corruptible by natural powers.

[4] And I also say the same thing about the partial diversity of worlds with respect to the most important parts, which are indeed the celestial bodies, because these [celestial bodies] are not naturally generable or corruptible.

[5] But in speaking of the partial diversity of worlds with respect to the least important parts [of a world], it must be said that from day to day the world is continually different because the many least important parts [of a world] are corrupted

---

[10] Aristotle, Metaphysics, 12.10.1075b.38–1076a.6.

[11] Buridan knew, of course, that Aristotle denied the existence of a creator God. But here, according to Buridan, Aristotle assumes that if there were a single unique God, he could not create more than one thing directly. Thus, if a plurality of things exist, this could occur only because the one thing that God created was able, in turn, to create something else, and so on, a process that presupposes differences between successively created things. But a plurality of identical worlds could not be created in this manner because if God created one world initially, that world could not give rise to a second world identical to itself, which then would give rise to a third identical world, and so on. To understand why, we must turn to Buridan's response to the second principal argument (later). The world as we know it is predominantly a perfect, incorruptible thing by virtue of its celestial bodies. But an incorruptible perfect thing cannot, as Buridan implies below (in his response to the second principal argument), generate a likeness of itself because that would cause a change, and therefore an imperfection, within itself (this is Albert of Saxony's argument in his *Questions on De celo*, ed. cit., bk.1, qu. 11, fols. 95r, col.2–95v, col. 1). Likenesses can only be produced by "perfect things that are in the genus of generable and corruptible things" (by "perfect things" Buridan means the fully developed members of a species, whether the latter is corruptible or incorruptible).

[12] Here Buridan may have had in mind the Condemnation of 1277, in which article 34 condemned the opinion "That the first cause [i.e., God] could not make several worlds." For a translation of the 219 condemned articles by Ernest Fortin and Peter D. O'Neill, see *Medieval Political Philosophy: A Sourcebook*, edited by Ralph Lerner and Muhsin Mahdi (New York, 1963), 337–354. The significance of article 34 and other articles is discussed by Edward Grant, "The Condemnation of 1277, God's Absolute Power, and Physical Thought in the Late Middle Ages," *Viator*, vol. 10 (1979), 211–244.

and many others are generated.[13] Thus the world does not remain wholly the same. But neither is the difference [between such successive worlds] made totally different, but the world remains the same with respect to the most important major parts and becomes only partially different with respect to the lesser and least important parts. But because the denomination [or signification] of a name ought to be made more from its most important parts [rather than from its least important parts], we say that the world remains the same rather than [say] that it becomes different from day to day.

[Response] to the [principal] arguments

(1) To the first [argument], it is easy to respond by saying that not only this term "world" but also the term "god" are called common and specific terms not because they actually stand for several things, nor because there could be several things for which they stand, but because it is not inconsistent (*repugnat*) for these terms to stand for several things from the mode of their signification and imposition. But it is inconsistent on the part of the things signified. Indeed the term "chimera" is also a common term.[14]

(2) To another [i.e., the second principal argument], I say that it is not universally the nature of every perfect thing to generate a likeness of itself, but this [characteristic] belongs only to the nature of perfect things that are in the genus of generable and corruptible things.[15]

(3) To another [i.e., the third principal argument], I concede that God can make several other worlds.[16]

---

[13] By contrast with the "most important parts" (partes principaliores) of a world, which are the incorruptible celestial bodies, the "least important parts" (partes minus principales) of a world are the continually generable and corruptible inferior, or sublunar, bodies composed of compounds comprised of different proportions of the four elements.

[14] Here Buridan rejects the first argument, which inferred the actual existence of a plurality of worlds from the fact that a term like "world" could be predicated of several things. Although he concedes that a common term like "world" is indeed predicable of more than one thing, just as is the term "god," it does not follow that more than one world, or more than one god, exists. Otherwise, the common term "chimera," which is predicable of many fanciful and imaginary beings, would signify a real entity every time it was predicated. Indeed, it is improper to believe that more than one world actually exists and impious to believe that more than one god exists.

[15] Nothing in the celestial region can generate its own kind because generation and corruption, that is, change of any kind, cannot occur in that incorruptible realm. Only in the sublunar, or terrestrial, region of the world can the members of virtually every species (except those that are spontaneously generated) produce their own kind.

[16] As a consequence of the Condemnation of 1277, which, among other things, laid emphasis on God's absolute power to do anything short of a logical contradiction, scholastic authors such as Buridan routinely conceded that God could do things that were considered naturally impossible in the Aristotelian world system to which they subscribed. The existence of a single, unique world was a fundamental feature of Aristotelian cosmology and Christian theology and faith. But it was essential to concede that God could create other worlds if He wished. Indeed, the contemplation of the hypothetical consequences of hypothetical natural impossibilities that could be affected by divine power were much discussed during

(4) To another [i.e., the fourth principal argument], I say that the possibility for good or better ought not to be denied because every good or better thing is possible and no impossible thing is good or better. But if you say that a plurality of gods would be better than one, I reply, speaking categorically, that there are not, nor will there be, nor was there ever a plurality of gods; nor can it be that more gods are better than one god because it is impossible that there be more gods; therefore it [i.e., a plurality of gods] can be neither better nor worse. But [now] you ask, if there were more gods, would they not be better than one god and would it not be better that there be more gods than one [only]? I reply that from an impossible [proposition], contradictories follow, so that it is conceded that if there were more gods, they would be better than one and [also] not better than one.[17]

(5) The final argument, about a plurality of worlds with respect to a partial diversity, is well argued. And so the question is obvious.[18]

Buridan has followed the six-step outline described earlier. He enunciates the question (step 1), presents the principal arguments in favor of the position he will eventually oppose (step 2), and then introduces the opposite opinion, which Aristotle supports (step 3). Buridan now reaches step 4 where he discourses on the word "world" (*mundus*). How is that term to be understood and what do we mean by other worlds? In step 5, Buridan presents his own interpretations of the question and does so without identifying them as numbered conclusions, as was often done. Finally, Buridan concludes with the sixth step by rejecting, in turn, each of the five principal arguments enunciated in the second step.

Indeed, the six-step format was used in the formulation of hundreds of questions during the course of the late Middle Ages. In every question, the objective was to present the affirmative and negative arguments and to choose, or "determine," the correct response. Aristotelian natural philosophy was divided into hundreds of questions, as can be seen by simply counting the number that various scholars included in their questions treatises on Aristotle's major texts. Thus in his *Questions on De caelo*, John Buridan included fifty-nine questions; Albert of Saxony treated 107 questions in his *Questions on the Physics* and thirty-five questions in his *Questions on*

the fourteenth century. The possibility of a plurality of worlds was one such problem (see Grant, "The Condemnation of 1277," *Viator*, 10, 239–242).

[17] Of course, this argument also applies to a plurality of worlds.

[18] Buridan had himself argued this position, namely, that daily changes in the sublunar region of the world could be construed as a continuous sequence of different partial worlds. It was only because the most important parts of the world – that is, the celestial bodies – remained constant that our cosmos was assumed to be a single, enduring world rather than an endless succession of different worlds. Thus, "the question is obvious," by which Buridan means *the answer* is obvious: there is a succession of different partial worlds which produces a plurality of worlds in a trivial sense. Because of the incorruptible celestial region, however, the world remains a single, identifiable entity from which one may conclude that there is only one world.

Generation and Corruption; Nicole Oresme included forty-four questions in his *Questions on De anima*; and Themon Judaeus presented sixty-five questions in his *Questions on the Meteorologica*. The total number of questions on these five basic Aristotelian treatises is 310. Because the five treatises form the heart of Aristotle's natural philosophy, we realize that the subject of natural philosophy was fragmented into hundreds of independent, though largely, unrelated questions. Occasionally, an author would refer from one question to another, and thereby link one or more questions. Most questions, however, were left isolated and unconnected.

Although the questions arrangement was used primarily for Aristotle's treatises on natural philosophy, it was also used in other contexts. Indeed, there were even questions on Euclid's geometry and Aristotle's logic.

The themes and topics of medieval questions in natural philosophy were far-ranging. The question was the basic vehicle for the analysis of problems in natural philosophy and theology. Numerous questions were frequently proposed for a given problem or topic of interest. Many of the questions appear strange, and even bizarre. But they all form part of the medieval concept of doing natural philosophy, and we shall have to consider these various aspects later in this study. At this point, however, I shall move on and mention one other kind of literature in which natural philosophy was often the dominant or sole theme: the treatise, or tractate (*tractatus*).

The Thematic Treatise, or "Tractatus"

A genre of treatise in which much natural philosophy appeared is the *Tractatus*, or Treatise. These were almost always thematic works concerned with a subject relevant to natural philosophy. Works in this category often included the term *Tractatus* in the title, thus informing the reader that the treatise in question was focused on some particular subject. To illustrate the medieval tractate, I shall cite the introduction of the *Treatise on Proportions, or on the Proportions of the Speeds of Motion* composed by Thomas Bradwardine (ca. 1290–1349) in 1328. In his Introduction, Bradwardine explains his objective in the four chapters of his work and thereby gives us a good idea of how a tractate could be used to pursue a subject independently of Aristotle's texts.

INTRODUCTION

Since each successive motion is proportionable to another with respect to speed, natural philosophy, which studies motion, ought not to ignore the proportion of motions and their speeds, and, because an understanding of this is both necessary and extremely difficult, nor has as yet been treated fully in any branch of philosophy, we have accordingly composed the following work on the subject. Since, moreover (as Boethius points out in Book I of his *Arithmetic*), it is agreed that whoever omits mathematical studies has destroyed the whole of philosophic knowledge, we have commenced by setting forth the mathematics needed for the task in hand, in order to make the subject easier and more accessible to the student. For the sake of this same ease and accessibility, the work is also divided into four sections, or chapters.

The first of these, setting forth the necessary mathematics, is subdivided into three parts of which the first takes up the definitions, types and other properties of proportion. The second deals with proportionality in a similar fashion. The third adds certain axioms, from which several mathematical conclusions are drawn.

Chapter 2, on the other hand, argues against four opinions, or schools of thought, which have arisen concerning the proportion between the speeds of motions and, following the number of those opinions, is divided into four parts.

Chapter 3 makes clear the correct understanding of the proportion between the speeds of motions, with respect to both moving and resisting powers, and this is also divided into two parts. The first of these develops several theorems concerning the proportion between the speeds of motions, and the second raises and settles objections to them.

Chapter 4 treats of the proportion between the speeds of motions with respect to the quantities of the moved body and the interval traversed, and includes a special discussion of circular motion. It is divided into three parts, the first of which commences by establishing the requisite mathematical material. Part two undertakes the refutation of several opinions concerning the proportion between the speeds of motions, with respect to the magnitudes both of moved bodies and of intervals traversed, and sets forth the correct account. The third, finally, discloses certain hidden truths concerning the proportions between the elements.

Let us then pass on to the task in hand.[19]

Although his treatise used mathematics, Bradwardine had in fact composed a treatise on problems of motion that were derived ultimately from the seventh book of Aristotle's *Physics*. The problems of motion in Aristotle's seventh book of the *Physics* were regarded as quintessentially in the domain of natural philosophy, which all agreed was mostly about motion and change. Bradwardine obviously wished to pursue this topic at some length, and in considerable detail, for which reason he could not use the questions format, which was only suitable for problems that were resolvable in relatively brief compass.

Bradwardine's treatise on proportions led other scholastic natural philosophers with some degree of ability in mathematics to consider the same problems, as well as additional ones in treatises with similar sounding titles. The most significant author of tractates in the Middle Ages was Nicole Oresme, who applied mathematics to various problems in natural philosophy, as we shall see later, in this chapter (in the section "Beyond Aristotle").

The treatise format also was useful for pedagogical purposes, as was the case for the lengthy, anonymous, untitled treatise on natural philosophy described in the preceding chapter. The opening words of the treatise reveal

---

[19] H. Lamar Crosby, Jr., ed. and tr., *Thomas of Bradwardine His "Tractatus de Proportionibus": Its Significance for the Development of Mathematical Physics* (Madison: University of Wisconsin Press, 1955), 65.

its pedagogical intent. The author wished to present Aristotle's opinions, "because," as he explains,

the texts of Aristotle are quite prolix and verbose and often filled with difficult words, making their study for young people difficult and time-consuming, it thus seems appropriate to summarily collect the opinions of Aristotle and other philosophers and bring them under brief compass so they are more easily understood.[20]

Although other forms of literature relevant to natural philosophy were composed in the late Middle Ages – encyclopedias immediately come to mind, as well as scientific texts in the exact sciences that contained relevant discussions of natural philosophy – those that I have described comprise the basic core of medieval natural philosophy. Of these types, there is no doubt that questions on the various natural books of Aristotle constituted the most commonly used format for the presentation of natural philosophy at universities. The great variety of questions on Aristotle's natural books was representative of what medieval natural philosophy was about.

THE SUBSTANTIVE NATURE OF NATURAL PHILOSOPHY IN THE LATE MIDDLE AGES

To appreciate the enormous range and variety of medieval natural philosophy, it is essential to gain a sense of the questions posed by scholastic natural philosophers. Those questions were largely incited by the themes and topics Aristotle had included in his corpus of natural philosophical texts. But circumstances of culture and religion led medieval scholars to venture considerably beyond Aristotle's limits. Despite its wider range and greater adventurousness, medieval natural philosophy remained solidly within Aristotle's rationalistic approach. Buridan's question on the possible plurality of worlds shows the way scholastic natural philosophers analyzed and resolved questions about nature.

We can discern the essence of medieval natural philosophy in the range of questions that medieval natural philosophers regularly posed and which they obviously regarded as important for a proper understanding of nature. In the fourteenth century, scholastic natural philosophers considered four of Aristotle's treatises as constituting the core of natural philosophy, namely, the *Physics*, *On the Heavens*, *On Generation and Corruption*, and the *Meteorology*.[21] Indeed, Aristotle himself regarded physics in the broadest sense

---

[20] I have here repeated the passage from Chapter 7, in which I discussed the treatise.

[21] I have translated the enunciations of all the questions in these treatises drawn from the following authors: Albert of Saxony (*Physics*); John Buridan (*On the Heavens*); Albert of Saxony (*On Generation and Corruption*); and Themon Judaeus (*Meteorology*). There are 266 questions in the four treatises. For the translations, see Edward Grant, ed., *A Source Book in Medieval Science*, 199–210.

as comprised of the subject matters of these four treatises, as is evident from his opening remarks in the *Meteorology*:

We have already discussed the first causes of nature, and all natural motion,[22] also the stars ordered in the motion of the heavens,[23] and the corporeal elements – enumerating and specifying them and showing how they change into one another – and becoming and perishing in general.[24] There remains for consideration a part of this inquiry which all our predecessors called meteorology.[25] It is concerned with events that are natural, though their order is less perfect than that of the first of the elements of bodies. They take place in the region nearest to the motion of the stars. Such are the milky way, and comets, and the movements of meteors. It studies also all the affections we may call common to air and water, and the kinds and parts of the earth and the affections of its parts. These throw light on the causes of winds and earthquakes and all the consequences of their motions. Of these things some puzzle us, while others admit of explanation in some degree. Further, the inquiry is concerned with the fall of thunderbolts and with whirlwinds and fire-winds, and further, the recurrent affections produced in these same bodies by concretion. When the inquiry into these matters is concluded let us consider what account we can give, in accordance with the method we have followed, of animals and plants, both generally and in details.[26] When that has been done we may say that the whole of our original undertaking will have been carried out.[27]

To these four treatises, we should add *On the Soul* (*De anima*) and the *Parva Naturalia*. Under these two titles, Aristotle considers "the cognitive faculties, sensation and reason, but much space is devoted, particularly in the *Parva Naturalia*, to what we should consider physiological questions, as for example, sleep and waking, and even respiration. These treatises lead on to the biology proper."[28] Because medieval natural philosophers rarely commented upon the *Parva Naturalia* (or *Short Physical Treatises*) and the biological treatises, I have paid little attention to them, as well as Aristotle's *On the Soul*, although all of these works were certainly regarded as part of natural philosophy.

The questions routinely posed by medieval natural philosophers in their *Questions* on any one of Aristotle's treatises usually, but not always, reflected problems that Aristotle considered in the treatise being commented upon.

---

[22] Aristotle obviously intends his *Physics*.

[23] Here Aristotle refers to *On the Heavens* (*De caelo*).

[24] A clear reference to *On Generation and Corruption* (*De generatione et corruptione*).

[25] This is, of course, the *Meteorology*, or *Meteors*, as it is also called.

[26] Aristotle wrote a number of works on biology, where he discussed animals in great detail. He probably wrote works on botany, but a work often attributed to Aristotle, titled *On Plants*, is spurious.

[27] Drawn from Aristotle, *Meteorology*, 1.338.20–339a.9. The translation is by E. W. Webster in *The Complete Works of Aristotle, The Revised Oxford Translation*, edited by Jonathan Barnes, 2 vols. (Princeton, NJ: Princeton University Press, 1984).

[28] From G. E. R. Lloyd, *Aristotle: The Growth and Structure of His Thought* (Cambridge: Cambridge University Press, 1968), 182.

This is evident from an inspection of Albert of Saxony's questions on Aristotle's *Physics*. Here we find questions on Aristotle's major themes in the *Physics*. Albert's questions in the first book are based on rather minimal discussions by Aristotle, sometimes but a mere mention of the subject of a given question. Occasionally, the subject of the question has no counterpart in Aristotle's treatise. For example, in book 1, question 15, Albert asks "whether prime matter could be knowable per se," a subject Aristotle does not discuss or mention in book 1. For the most part, however, there is some basis, tenuous though it sometimes is, for each question in each book, including book 1. For all the major themes Aristotle discussed in each book, Albert, and all other scholastic commentators, usually included two or more questions. These are questions that emerged in the evolution of medieval natural philosophy and illustrate the kinds of themes that medieval natural philosophers deemed most important and interesting. In his second book of the *Physics*, Albert considers how the term "nature" is used and its definition and a few subsidiary questions (questions 1–6),[29] the relationship between natural science and mathematics (questions 7 and 8), the four different causes and their effects (questions 9–12), the role of chance (questions 13–14), the possibility that nature can produce a monster (question 15) and the concept of necessity (question 16). In the remaining books, Aristotle considered a number of themes that were focal points of medieval natural philosophy. Among these topics are what is the nature of motion (questions 1–7) and the infinite (questions 9–13) (book 3); what are place (questions 1–7), void (questions 8–12), and time (questions 14–17) (book 4); numerous aspects of motion (questions 1–8) (book 5); the continuum and whether it is composed of divisible quantities or indivisible magnitudes (questions 1–3) (book 6); on the relationship of mover and moved (questions 1–2), including a discussion of how motions or velocities can be represented mathematically (questions 6–7) (book 7); whether motion is eternal (questions 1–3) and discussions of rectilinear and circular motion that are background for Aristotle's treatment of the first mover (or unmoved mover) (questions 4–13) (book 8). Not all major topics were discussed. For example, in book 1, chapter 2, of his *Physics*, Aristotle inquires whether all things are one. Albert of Saxony includes no question on this theme in his first book, or anywhere else in his treatise. What is true for questions in the *Physics*, is also true for Aristotle's other treatises on which medieval natural philosophers posed questions.

Departures from Aristotle

Although some discussion by Aristotle, or some remark he made, served as the basis for most questions in natural philosophy, the responses, surprisingly, often disagreed with Aristotle's position, or qualified it in some

---

[29] English translations of all the questions cited here can be found in Grant, *Source Book in Medieval Science*, 199–210.

significant way. Despite their great admiration for Aristotle the man and his natural philosophy, medieval natural philosophers did not regard him as infallible. As Albertus Magnus expressed it, "if . . . one believes him to be but a man, then without doubt he could err just as we can too."[30] Nicole Oresme reveals the ambivalence toward Aristotle even when he had reason to believe that Aristotle had erred. In the first proposition of the fourth book of his *Treatise on Ratios of Ratios* (*Tractatus de proportionibus proportionum*), Oresme declares:

That the following rules are false: If a power moves a mobile with a certain velocity, double the power will move the same mobile twice as quickly. And this [rule]: If a power moves a mobile, the same power can move half the mobile twice as quickly.

Thus, the first rule may be represented as follows: if $F/R \propto V$, then $2F/R \propto 2V$; and the second rule: if $F/R \propto V$, then $F/(R:2) \propto 2V$. In the final paragraph of the proposition, Oresme indicates that these false rules appear to have been proclaimed by Aristotle. "What, then," he asks, "should we say to Aristotle, who seems to enunciate the repudiated rules in the seventh [book] of the *Physics*?" Oresme explains that unless certain qualifications are added, the rules are certainly false. Perhaps we should read Aristotle as really intending these qualifications, which would make his rules correct for those instances. "Perhaps Aristotle said this," Oresme concludes, "but has been poorly translated. But if he did not say it, perhaps he failed to understand [the rules] properly."[31]

In this proposition, Oresme is obviously trying to save Aristotle. Perhaps Aristotle really understood the rules correctly but was poorly translated. In his last extant work, *Le Livre du ciel et du monde*, dated in 1377, Oresme abandons his earlier attempt to excuse Aristotle and assumes instead that Aristotle failed to understand the rules properly. This certainly seems to be the import of his assertion that "saving his reverence [i.e, Aristotle], it [that is, the false rules] is not well stated."[32] In truth, Oresme actually misinterpreted Aristotle,[33] but that is of no relevance here. Of significance, is his obvious disagreement with Aristotle. Over the next few centuries, many other natural philosophers also disagreed with one or more aspects of Aristotle's natural philosophy. In what follows, I shall present a number of these departures, not only to show that medieval natural philosophers were anything but slavish followers of Aristotle, but also because the disagreements reveal the kinds of issues that were regarded as controversial and with respect to which Aristotle's positions were found wanting or judged erroneous. The disagreements described in what follows are not the kind that arose from difficulties

---

[30] From Albertus's *Commentary on Aristotle's Physics*, bk. 8, tract 1, ch. 14. Cited from Grant, *The Foundations of Modern Science in the Middle Ages*, 164.

[31] See Grant, ed. and tr., *Nicole Oresme "De proportionibus proportionum" and "Ad pauca respicientes,"* 275.

[32] See ibid., 368–369.   [33] See ibid., 369, where I point out Oresme's misinterpretation.

as to Aristotle's real meaning, or ambiguities resulting from poor translations, or even errors in copying and transmission of manuscript versions. They are genuine criticisms of Aristotle's interpretation and understanding of various aspects of his natural philosophy.

Impetus Theory

One of the most profound departures from Aristotle's ideas occurred in a basic problem of motion: what enables a body to keep moving after its initial external mover has lost contact with it? Aristotle had argued (in *Physics*, book 8, chapter 10) that after a body loses contact with its initial mover, it is propelled forward by the air, which has been activated by the action of the initial mover, an explanation that was known as *antiperistasis*. As successive segments of air are activated, they gradually lose some force until eventually the last segment of air is incapable of causing motion and the body comes to rest, or, as Aristotle described the action of the air as motive power: "The motion ceases when the motive force produced in one member of the consecutive series is at each stage less, and it finally ceases when one member no longer causes the next member to be a mover but only causes it to be in motion The motion of these last two – of the one as mover and the other as moved – must cease simultaneously, and with this the whole motion ceases."[34] Thus did Aristotle use air as both a motive power and a resistance to motion, arguing that without the air to resist and slow the motion of bodies, motions would be instantaneous, that is, of infinite speed.

In the late fifth and early sixth centuries AD, John Philoponus, a Greek convert to Christianity, rejected Aristotle's explanation by arguing that if air were the motive force in continuous forced motion, one ought to be able to cause a stone to move by agitating the air behind it.[35] Philoponus replaced Aristotle's explanation with one that invoked an incorporeal impressed force, which he believed was imparted to the projectile by the motive force, when the latter was brought into contact with the former. Thus did Philoponus seek to account for forced motion in Aristotle's physical system. Islamic natural philosophers, who were probably familiar with Philoponus's arguments, continued on the path that Philoponus pioneered and added some new dimensions to his impressed force concept, which they called *mail*. Avicenna, for example, regarded *mail* as a permanent quality that would endure in a moving body forever if there were no external resistances to the motion. Another Islamic author, Abu'l Barakat (d. ca. 1164), described the impressed force, or *mail*, as a nonpermanent force that was self-dissipating. These arguments were important for discussions about the possibility of motion in a vacuum, the existence of which Aristotle denied. A permanent impressed

[34] Aristotle, *Physics*, 8.10.267a.8–12.
[35] For a translation of Philoponus's argument, see Morris R. Cohen and I. E. Drabkin, eds. *A Source Book in Greek Science* (New York: McGraw-Hill Book Co., 1948), 223.

force would cause a body in a vacuum to move forever, and therefore would tend to reinforce the Aristotelian idea that a vacuum was impossible. By contrast, a self-dissipating impressed force would dissipate in a void, so that a body moving in the void would come to rest immediately thereafter. Both of these kinds of impressed force would find counterparts in the Latin West.

As a result of the translations of Greek and Arabic texts in natural philosophy, the impressed force theory was already known in the thirteenth century but was rejected. In the fourteenth century, however, many natural philosophers found the arguments in favor of impressed force persuasive. As early as 1323, Franciscus de Marchia, a theologian, accepted a self-dissipating kind of impressed force, and also assigned a subsidiary role to the air. The most famous account of the impressed force theory, however, was given by John Buridan, one of the most famous natural philosophers of the Middle Ages.

Buridan may have been the first to use the Latin term *impetus* to describe the impressed force theory. He dealt with the problem at great length in his *Questions on the Physics*, book 8, question 12.[36] In this question, he asks "whether a projectile after leaving the hand of the projector is moved by air, or by what it is moved." He judged "the question to be very difficult, because Aristotle...has not solved it well."[37] He first brings three experiences to bear on Aristotle's position, showing that certain motions could not be explained by invoking the air, namely, a smith's wheel, a lance that is pointy at both ends, and the movement of a ship under certain circumstances. After mustering other arguments against air as the mover of projectiles, Buridan declares:

it seems to me that it ought to be said that the motor in moving a moving body impresses (*imprimit*) in it a certain impetus (*impetus*) or a certain motive force (*vis motiva*) of the moving body, [which impetus acts] in the direction toward which the mover was moving the moving body, either up or down, or laterally, or circularly. *And by the amount the motor moves that moving body more swiftly, by the same amount it will impress in it a stronger impetus.*[38] It is by that impetus that the stone is moved after the projector ceases to move. But that impetus is continually decreased (*remittitur*) by the resisting air and by the gravity of the stone, which inclines it in a direction contrary to that in which the impetus was naturally predisposed to move it. Thus the movement of the stone continually becomes slower, and finally that impetus is so diminished or corrupted that the gravity of the stone wins out over it and moves the stone down to its natural place.[39]

[36] The most significant parts of Buridan's question are translated and commented on in Marshall Clagett, *The Science of Mechanics in the Middle Ages* (Madison: University of Wisconsin Press, 1959), 532–540. Clagett's translation and commentary are reprinted in Grant, *A Source Book in Medieval Science*, 275–280.

[37] Clagett, ibid., 532.   [38] The italics are Clagett's.

[39] Clagett, *The Science of Mechanics*, 534–535.

Buridan used the speed of a body and its quantity of matter to measure the amount of impetus it had. These were the same quantities that measured momentum in Newtonian physics, although momentum for Newton was an effect of motion, whereas impetus was a cause of motion for Buridan.[40]

Impetus and the Acceleration of Falling Bodies

Although Aristotle was aware that freely falling bodies accelerated, he paid little attention to acceleration. Buridan, however, recognized that bodies fall with a continuous acceleration and used his impetus theory to explain this phenomenon. He assumed that a body's heaviness was the cause of its downward motion. To explain its acceleration, however, he assumed that the body's heaviness produced successive increments of impetus, or "accidental heaviness," as it was occasionally called. At each moment, the body's heaviness produces an increment of impetus. Therefore, in successive moments, successive increments of impetus are produced and the body moves down with a continuous acceleration. Impressed force theories were influential into the sixteenth century. To explain the downward acceleration of bodies, Galileo ultimately adopted a version of the impetus theory that was quite similar to Buridan's.

Is There a Moment of Rest between the End of an upward Motion and Its Descent?

In book 8, chapter 8 of his *Physics*, Aristotle argues that if something is thrown upward and then descends to the ground, that motion cannot be regarded as a continuous motion, because up and down are contrary to each other. There has to be at least a momentary temporal break between the upward motion and the downward motion. This came to known as the "moment of rest," or *quies media* (in medieval Latin texts). In their questions on Aristotle's *Physics*, scholastic natural philosophers, including John Buridan, Albert of Saxony, and Marsilius of Inghen, vigorously disagreed with Aristotle, using a series of imaginary experiential illustrations to subvert his position. A popular "experience" was one in which a millstone descends and strikes an upward moving bean. As Marsilius of Inghen describes it:

The proof is that if a bean were projected upward against a millstone which is descending, it does not appear probable that the bean could rest before descending, for if it did rest through some time it would stop the millstone from descending, which seems impossible.[41]

---

[40] For a more extended treatment of impetus, see my *Physical Science in the Middle Ages* (Cambridge: Cambridge University Press, 1977), 48–54, and my *Foundations of Modern Science in the Middle Ages* (Cambridge: Cambridge University Press, 1996), 93–98.

[41] My translation from Grant, *A Source Book in Medieval Science*, 286. I also cite this passage in my *God and Reason in the Middle Ages* (Cambridge: Cambridge University Press, 2001), 170.

There were at least two other arguments that were introduced as counterinstances to Aristotle's position. One involves a ship on which Socrates is imagined to be walking in a direction opposite to that in which the ship is moving; the other assumes that a fly is walking up a lance that is moving in the opposite direction.[42] Galileo mentions a version of the "millstone-bean" argument in his early work *On Motion* (*De motu*), in which he refers to it "as the well-known one about a large stone falling from a tower" and descending on a pebble thrown up from below.[43]

Is the Earth at Rest in the Center of the Universe?
In his *Questions on On the Heavens* (*De caelo*), book 2, question 22, John Buridan asks "whether the earth always is at rest in the center of the universe."[44] He was reacting to Aristotle's arguments in book 2, chapter 14 of *On the Heavens*, in which Aristotle concludes that "it is clear that the earth does not move and does not lie elsewhere than at the centre."[45] To counter Aristotle, Buridan considers two kinds of motions that the earth might possess: a rotational motion and a slight rectilinear motion around the center.

Does the Earth Have a Rotational Motion at the Center of the World?
In treating the first case, Buridan asserts that the earth's daily rotation from west to east is compatible with astronomical phenomena. If we assume that the stellar sphere is at rest and the earth rotates daily on its axis, the earth's rotation would produce night and day. To show that this arrangement would produce the proper results, Buridan resorts to an example involving relative motion. "If anyone is moved in a ship," he declares,

and he imagines that he is at rest, then, should he see another ship which is truly at rest, it will appear to him that the other ship is moved. This is so because his eye would be completely in the same relationship to the other ship regardless of whether his own ship is at rest and the other moved, or the contrary situation prevailed. And so we also posit that the sphere of the sun is everywhere at rest and the earth in carrying us would be rotated. Since, however, we imagine that we are at rest, just as the man located on the ship which is moving swiftly does not perceive his own motion nor the motion of the ship, then it is certain that the sun would appear to us to rise and then to set, just as it does when it is moved and we are at rest.[46]

[42] For a description of these two counter examples, see my *God and Reason in the Middle Ages*, 171.

[43] For the reference, see Grant, ibid., 170.

[44] Most of the question is translated and annotated by Marshall Clagett, *The Science of Mechanics in the Middle Ages*, 594–599. It is reprinted in my *Source Book in Medieval Science*, 500–503.

[45] Aristotle, *On the Heavens*, 2.14.296b.23–24. The translation is by J. L. Stocks in the Oxford English Translation.

[46] Clagett, *The Science of Mechanics in the Middle Ages*, 501.

Although Buridan offered other arguments in favor of the earth's daily rotation, he ultimately found the arguments against rotation more persuasive than the favorable arguments.

In both his *Questions on On the Heavens* and in his French commentary on *On the Heavens*, known as Le *Livre du ciel*, Nicole Oresme considers whether the earth rotates daily on its axis. In his French commentary, where Oresme presents the more powerful arguments, Oresme ultimately agreed with Aristotle and Buridan that the heaven rather than the earth rotates daily. But Oresme departed from both Aristotle and Buridan when he insisted that the arguments in favor of a daily rotation were just as convincing as those that denied rotation. Oresme emphasized that neither experience nor reason could conclusively demonstrate either alternative. Indeed, Oresme advanced beyond the earlier arguments of John Buridan, formulating an impressive array of arguments in favor of the earth's rotation that Nicholas Copernicus himself did not surpass in his *On the Revolutions of the Heavenly Orbs* (1543). Like Buridan, Oresme relied on relativity of motion to show the plausibility of the earth's daily rotation. He observed that we only perceive the local motion of a body when it assumes a different position relative to another body. If a man were carried around by the heavens and could see the earth in some detail, it would seem to him that the earth moved with a daily motion, just as we, on earth, viewing the rotating heavens, attribute the daily motion to the heavens. To the argument that if the earth rotated from west to east, a noticeable wind should blow constantly from the east, Oresme counters that the air rotates with the earth and therefore would not blow from the east. Similarly, some argued that because an arrow shot into the air falls back approximately to the place from whence it was shot, and does not fall to the west, it follows that the earth does not rotate. Oresme countered that this argument was inconclusive because if the earth rotated, the arrow would share the earth's rotation and fall back to the place from which it was launched. Thus, the same effects would occur whether or not the earth had a daily axial rotation. Oresme offered additional plausible, although nondemonstrative, arguments in favor of the earth's axial rotation. For example, it would be simpler if God caused the daily rotation of the heavens by causing the small earth to rotate, rather than making the monumentally large heavens revolve at enormously higher speeds. To the biblical argument that God aided the army of Joshua by making the sun stand still over Gibeon (Joshua 10:12–14), thus demonstrating that the heavens rotate and the earth is at rest, Oresme suggests that God could have achieved the same effect in performing His miracle by temporarily halting the earth's rotation. Both Galileo and Kepler presented explanations of the Joshua miracle, with Kepler's arguments resembling Oresme's.

At the conclusion of his lengthy discussion, however, Oresme opts for Aristotle's traditional opinion. He was apparently convinced that the earth's daily rotation "seems as much against natural reason as, or more against natural reason than, all or many of the articles of our faith." Oresme adopted the customary position despite the fact that his arguments in favor of the earth's rotation were more plausible and powerful than the universally assumed alternative.[47]

The arguments in favor of rotation mark a significant departure from Aristotle on the important issue of the earth's status at the center of the world. Both Buridan and Oresme marshaled powerful arguments in favor of the earth's daily rotation, a number of which were repeated in Copernicus's revolutionary treatise of 1543. Oresme went further than Buridan and argued that on the basis of reason and evidence there was no basis for choosing one hypothesis over the other. Both Buridan and Oresme seriously challenged Aristotle's idea of a resting nonrotating earth and marked a departure because they both assumed that the earth's daily rotation was not impossible or absurd but a viable hypothesis.[48]

Does the Earth Move Rectilinearly?

If Buridan ultimately sided with Aristotle in denying the earth's diurnal rotation, he definitely broke with him on the issue of the earth's possible rectilinear motion. In Aristotle's view not only did the earth not rotate, but neither did it move rectilinearly in any way whatever. It was wholly at rest in the center of a spherical universe. In the latter part of the same question in which he discusses diurnal rotation, Buridan vigorously challenges that dogma, using the concepts of "center of gravity" and "center of magnitude." The earth undergoes incessant geological changes that at every moment alter the rarity and density of its different parts, which, in turn, continually change the earth's center of magnitude. As a result, the center of magnitude does not coincide with the earth's center. Rather, it is the earth's center of gravity that coincides with the earth's geometric center. As Buridan explains, "the center of the universe is the center of gravity of the earth." Because the earth is constantly changing its density in its different parts, the center of magnitude is also changing. As the earth's different parts undergo incessant change, the magnitude of the earth is constantly changing and therefore its center of magnitude is also changing. Because the center of the earth's magnitude is constantly shifting, Buridan argues that it does not coincide with the center of the universe. Rather, the earth's center of gravity coincides with the center of the universe. In sum, Buridan explains that "the earth, with respect to its magnitude, is not directly in the center of the universe. We commonly say,

---

[47] For Oresme's arguments, see Grant, *A Source Book in Medieval Science*, 503–510. For the brief quotation, see p. 510, and also n. 61 on that page.

[48] For further elaboration on the theme of the daily rotation, see Grant, *The Foundation of Modern Science in the Middle Ages*, 112–116, and for translated selections on this theme, also see my *Source Book in Medieval Science*, 494–516.

however, that it is in the center of the universe, because its center of gravity is the center of the universe."[49]

As the earth's center of magnitude constantly shifts, so does its center of gravity. Does this mean, Buridan inquires, that "the earth is sometimes moved according to its whole in a straight line"?[50] He replies in the affirmative. As the earth's center of gravity constantly shifts in accordance with the continually changing center of magnitude, each new center of gravity moves sufficiently to coincide with the geometric center of the universe. To achieve this, the earth actually moves a short rectilinear distance. Therefore, contrary to what Aristotle declared, the earth is not at rest in the center of the universe, but constantly moves short rectilinear distances to bring its center of gravity into coincidence with the center of the universe.

THOUGHT EXPERIMENTS AND THE ROLE OF THE IMAGINATION

In the topics just described, scholastic natural philosophers departed from a number of Aristotle's ideas about the operation of the physical world. In doing so they undoubtedly used their imaginations to consider the alternatives many of them adopted. But the new emphasis on imagination by medieval natural philosophers went far beyond the realm of reality and focused on hypothetical problems relevant to imaginary worlds.

Perhaps the greatest difference between Aristotle's natural philosophy and that of his medieval Latin followers lies in the role played by the imagination. Although Aristotle ostensibly based his natural philosophy on sense perception and empiricism, he relied most heavily on theory and his sense of how things had to be in order to produce the operations of our divine world, as he understood them. Despite his emphasis on observation and sense perception, Aristotle recognized their inadequacy when he declared that although the senses "give the most authoritative knowledge of particulars ... they do not tell us the 'why' of anything – e.g. why fire is hot; they only say that it is hot."[51] During the late Middle Ages, much lip service was paid to observation and experience, but direct observation played a small role in most of the examples and illustrations cited in behalf of this or that claim. Medieval empiricism was largely an "empiricism without observation."[52]

If genuine observation and empiricism played a minor role in medieval natural philosophy, medieval natural philosophers departed radically from Aristotle by virtue of their heavy emphasis on the imagination. For the first time in the history of science, the imagination came to play a major role in the analysis of problems in natural philosophy. Of course, use of the imagination

---

[49] Clagett, *The Science of Mechanics in the Middle Ages*, 597; Grant, *Source Book*, 502.
[50] Clagett, ibid.; Grant, ibid. [51] Aristotle, *Metaphysics*, 1.1.981b10–11.
[52] For more on "empiricism without observation," see the latter part of this chapter.

in the study and resolution of problems about nature was unavoidable. It had always played a role from the very beginning of human efforts to understand nature's operations. But, during the late Middle Ages, the imagination became a formidable instrument in natural philosophy and theology in ways that would have astonished Aristotle. It became a common practice to assume hypothetical conditions relevant to other worlds and empty spaces and then to imagine how various problems could be resolved within the boundary conditions of these hypothetical constructs. When engaged in this kind of activity, scholastics used the expression *secundum imaginationem*, "according to the imagination." They fully recognized that they were engaged in hypothetical activities that relied heavily on their imaginations, as guided by reason and logical analysis.

With a few exceptions that involved problems of motion, they made no meaningful effort to transform their hypothetical conclusions into specific knowledge about the real physical world. They did, however, assume that, although these hypothetical conclusions were naturally impossible, God could produce them supernaturally if He wished. Indeed, the most interesting cases involved ways in which God was first assumed to produce counterfactual cosmic creations, after which it was imagined how basic features of Aristotle's natural philosophy would be altered, as will be evident in the following illustrations.

Departures from Aristotle Based on Appeals to God's Absolute Power

Significant departures from Aristotle's natural philosophy occurred not only directly, but also indirectly by way of counterfactuals, or hypothetical assumptions about the world. As we saw, such departures were said to occur "according to the imagination" (*secundum imaginationem*). The urge to invoke counterfactuals in natural philosophy was considerably stimulated by a theological condemnation in 1277, known as "The Condemnation of 1277," which was triggered by Aristotle's natural philosophy. The teaching of Aristotle's natural philosophy was officially banned in Paris, and therefore banned at the University of Paris, from 1210 until the 1240s. In the decade of the 1240s, the authorities apparently relented and permitted lectures on Aristotle's natural philosophy in the faculty of arts, where Roger Bacon gave the first lectures on that subject. Despite its acceptance, an uneasiness about Aristotle's interpretation of the world continued among some influential theologians and Church officials in Paris into the 1270s.

The cause of their apprehension was the simple fact that a number of Aristotle's doctrines clashed with revelation and biblical texts. Aristotle's arguments for an eternal world that had neither beginning nor end conflicted with the Christian doctrine of creation from nothing. Aristotle believed that only the rational part of the human soul is immortal, whereas Christians made no distinctions between different levels of soul, regarding the entire soul as immortal. Aristotle also insisted that all accidental qualities necessarily inhered in a substance, a view that conflicted with the Christian doctrine of the Eucharist, or Mass, in which, after conversion to the body and blood of Christ, the bread and wine continued to exist without inhering in any substance.

If these were the only differences, Church officials might have simply informed university teachers to point out Aristotle's doctrinal errors and then move on to the greater part of his natural philosophy that did not conflict with church doctrine. There was, however, one other aspect of Aristotle's approach to nature that caused great concern in the 1260s and 1270s and led to the Condemnation of 1277. Aristotle often presented his conclusions about the world as absolutely necessary and impossible to be otherwise. For example, Aristotle argued that there is only one world, and that it is impossible for other worlds to exist. He also argued that the existence of a vacuum is impossible, because finite motion could not occur in a vacuum and all bodies, whatever their weight, would fall with equal speeds in a vacuum, which Aristotle thought absurd. Did this mean that even if God wished to create other worlds, or create a vacuum, that He could not do so, because such things were shown to be impossible in Aristotle's natural philosophy? For a true follower of Aristotle, the answer is no, God could not perform such actions.

Because of their concern about such matters, a group of conservative theologians prevailed upon the bishop of Paris, Etienne Tempier, to condemn thirteen propositions in 1270, and, when that was seemingly ineffective, in 1277, Pope John X, hearing of the continuing controversies in Paris, instructed bishop Tempier to investigate and resolve the issue. Acting on the advice of his theological advisors, bishop Tempier responded within three weeks (in March 1277) with a massive condemnation of 219 propositions.[53]

The Possibility of Other Worlds[54]

Many of the condemned articles were relevant to natural philosophy. One of the most important involving God's absolute power is article 34, which declares "That the first cause [that is, God] could not make several worlds."[55] This is condemned because it denies to God the ability to make more worlds if He wishes. Although scholastic natural philosophers and theologians did not really believe that God had made other worlds, or would ever make

---

[53] For a translation of the condemned articles by Ernest L. Fortin and Peter D. O'Neill, see Ralph Lerner and Muhsin Mahdi, *Medieval Political Philosophy: A Sourcebook*, 337–354. For a translation of the articles relevant to science and natural philosophy, see my translation in Grant, *A Source Book in Medieval Science*, 45–50.

[54] For a more complete treatment of medieval discussions of the possible existence of other worlds, see Grant, *Planets, Stars, and Orbs: The Medieval Cosmos, 1200–1687* (Cambridge: Cambridge University Press, 1994), ch. 8 ("The Possibility of Other Worlds"), 150–168.

[55] See Grant, ibid., 48. I have added the bracketed qualification.

other worlds, they all conceded that God could make them if He wished. This was a significant concession, because it marked a major departure from Aristotle: although Aristotle regarded it as impossible for other worlds to exist, medieval natural philosophers allowed that the existence of other worlds was possible, albeit only by supernatural action.

Discussions of the possibility of other worlds were essentially hypothetical and counterfactual. But they are nonetheless important because once the question was posed and it was conceded that God could create other worlds, it was necessary to inquire about the properties and characteristics of those worlds. Would these worlds be successive, that is, would only one world exist at any time, or would they exist simultaneously, and if the latter, how would they be arranged? Would they be scattered in space; would they be concentric to one another; or would they be arranged eccentric to one another? If so, each world would come into being and pass away. It would then be replaced by another world that would also come into being and pass away, and so on, as long as it pleased God to create and destroy them.

Earlier in this chapter, I presented the full text of John Buridan's question on the possibility of other worlds, where he sought to respond to some of these hypothetical queries. Somewhat later in the century, in his *Questions on the Heavens*, Albert of Saxony (ca. 1316–1390) also devoted a question to the possibility of other worlds.[56] In his question, Albert sided as much as he could with Aristotle, whereas Buridan was much more critical of him. Both of course were compelled to concede that God could make other worlds, but Albert of Saxony chose to emphasize how such worlds would violate Aristotle's natural laws.

The departures from Aristotle are, however, of primary interest. One of Aristotle's major arguments against a plurality of worlds – worlds, by the way, that he assumed were identical in structure and operation – is that the earth, or parts of the earth, of one world would move toward the earth in the center of another world. In order to move out of its world, any piece of earth would have to rise upward and pass through the outermost surface of its own world and move toward the center of another world. This would mean that a heavy earth, or part of the earth, would have to move upward contrary to its natural tendency to move downward. Albert of Saxony and Buridan both denied this. The earth of each world would remain where it is, because each world is a self contained entity. Thus contrary to Aristotle, if God created simultaneous identical worlds, they would coexist without any difficulty.

---

[56] Albert of Saxony, *Questions on Aristotle's On the Heavens (De caelo)*, bk. 1, qu. 11 in *Questiones et decisiones physicales insignium virorum: Alberti de Saxonia in octo libros Physicorum; tres libros De celo et mundo; duos libros De generatione et corruptione; Thimonis in quatuor libros Meteororum; Buridani in tres libros De anima* ... (Paris, 1518), fols. 95r, col. 1–95v, col. 1.

The possibility of a plurality of worlds subverted some of Aristotle's foundational assumptions about our world, assumptions that were all based on his unqualified belief that there is, and can be, only one world. In that world Aristotle argued that the four elements – earth, water, air, and fire – each had a natural place that it would always move toward if otherwise unimpeded. These natural places were unique to our world, being located within the single center and circumference of our world. The possibility of other worlds, however, each with its own center and circumference and its own four elements destroys Aristotle's system. With a multiplicity of centers and circumferences, all would be equal and none unique. The mere possibility of a multiplicity of centers and circumferences each with its own four elements completely destroys Aristotle's notion of natural place.

The problem of the plurality of worlds is a good springboard for describing another great departure from Aristotle: the possible existence of void or empty space. Indeed, in his brilliant discussion of the plurality of worlds in his French commentary on Aristotle's *On the Heavens – Le Livre du ciel et du monde –* Nicole Oresme links the problem of void space to the problem of other worlds.[57] He observes that Aristotle argues "that outside this world there is no place or plenum, no void, and no time."[58] Oresme counters, however, that "if two worlds existed, one outside the other, there would have to be a vacuum between them for they would be spherical in shape." In support of this claim, Oresme declares:

It seems to me and I reply that, in the first place, the human mind consents naturally, as it were, to the idea that beyond the heavens and outside the world, which is not infinite, there exists some space whatever it may be, and we cannot easily conceive the contrary.

After imagining a few situations in which void spaces could occur, and conceding that such empty spaces could not occur from natural causes, Oresme concludes that:

outside the heavens, then, is an empty incorporeal space quite different from any other plenum or corporeal space.... Now this space of which we are talking is infinite and indivisible, and is the immensity of God and God Himself.[59]

Oresme explains that we cannot perceive this space by our corporeal senses, but "reason and truth, however, inform us that it exists." Because there is an infinite void space beyond our world, Oresme concludes that God could indeed make other worlds, and Aristotle cannot demonstrate the contrary.

---

[57] For Oresme's discussion, see *Nicole Oresme: Le Livre du ciel et du monde*, ed. Albert D. Menut and Alexander J. Denomy; translated with an Introduction by Albert D. Menut, 167–179. This section is reprinted in Grant, *A Source Book in Medieval Science*, 547–554.

[58] *Oresme: Le Live du ciel et du monde*, 177.

[59] Ibid. This citation includes the last two cited passages.

Oresme concludes, however, that "there has never been nor will there be more than one corporeal world."[60]

In this fascinating discussion, Oresme not only argues for the possible plurality of worlds by supernatural means, but he argues for the reality of an infinite extracosmic void space that lies beyond our finite cosmos and in which God could presumably create other worlds. In this Oresme was one of a number of theologian-natural philosophers who identified an infinite void space with God's infinite immensity.[61] One can hardly imagine a more dramatic departure from Aristotle, who denied the existence of anything outside of our finite cosmos.

The Rectilinear Motion of the Whole Cosmos

Article 49 of those condemned in 1277 caught the attention of numerous natural philosophers, among whom were Thomas Bradwardine, John Buridan, Nicole Oresme, Richard of Middleton, and Gaietanus of Thienis. It condemned the idea "That God could not move the heavens [that is, the world] with rectilinear motion; and the reason is that a vacuum would remain."[62] The implications of article 49 for natural philosophy were truly significant. It subverted Aristotle's concepts of vacuum, place, and motion. Nicole Oresme mentions the condemned article and uses it to counter those who insist that "to move with respect to place is to change one's position in relation to some other body which may, or may not, be in motion itself." In denying this claim about change of position, Oresme invokes article 49 and declares that "assuming such a motion, there would be no other body to which the world could be related with respect to place, and the description given above would be invalid."[63] For Oresme, the rectilinear motion of the cosmos in the infinite vacuum that lay outside of it represented an absolute motion that is unrelated to anything else, because there is nothing outside of the spherical world to which it can be related.

In proclaiming the plausibility of such a rectilinear motion, Oresme undermined Aristotle's notion of place, which assumed that place is the innermost immobile surface of the body that surrounds the body in place. Moreover, motion for Aristotle was from one place to another place in a space that is a plenum. In Oresme's hypothetical motion of the cosmos, the cosmos is not in a place, but in a vacuum; and yet it moves through space unrelated to any

---

[60] Oresme: *Le Livre du ciel*, 179.

[61] For a lengthy discussion of the concept of an extracosmic infinite void space in medieval natural philosophy and theology, see my book *Much Ado about Nothing: Theories of Space and Vacuum from the Middle Ages to the Scientific Revolution* (Cambridge: Cambridge University Press, 1981), Part II ("Infinite Void Space Beyond the World"), 103–255. My account includes the discussion of this concept in the seventeenth century.

[62] Drawn from my translation in Grant, *A Source Book in Medieval Science*, 48.

[63] My translation from Menut's French text in *Le Livre du ciel et du monde*, 368, 370. I draw on my translation in Grant, *A Source Book in Medieval Science*, 553, n. 25.

other body. It matters but little that the motion Oresme describes is hypothetical and imaginary. It is of major significance because Oresme thought the motion he describes is plausible and possible, if the conditions of article 49 are realized.

Strange as it may seem at first glance, the idea of God moving the entire world with a rectilinear motion proved useful in the debates about space, void and the infinite in the seventeenth century. Samuel Clarke, Isaac Newton's spokesman, found occasion, in his famous controversy with Gottfried Leibniz (1646–1716), to invoke the concept of God moving the whole world with a rectilinear motion.

Void Spaces within the World

Although no article condemned in 1277 was concerned with the possibility of void spaces within our cosmos, medieval natural philosophers often assumed the existence of such vacua, occasionally hypothesizing that God had destroyed part of the plenum that fills our world and then describing the hypothetical motions of bodies in the vacuum left behind.

One of Aristotle's most important arguments against the possible existence of a vacuum was that the speed of any body in a vacuum would be instantaneous, that is, infinite, because a vacuum lacked a resisting medium. As Aristotle put it "the void can bear no ratio to the full, and therefore neither can movement through the one to movement through the other, but if a thing moves through the thinnest medium such and such a distance in such and such a time, it moves through the void with a speed beyond any ratio."[64] To counter Aristotle and make motion in a void intelligible, medieval natural philosophers had to show that such a motion would necessarily be both finite and successive. To show that bodies moving in a void space did so with a finite and successive motion, some natural philosophers argued that the vacuum itself served as a resistance because it was a three-dimensional extended space analogous to a material medium. And just as a material medium resisted the movement of bodies through it, so did a dimensional vacuum. The vacuum itself functioned as a resistance, because as a dimensional entity it was divisible into parts that were necessarily traversed in sequence. Hence, bodies moving through a vacuum would do so by successively traversing part after part, a process that would necessarily take time. Because the function of a resistance was to prevent a body from moving instantaneously, and because the vacuum performed that same function, the vacuum itself was regarded as a resistance. Scholastics designated it the *distantia terminorum*, or "the separation of the termini." The most famous names associated with this interpretation were Roger Bacon and Thomas Aquinas.[65]

---

[64] Aristotle, *Physics*, 4,8.215b.20–24; the Oxford translation of R. P. Hardie and R. K. Gaye.
[65] For further discussion, see Grant, *Much Ado about Nothing*, 27–30.

The identification of the void itself as a resistance, which meant that empty space was treated as if it were a material medium, left many natural philosophers dissatisfied.[66] Some felt it was more plausible to seek resistance not in the void itself, but in the body that was moving through it. A significant explanation was provided by Nicholas Bonetus (ca. 1280–1343) and Johannes Canonicus (fl. fourteenth century), who argued that a self-expending impressed force, or impetus, could produce a continuous, finite motion in a vacuum. Bonetus insists that a body could move in a void with a violent motion by virtue of a self-expending impressed force transmitted to it by the motive force that initially caused its motion. Moreover, because the impressed force is self-expending, the motion of the body would eventually terminate when the impressed force was completely dissipated. Thus, Bonetus argues that it was not necessary that an external agent be conjoined to a body in order to move it through the vacuum. "The reason for this," he explains, "is that in a violent motion some non-permanent and transient form is impressed in the mobile so that motion in a void is possible as long as this form endures; but when it disappears, the motion ceases."[67] Thus, Bonetus identified the agent that would act as a motive force in a vacuum and also accounted for the manner in which the body would cease its motion. By these mechanisms, he countered Aristotle's arguments that motion in a void would be instantaneous. It also subverted Aristotle's other contentions that motion in a vacuum was impossible because of the absence of an external medium to push the body forward; and that motion in a vacuum was impossible because the same absent material medium could not operate to cause the body to come to rest.

But what about natural motions of bodies in a void? Would they not move instantaneously? Scholastic natural philosophers sought to show that natural motion in a vacuum would be finite and successive for both elemental bodies – that is, bodies of pure earth, water, air, or fire – and bodies compounded of two or more elements, which were known as mixed bodies. In the natural motion of both elemental and mixed bodies, scholastics introduced the concept of internal resistance that produced finite natural motions. In the case of elemental bodies, the body's own weight, or heaviness, served to resist its own motion. It was assumed that the greater the weight of the body, the greater its velocity in the void.[68] But this implied that the weight, or heaviness of the body served as both motive force and resistance. In fact, no generally acceptable explanation for motion in a vacuum for a pure elemental body was devised in the Middle Ages. They could not convincingly assign a proper resistance that was associated solely with the body and not

[66] For criticisms of the *distantia terminorum* arguments, see Grant, ibid., 31–38.

[67] See Grant, ibid., 43 and n. 81 on p. 291.

[68] A detailed description of natural elemental motion in a vacuum appears in Grant, ibid., 45–49.

the vacuum. By default, the most successful explanation was that which assigned the function of resistance to the vacuum itself, which as we saw, they called *distantia terminorum*, or "separation of the termini."

By contrast with elemental bodies, the attempt to explain the natural motion of mixed bodies in a void can be regarded as successful within the framework of Aristotelian physics. To account for the natural motion of mixed bodies, scholastics in the fourteenth century abandoned Aristotle's concept that the natural motion of a mixed body was determined by the dominant element in it. They replaced it with a concept in which the direction of motion of a body in natural motion was determined by the relationship between the heavy and light elements within the body. If the light elements prevailed the body was considered a "light mixed body"; if the heavy elements prevailed, it was regarded as a "heavy mixed body." By this device, every body had within itself contrary light and heavy elements with one contrary functioning as an internal motive force (the more powerful predominant element or elements), and the other opposing it as an internal resistance. Because this met Aristotle's conditions for motion, namely, a motive force operating against a resistance, natural motion in a vacuum became feasible.

Certain of Aristotle's basic ideas about the natural places of the elements were associated with this approach. Within any given body, the sum total of lightness that was opposed by the sum total of heaviness was determined not only by the number of elements in the body and their relative proportions, but also on the location of the body, that is, whether the body was in the sphere of fire, air, water, or earth. It became customary to assign arbitrary numerical values to represent the degree of heaviness or lightness that an element possessed. The examples imagined often were of the kind where God was assumed to have annihilated all the matter below the moon or below the sphere of fire. Despite the fact that the vacuum God created should have been uniformly identical throughout its extent, this was not the case in the examples Albert of Saxony presented in his *Questions on the Physics*, book 4, question 11. Albert assumes that when an element in a mixed body is in that part of the void that was formerly the element's natural place – that is, the natural place of that element before God destroyed the plenum and created the void – that element will resist the downward movement of its mixed body.

In an example,[69] Albert assumes a mixed body consisting of earth, to which he assigns a degree of 3, and air, to which he assigns a value of 2. Albert then imagines that everything below the sphere of fire is annihilated,[70] leaving behind a vast vacuum. Because the body will begin its fall in the sphere

---

[69] In this example, I rely on my account in Grant, ibid., 53–54.
[70] Interestingly, Albert does not attribute the annihilation of the matter to God, but simply assumes it for the sake of the illustration.

of fire, which is not void and is also not an element in the mixed body, Albert arbitrarily assumes that the elemental fire will resist the descent of the mixed body with a resistance of 1. Thus, when the body begins its descent in the sphere of fire, it falls with a speed that results from a ratio of force to resistance that is as 5/1 (that is, F/R = 5/1, where F is force, consisting of the combined effort of earth and air; and R is the resistance of the fire). When the body reaches the natural place of air, which is now void, the air acts as if it was in its natural place and resists the force of the earth. Hence, the ratio of force to resistance is now F/R = 3/2 and produces a speed in the natural place of air that is less than the speed of the same mixed body as it fell through the sphere of fire. Albert observes that in this instance, "the same mixed body has a much slower motion in a vacuum than in a plenum, namely fire."[71] This consequence is a serious violation of Aristotle's physics in which the motion of any body imagined in a vacuum would have been instantaneous. Indeed, that is one of the major reasons Aristotle rejected the possibility of the existence of void spaces.

That Homogeneous Mixed Bodies of Unequal Weight Would Fall with the Same Speed in a Void

One of Aristotle's arguments to show the absurdity of void space involved his conclusion that in a void, bodies of unequal weight would fall with the same speed. In a plenum, bodies fall with velocities proportional to their weights. A heavier body cleaves through the medium more easily than a less heavy body; the greater the weight, the greater the speed. In the absence of a resistant medium – that is, in a void – Aristotle could see no reason why a heavy body should fall more quickly than a less heavy body. He concluded that they must fall with the same speed in a void, a consequence he regarded as absurd and therefore serving as one more good reason to deny the possibility of the existence of any void spaces.

The doctrine of mixed bodies changed all this. Fourteenth-century scholastic natural philosophers such as Walter Burley, Thomas Bradwardine, and Albert of Saxony, who accepted the existence of mixed bodies, thought it made perfectly good sense to assume that homogeneous mixed bodies of unequal weight would fall with the same speed in a vacuum. If the mixed bodies were not homogeneous and had different proportions of light and heavy, they would fall at different speeds. As the eighth conclusion of book 4, question 12, of his *Questions on the Physics*, Albert of Saxony asserts that "Mixed [or compound] bodies of homogeneous composition are moved with equal velocity in a vacuum but not in a plenum." He insists that in homogenous bodies, "the ratio of motive power to total resistance in one body is the

---

[71] I quote from my translation of Albert's question in Grant, *A Source Book in Medieval Science*, 337.

same as in another homogenous body, because they both have only internal resistance."[72] This would not be true for homogeneous bodies falling in a plenum, where an external resistance is added to the internal resistance and causes the bodies to acquire different speeds.

More than two centuries later, Galileo, in his early treatise *De motu* (ca. 1590), arrived at the same conclusion as did Burley, Bradwardine, and Albert of Saxony. By the time he wrote his *Discourses and Mathematical Demonstrations Concerning Two New Sciences* in 1638, he had expanded the medieval conclusion about homogeneous bodies falling with equal speed in the void to a generalization that proclaimed that all bodies of whatever weight and composition – homogeneity was not a requirement – would fall in a vacuum with equal speed.

There were many other departures from Aristotle in questions and commentaries on his treatises, especially in cosmology. For example, many abandoned Aristotle's system of concentric orbs in favor of epicycles and solid eccentric orbs based on Ptolemy's *Hypotheses of the Planets*. The reason was simple: Aristotle's concentric orbs could not save the astronomical phenomena, whereas Ptolemy's epicycles and eccentrics could. This had enormous consequences. The system of eccentrics and epicycles permitted motion around a geometric point as well as a physical body, a move that undermined Aristotle's claim (and that of his followers Averroes and Maimonides) that celestial motions could only occur around a physical body, namely, the earth. Aristotle believed that the celestial orbs were nested one within another and that neither matter nor empty spaces could lie between them. But others, Albertus Magnus among them, assumed that matter lay between the orbs, a matter that was radically different than the incorruptible and indivisible, matter that composed the orbs. This interorbicular matter was capable of expansion and contraction and thereby capable of change. This departed from Aristotle in other dramatic ways: it shattered Aristotle's idea of a completely homogeneous celestial matter and it denied that celestial orbs were in direct contact. Another significant departure involved the planets, which Aristotle believed were simply carried around by their respective orbs and had no motion of their own. Many abandoned this opinion and insisted that at least the Moon had a rotatory motion of its own, and perhaps other planets.[73]

I have now described or mentioned a sufficient number of departures from Aristotle's natural philosophy to show that while Aristotle's interpretations lay at the heart of medieval natural philosophy, medieval natural

---

[72] Albert of Saxony, *Questions on the Physics*, bk. 4, qu. 12 in Grant, *A Source Book in Medieval Science*, 341.

[73] I have drawn the examples in this paragraph from my article, "Ways to Interpret the Terms 'Aristotelian' and 'Aristotelianism' in Medieval and Renaissance Natural Philosophy," in *History of Science* 25 (1987), 338–339.

philosophers altered and subverted his interpretations in many ways. This is true for the whole of his natural philosophy. From what has been said here, however, I would emphasize that a major mode of departing from Aristotle was by way of hypothetical or counterfactual assertions about the world, whether these were by the assumption of some naturally impossible action by way of God's absolute power, or simply the assertion of a counterfactual by way of the imagination, that is, *secundum imaginationem*. In most of these departures, the object was to show that what Aristotle may have thought was naturally impossible was indeed possible and intelligible. And yet, none of the departures mentioned here were accepted as representative of the natural physical world. Aristotle's many opinions on a host of topics remained the basis of medieval natural philosophy. What scholastic natural philosophers showed is that one could easily imagine a world that was radically different than Aristotle's conception of it.

BEYOND ARISTOTLE

The departures from Aristotle mentioned thus far were made in treatises in which authors had little opportunity to develop this or that theme, or a plurality of themes. Particular questions were too brief to allow for extended argument and development. Moreover, most questions were self-contained and generally unrelated to one another. They were independent units, each of which resolved a particular problem that was only rarely related to what went before or what was to come. That format was characteristic of questions treatises. By contrast, the tractates, or treatises, in natural philosophy provided natural philosophers an opportunity to develop any theme in any manner they wished and at any length. It was in the tractates that themes were developed that Aristotle had barely mentioned, or not mentioned at all.

Most medieval natural philosophers chose to confine themselves to questions on various works of Aristotle. One of those who, as already mentioned, found the tractate, or treatise, a convenient format, was Nicole Oresme, who may have composed more independent treatises than any other medieval scholastic author. Not only did he write numerous treatises, but most of them had the stamp of originality and were of high intellectual quality. Oresme wrote extensively on a number of important themes. He inquired at great length whether the celestial motions are commensurable or incommensurable and offers brilliant arguments to show that it is probable that any two celestial motions are probably incommensurable, from which he concludes that astrological predictions are by the very nature of things imprecise and unreliable.[74]

---

[74] On this theme, see Edward Grant, ed. and tr. *Nicole Oresme and the Kinematics of Circular Motion: Tractatus de commensurabilitate vel incommensurabilitate motuum celi* (Madison: University of Wisconsin Press, 1971).

Sometime around 1350, Oresme wrote the most important medieval treatise on the intension and remission of forms or qualities titled *Treatise on the Configuration of Qualities and Motions*.[75] In the intension and remission of forms or qualities, medieval natural philosophers assumed that almost any variable quality could be increased or decreased by the addition or subtraction of identical parts. Of course, qualities cannot be meaningfully varied by the addition or subtraction of parts. By assuming that such operations were conceptually meaningful, however, Oresme and others gave to the intension and remission of qualities a quantitative and mathematical character. And so it was that Oresme represented variable qualities by geometric figures, thus marking a departure from his predecessors at Merton College, Oxford University, who had used arithmetic to represent variable qualities.[76] His most important contribution, however, was a geometric proof of the mean speed theorem (in modern symbols: $s = (1/2)at^2$, where s is distance, a is acceleration, and t is time) in which velocity is treated as a variable quality. Thus, Oresme represented uniformly accelerated motion from rest by a right triangle, and employed a rectangle to represent uniform motion, which was assigned the speed acquired at the middle instant of the uniform acceleration. Oresme showed that the triangle and rectangle are equal in area and therefore a body moving with a velocity that is uniformly accelerated from rest would traverse the same distance as a body moving during the same time with a uniform speed equal to the middle instant of the uniform acceleration. Oresme's proof and diagram were printed in numerous editions of the sixteenth century and probably influenced Galileo in his *Two New Sciences* (1638), where in the Third Day, Theorem I, Galileo gives essentially the same geometric proof.[77]

In addition to his extensive use of mathematics in the *De configurationibus* and his use of it in his discussions of commensurability and incommensurability of the celestial motions, Oresme used mathematics to resolve physical problems of motion. In his *On Ratios of Ratios* (*De proportionibus proportionum*), Oresme used rational and irrational exponents and also introduced probability considerations to determine that given any two ratios, it was probable that they would be related by an irrational, rather than a rational,

---

[75] For the edition of the Latin text and an English translation, see Marshall Clagett, ed. and tr., *Nicole Oresme and the Medieval Geometry of Qualities and Motions: A Treatise on the Uniformity and Difformity of Intensities Known as "Tractatus de configurationibus et qualitatibus et motuum"* (Madison: University of Wisconsin Press, 1968).

[76] In his tractate titled *Rules for Solving Sophisms*, the Oxford scholar, William Heytesbury, gives an arithmetic proof of the mean speed theorem. For the Latin text and translation, see Clagett, *The Science of Mechanics in the Middle Ages*, 270–283.

[77] For the Latin text and translation of Oresme's mean speed theorem, see Clagett, ed. and tr., *Nicole Oresme and the Medieval Geometry of Qualities and Motions*, 408–411. For a more detailed discussion, see two of my books: *The Foundations of Modern Science in the Middle Ages*, 98–104, and *God and Reason in the Middle Ages*, 172–175.

exponent. Using Euclid's theory of proportionality in the fifth book of the *Elements*, Oresme develops the concept of a "ratio of ratios," which is actually what we, although not Oresme, call the exponent. For example, in the relationship A/B = (C/D)$^{p/q}$, the exponent, p/q, is a "ratio of ratios" that relates the two ratios A/B and C/D. If p/q is rational, then ratios A/B and C/D, which may be rational or irrational, or one rational, the other irrational, must be commensurable and represent a "rational ratio of ratios." For example, (2/1)$^{1/2}$ and 2/1 are commensurable and form a rational ratio of ratios because 2/1 = [(2/1)$^{1/2}$]$^{2/1}$; similarly, 27/1 = (3/1)$^{3/1}$ is also a rational ratio of ratios because the rational exponent 3/1 relates the two ratios 27/1 and 3/1. But 6/1 and 3/1 form an "irrational ratio of ratios," because 6/1 ≠ (3/1)$^{p/q}$, where p/q is rational. In fact, p/q must be irrational.

Oresme also introduces probability considerations in demonstrating that any two given unknown ratios are more likely to be incommensurable than commensurable. He takes 100 rational ratios from 2/1 to 101/1 and relates them two at a time. For example, 2/1 and 3/1 form an irrational ratio of ratios, namely 3/1 ≠ (2/1)$^{p/q}$, whereas 2/1 and 4/1 form a rational ratio of ratios, namely 4/1 = (2/1)$^{p/q}$, where p/q = 2/1. These 100 ratios taken two at a time form 9,900 possible ratios of ratios. Because Oresme is only interested in ratios of greater inequality, where the numerator is greater than the denominator, only half of the ratios of ratios are relevant, namely 4,950. Of these ratios, Oresme shows that only 25 are rational, while the other 4,925 are irrational. Thus, when a set of 100 rational ratios are posited, the ratio of irrational to rational ratios is 4925/25, or 197/1. As one takes more and more rational ratios, the odds that any given "ratio of ratios" is irrational increases.[78]

In *On Seeing the Stars* (*De visione stellarum*),[79] Oresme considers atmospheric refraction and makes a spectacular contribution to the history of science. He arrives at innovative conclusions "in at least three separable areas: (1) in *optics*, he argues that light travels on a curved path in a medium of uniformly varying density and that refraction does not require a single, specific refracting surface; (2) in *mathematics*, he contends that convergent infinite series may be used to equate infinitely small straight lines with a curve; and (3) in *astronomy*, he asserts that atmospheric refraction occurs along a

---

[78] Oresme developed his concept of "ratio of ratios" and his probability considerations in his Treatise *on Ratios of Ratios* (*Tractatus de proportionibus proportionum*). See my edition and translation in *Nicole Oresme: "De proportionibus proportionum" and "Ad pauca respicientes,"* edited with Introductions, English translations and Critical Notes by Edward Grant (Madison: University of Wisconsin Press, 1966).

[79] Recently edited by Danny Ethus Burton under the title *Nicole Oresme's On Seeing the Stars (De visione stellarum): A Critical Edition of Oresme's Treatise on Optics and Atmospheric Refraction, with an Introduction, Commentary, and English Translation* (Ph.D. diss., Indiana University, Feb. 2000).

curved path, as Hooke and Newton later confirmed."[80] In this significant treatise, Oresme rejects two traditional views in optics – held by Ptolemy, Alhazen, Roger Bacon, and Witelo – namely, that a light ray is refracted only at the interface of two media of differing densities *and* that no refraction could occur in a single medium whose density varies uniformly. By contrast, Oresme argues that refraction of light does not require a single refracting interface between two media of differing densities. It will be refracted along a curved path when it is in a single medium of uniformly varying density. For example, if air increases in rarity as its distance from the earth increases, light would pass through it along a curved path. To deduce a curved path, Oresme used his knowledge of convergent infinite series, assuming that successive refractions produced successively smaller line segments and that as the line segments increased to infinity they would form a curved line.

In his analysis of Oresme's text, Danny Burton, who first discovered Oresme's contribution to our understanding of atmospheric refraction, observes that Robert Hooke and Isaac Newton were previously thought to have been the first to argue that light is continuously refracted as it moves along a curved path through a uniformly decreasing medium. However, "while the definitive demonstration of the curvature of light in the atmosphere was Hooke's and Newton's, the original argument for such curvature was Oresme's."[81]

Oresme wrote other important treatises, including one on money, one against astrology, and one that was essentially against magic. His writings reveal how the treatise format could be put to the most telling and effective use.

In describing medieval departures from Aristotle's natural philosophy, I have presented two kinds of disagreements. In the first category, examples are based on phenomena in the real physical world, as, for example, impetus theory, and the possible motions of the earth. In the second category, the conditions and circumstances that were assumed are not about the real physical world, but about imaginary conditions that were created by God as a consequence of His absolute power and all of which were naturally impossible in Aristotle's real world; or they were simply hypothetical imaginings proposed in order to test whether such conditions might be intelligible if perchance they were brought into being. Medieval natural philosophy was about both of these categories: phenomena in the real world and imaginary phenomena in one or more hypothetical worlds. We must now inquire about a vital aspect of Aristotelian natural philosophy as it pertained to these two categories and ostensibly underlies all of natural philosophy. I refer to the role of empiricism, that is, experience and observation. Twice previously, I have written on this theme and in both instances have characterized medieval

---

[80] See ibid., 40. For the detailed arguments, see pp. 50–55.
[81] Ibid., 53.

empiricism as an "empiricism without observation."[82] Let me now explain what I understand by this.

The Role of Experience in Medieval Natural Philosophy

Aristotle regarded experience as the basis for science, expressing the conviction in one place that "science and art come to men *through experience*."[83] Elsewhere, Aristotle uttered a similar sentiment when he declared that "it is the business of experience to give the principles which belong to each subject. I mean for example that astronomical experience supplies the principles of astronomical science; for once the phenomena were adequately apprehended, the demonstrations of astronomy were discovered. Similarly with any other art or science."[84] Experience was acquired through the senses, so that Aristotle regarded sense perception as the basis of all knowledge. But if Aristotle believed that the senses "give us the most authoritative knowledge of particulars," he also explains that "they do not tell us the 'why' of anything – e.g. why fire is hot; they only say that it is hot."[85] The "why" of things involves theoretical and causal explanations and thus goes beyond empirical evidence. Aristotle's books on natural philosophy, mentioned earlier in this chapter, are largely concerned with theoretical accounts of their respective subject matters and often seem far removed from observational experience. They often seem to describe a world that conforms to Aristotle's preconceived ideas of what is fit and proper for our material universe.

At best, Aristotle left an unclear and uncertain picture about the relationship between the empirical and the theoretical. His biological treatises were heavily empirical, filled with descriptions of animal behavior. But his books about the inanimate physical world were heavily theoretical and much less empirical. Apart from brief remarks here and there, Aristotle left no major discussion about the role of experience and observation in arriving at scientific knowledge about the world. As the inheritors of Aristotle's natural philosophy, medieval natural philosophers were largely left to find their own way in this matter. As did Aristotle, they paid little attention to the role of experience in acquiring knowledge about the physical world. Indeed, it is often difficult to determine what they regarded as experiential in their treatment of various problems in natural philosophy. They also failed to

---

[82] The first account is in *God and Reason in the Middle Ages*, 160–182, in which the relevant section is titled "Reason and the Senses in Natural Philosophy: Empiricism without Observation." The second discussion appears in an article titled "Medieval Natural Philosophy: Empiricism without Observation," and appears in a book published by Cees Leijenhorst, Christoph Lüthy, and Johannes M. M. H. Thijssen, eds., *The Dynamics of Aristotelian Natural Philosophy from Antiquity to the Seventeenth Century* (Leiden: Brill, 2002), 141–168.

[83] *Metaphysics*, 1.1.981a.4–5.   [84] *Prior Analytics*, 1.30.46a.18–21.

[85] *Metaphysics*, 1.1.981b.10–12.

distinguish between different kinds of experiences, the kind that might turn up in a real-world situation, or one that might be injected into a counterfactual discussion. Because of these difficulties, I shall proceed by way of the analysis of examples in the two different categories that I distinguished earlier: the phenomena of the real world and the phenomena of imaginary worlds.

That medieval natural philosophers were well aware of Aristotle's emphasis on empiricism and sense data, and that they thoroughly approved, is made evident by explicit statements of such eminent natural philosophers as Albertus Magnus, Roger Bacon, and John Buridan. For example, Albertus Magnus declared that "anything that is taken on the evidence of the senses is superior to that which is opposed to sense observation; a conclusion that is inconsistent with the evidence of the sense is not to be believed; and a principle that does not accord with the experimental knowledge of the senses is not a principle but rather its opposite."[86] Roger Bacon insisted that "reasoning does not suffice, but experience does";[87] Buridan upholds experience by invoking Aristotle's opinion, declaring that "Aristotle puts it very well [when he says] that many principles must be accepted and known by sense, memory, and experience. Indeed, at some time or other, we could not know that every fire is hot [except in this way]."[88]

Despite such positive statements in favor of experience and sense perception, and the reasonable assumption that most medieval scholars would have agreed with those sentiments, the role that experience played in medieval natural philosophy is puzzling. It is one thing to proclaim the importance of experience and quite another to use it frequently in the resolution of questions and problems. The greatest obstacle to assessing the role of experience and sense perception in medieval natural philosophy is that we are largely ignorant of the way natural philosophers used experience to resolve questions and problems. Medieval scholars were no more forthcoming about their use of experience, or what they understood by the term experience (*experientia*), than was Aristotle. The sentiments in favor of experience I quoted from Albertus Magnus, Roger Bacon, and John Buridan are strong declarations in favor of experience, but they are quite abstract and tell us nothing about what they understand by experience. To gain some insight into medieval conceptions of experience and how it was used in various contexts and questions, I deem it essential to analyze a variety of instances where experience appears to play a role. I

---

[86] Translated by William A. Wallace, *Causality and Scientific Explanation*, 2 vols. (Ann Arbor: University of Michigan Press, vol. 1, 1972; vol. 2, 1974), vol. 1, 70.

[87] For the full passage and reference, see Grant, *God and Reason in the Middle Ages*, 161. The passage occurs in Bacon's *Opus Majus*, Part VI ("On Experimental Science").

[88] Buridan's remark appears in his *Questions on the Physics*, bk. 4, qu. 7 and is quoted more fully in Grant, *God and Reason*, 161.

shall first consider examples from problems relevant to the real physical world.

The Role of Experience in the Real Physical World

The most common and obvious use of experience occurs in situations where an author has occasion to mention an observed or perceived phenomenon that is known to everyone. Included in this category are experiences that produce observations such as: snow and ice are cold; the sun and fire give heat, or, the heavens move, as John Buridan observed.[89] Observations of this kind play no role here.

In the course of their discussions, scholastic natural philosophers found reasons to report experiences, or even occasional experiments. In all such instances, it is important to know whether the experience was personally observed by the author, or was gleaned from another treatise, or was, perhaps, received by word-of-mouth from a friend or colleague. Finally, we should attempt to determine the role a particular experience plays in an author's judgment.

THE IMPETUS THEORY OF PROJECTILE MOTION Among the examples of departures from Aristotle's interpretations of the physical world described earlier in this chapter was the impetus theory of impressed force to account for projectile motion. Impetus theory replaced Aristotle's theory that air, an external agent, propelled a body forward after it lost contact with its motive force. To bolster his arguments against Aristotle's explanation, Buridan introduces three experiences (*experientie*). The first of these, as he explains,

concerns the top (*trocus*) and the smith's mill (i.e. wheel – *mola fabri*) which are moved for a long time and yet do not leave their places. Hence, it is not necessary for the air to follow along to fill up the place of departure of a top of this kind and a smith's mill. So it cannot be said [that the top and smith's mill are moved by the air] in this manner.[90]

The second experience is this: A lance having a conical posterior as sharp as its anterior would be moved after projection just as swiftly as it would be without a sharp conical posterior. But surely the air following could not push a sharp end in this way, because the air would be easily divided by the sharpness.

---

[89] Buridan mentions this obvious fact in his *Questions on the Metaphysics*, bk. 2, qu. 1. For the full reference, see Grant, *God and Reason*, 164, n. 37.

[90] In the final question of his *Questions on the Physics*, bk. 8, qu. 13, Albert of Saxony cites the same experience involving the smith's wheel and the top in a question "by what is a projectile moved upward after its separation from the projector?" ("Ultimo quaeritur a qua moveatur proiectum sursum post separationem illius a qua proiicit.") Here is what Albert says: "Similiter ista opinio non habet locum in motu mole fabri; similiter in motu troci. Vidimus enim quod trocus post exitum eius a manu proiicientis diu movetur circulariter absque hoc quod aliquis aer ipsum insequatur, movet enim super eodem puncto spatii." See *Questiones et decisiones physicales insignium virorum. Alberti de Saxonia in octo libros Physicorum; tres libros De celo et mundo; duos libros De generatione et corruptione* ... (Paris, 1518), fol. 83v, col. 1.

The third experience is this: a ship drawn swiftly in the river even against the flow of the river, after the drawing has ceased, cannot be stopped quickly, but continues to move for a long time. And yet a sailor on deck does not feel any air from behind pushing him. He feels only the air from the front resisting [him]. Again, suppose that the said ship were loaded with grain or wood and a man were situated to the rear of the cargo. Then if the air were of such an impetus that it could push the ship along so strongly, the man would be pressed very violently between that cargo and the air following it. Experience shows this to be false. Or, at least, if the ship were loaded with grain or straw, the air following and pushing would fold over (*plico*) the stalks which were in the rear. This is all false.[91]

Buridan's first and third experiences were probably based on direct observation. In the first, Buridan trades on the fact that the top and the smith's mill wheel move with rotational motions and are therefore not pushed from behind by the air. Indeed, Buridan would argue that they move around by virtue of an impressed force. By contrast, the rectilinear motions Aristotle had in mind occupied a succession of places with air pushing the moving bodies from behind. The implication of this experience is that Aristotle's theory of motion cannot explain the motion of a top and smith's wheel.

It is quite likely that Buridan directly observed the rotational motions of a top and a smith's wheel. Most of his contemporaries also would have had occasion to see them. Buridan was clever enough to realize the value of these devices in refuting Aristotle's theory.

In the third experience, Buridan capitalizes on the fact that when shiphaulers cease to pull a ship, the ship's momentum will carry it forward for some distance. In the course of his life, he was very likely to have witnessed this phenomenon. Whether he also observed the other conditions he mentions is more problematic, but it is certainly possible.

The second experience is of great importance, because it reveals a significant characteristic about medieval appeals to experience. It is a "reasoned" experience, which Buridan conjured up for the occasion. In this experience, Buridan criticizes Aristotle's theory by showing the implausibility that a lance with a posterior that is as sharp as the pointy front end could be moved as quickly as a lance that did not have a pointy posterior but presumably had a greater surface area. Reason tells us, Buridan means to inform us, that air cannot push a lance that has a pointy end. It can only push it forward if the posterior of the lance had a sufficiently broad surface. And yet, a lance that had a posterior as pointy as its anterior can be hurled through the air for

91 Clagett, *The Science of Mechanics*, 533; Grant, *Source Book*, 275–276. For the Latin text, see *Acutissimi philosophi reverendi Magistri Johannis Buridani subtilissime Questiones super octo phisicorum libros Aristotelis* (Paris, 1509), bk. 8, qu. 12, fols. 120r, col. 2–120v, col. 1.

some distance. But it could not be moved by the air pressing on its pointy end. It would be moved by the impetus, or force, that had been impressed in it by the one who hurled the lance.

Buridan makes other experiential claims in the same question. If air were a motive force, "it follows," he argues,

that you would throw a feather farther than a stone and something less heavy farther than something heavier, assuming equal magnitudes and shapes. Experience shows this to be false. The consequence is manifest, for the air having been moved ought to sustain or carry or move a feather more easily than something heavier.[92]

We may rightly assume that Buridan had observed the conditions he describes, or something analogous to them. A stone can easily be thrown farther than a feather, or some other very light object. But if Aristotle's theory is correct, the air should carry a feather farther than a stone.

Buridan also considered the impetus theory in his *Questions on On the Heavens*, book 2, question 12, where he asks "whether natural motion ought to be swifter in the end than the beginning."[93] Buridan agrees with Aristotle that a falling body has a greater speed at the end of its motion than at the beginning. Indeed, they both assumed that heavy bodies accelerate when they fall naturally. Their causal explanations for this phenomenon differ radically. Aristotle held that heavy falling bodies accelerate because they are attracted to their natural place. Buridan refutes Aristotle's position by three experiences of which the second and third will be cited here.

In the second experience, Buridan assumes that there are two stones, one of which will fall to the earth from a high place and the other from a low place. Now if the velocities of the stones are affected only by proximity to their natural place, their speeds should be equal when they are both one-foot from the earth, even though they began their respective falls from radically different heights. "Yet it is manifest to the senses," Buridan argues,

that the body which should fall from the high point would be moved much more quickly than that which should fall from the low point, and it would kill a man while the other stone [falling from the low point] would not hurt him.[94]

From this example, which is a reasoned argument and not an experiment that was carried out, Buridan considers it evident that the heights of the stones determines their respective speeds and not proximity to their natural

---

[92] Clagett, ibid.; Grant, ibid.

[93] This question is translated and annotated by Marshall Clagett in Clagett, *The Science of Mechanics*, 557–564. The translation and annotation are reproduced in Grant, *A Source Book*, 280–284.

[94] Clagett, ibid., 559; Grant, ibid., 281.

place. The next experience leads us to the same conclusion, when Buridan declares:

Again, if a stone falls from an exceedingly high place through a space of ten feet and then encountering there an obstacle comes to rest, and if a similar stone descends from a low point to the earth, also through a distance of ten feet, neither of these movements will appear to be any swifter than the other, even though one is nearer to the natural place of earth than the other.[95]

If Buridan had carried out such an experiment, he could not have determined by direct observation that the speeds of the two stones are equal, although he could have arrived at a rough estimate. If Buridan had carried out the first experiment, he would have risked killing someone and obviously could not have performed it.

In the same question on whether natural motion is faster at the beginning than at the end, Buridan refutes another theory attributed to Aristotle. This theory held that a naturally falling body accelerates because as it falls and nears the ground there is less and less air to resist its fall; thus its speed will continually accelerate. Buridan uses a similar argument against this theory, declaring that:

this opinion falls into the same inconsistency as the preceding one, because, as was said before, if two bodies similar throughout begin to fall, one from an exceedingly high place and the other from a low place such as a distance of ten feet from the earth, those bodies in the beginning of their motion are moved equally fast, notwithstanding the fact that one of them has a great deal of air beneath it and the other has only a little. Hence throughout, the greater velocity does not arise from a greater proximity to the earth or because the body has less air beneath it, but from the fact that that moving body is moved from a longer distance and through a longer space.[96]

Buridan believed that a body accelerates as it falls because at each moment the heaviness of the body creates an equal quantity of impetus in the body, thus causing it to accelerate throughout its fall. How do we know there is impetus in the body? Here Buridan again resorts to the smith's wheel, just as he did in his *Questions on the Physics*. He informs his audience that:

you have an experiment (*experimentum*) [to support this position]: if you cause a large and very heavy smith's mill [i.e., a wheel] to rotate and you then cease to move it, it will still move a while longer by this impetus it has acquired. Nay, you cannot immediately bring it to rest, but on account of the resistance from the gravity of the mill, the impetus would be continually diminished until the mill would cease to move.[97]

---

[95] Clagett, ibid.; Grant, ibid.   [96] Clagett, ibid., 559–560; Grant, ibid.

[97] Clagett, *The Science of Mechanics*, 561; Grant, *Source Book*, 282. In his *Questions on De celo*, bk. 2, qu. 7, Nicole Oresme gives a similar example. See Claudia Kren, ed. and tr. *The "Questiones super De celo" of Nicole Oresme* (Ph.D. diss., University of Wisconsin, 1965), 560–562.

Here Buridan is clearly drawing on an experience that was common in medieval Europe, where mill wheels were a familiar sight and the phenomenon Buridan describes was compatible with the actions of an impressed force, or impetus.

In his questions on impetus theory, Buridan made good use of experience to support his theory of an impressed motive force. His best observational experiences were associated with the two smith mill experiences and his use of the rotating top. They were drawn from life experiences, as he had almost certainly seen smith's mills and tops in action. The rest of the experiences were based on reason, and it is highly unlikely that Buridan actually performed any of them. If he thought them up, and did not learn about them from other sources, we may credit him with sufficient cleverness to have seen their direct relevance to his objectives. Most of the experiences and observations that medieval natural philosophers introduced into their arguments were rather remote and were more like Buridan's second experience with the lance. They were formulated for the purpose and were not performed or even observed. But were they regarded as legitimate experiences? Medieval natural philosophers did not concern themselves with such matters. They apparently did not believe it necessary to do so, choosing to regard experiences simply as examples and counterexamples. Did they distinguish between personally observed experiences and examples formulated for the occasion? There is no evidence that they did. We can see this in other examples about the real world.

Earlier in this chapter, I cited a passage about the moment of rest that Aristotle had insisted was necessary between an upward and downward motion. Aristotle regarded them as two distinct motions with an interval of rest between. As we saw, scholastic natural philosophers disagreed and relied on the "experience" of the millstone falling downward and striking a bean that had been thrown upward. They concluded that when the millstone struck the bean, there could be no moment of rest. We have no reason to believe that anyone witnessed this experience and even less reason to suppose that anyone actually performed it. The millstone-bean example, which was first formulated by Islamic natural philosophers, was a clever counterargument to subvert Aristotle's position. It was a vivid counterexample that stood the test of reason.

In the same question, Marsilius of Inghen presents another example in which no moment of rest occurs. Instead of vertical motions, Marsilius employs contrary horizontal motions, imagining that:

Socrates (*Sortes*) is moved toward the west in a ship that is at rest. Then it is possible that Socrates might cease moving in any instant. Now let it be assumed that in the [very] same instant in which Socrates should cease to be moved [toward the west], the ship with all its contents, begins to be moved toward the east. Hence, immediately before, Socrates was moved to the west, and immediately after, will be moved toward the east. Therefore, previously he was moved with one motion and afterward with another, and contrary, motion without a moment of rest.[98]

Here again, Marsilius conjures a situation in which there can be no moment of rest. It is beyond plausibility to assume that Marsilius ever witnessed anything like the example he offers. He, or someone else, imagined the conditions on this imaginary ship. In treating the same question, John Buridan also imagined a ship example, but assumed somewhat different conditions.[99]

Despite his powerful arguments against the moment of rest, Marsilius accepted it under certain conditions,[100] whereas John Buridan rejected Aristotle's moment of rest without qualification. In the process, Buridan added to the collection of imaginary experiences by relating the actions of a lance and a fly. Buridan explains that

if a lance is hanging from a tree [and] a fly (*musca*) ascends on that lance and the cord by which the lance is hanging is broken, and then the lance and fly fall down, the motion of the fly will be contrary, from up to down, but there will be no moment of rest.[101]

By having the rope holding the lance to the tree break and carry the fly down with it, Buridan conferred an air of implausibility on his "experience" against the moment of rest. As soon as the lance began its downward motion, the fly would very likely have flown from the lance. There would have been no downward motion for the fly. Albert of Saxony made better use of the fly when he repeated the essential features of Buridan's example but abandoned the tree and the cord. Albert simply assumed that the fly ascends the lance quicker than the lance descends. He then assumes that the upward speed of the fly diminishes until it is less than the speed of the lance's descent. However, at the very instant when the fly's speed of ascent equals the lance's speed of descent,

it is true to say that immediately before [the speeds were equal] this fly was ascending; and it is [also] true to say that immediately after [the speeds were equal] it descends, because immediately after this the descent of the lance will be quicker, from which it again follows that between the ascent and descent of the fly there is no moment of rest.[102]

[98] Grant, *Source Book*, 287.

[99] Buridan, *Questions on the Physics*, bk. 8, qu. 8: "Whether it is necessary that in every contrary motion, the mobile rests in the turning point." See his *Questions on the Physics*, fols. 116r, col. 2–116v, col. 2.

[100] For Marsilius's qualifications, see my translation of his question in Grant, *Source Book*, 289, n. 31.

[101] My translation from Buridan, *Questions on the Physics*, bk. 8, qu. 8, fols. 116r, col. 2–116v, col. 1.

[102] My translation from Albert of Saxony, *Questions on the Physics*, bk. 8, qu. 12, fol. 82v, col. 2.

One of the most vital assumptions in Aristotle's physics was his absolute conviction that the world is everywhere filled with matter and that it is impossible for a vacuum to exist or to be created. Experiences were invoked to support Aristotle's arguments. With the possible exception of Nicholas of Autrecourt, all were in agreement with Aristotle that the existence of a vacuum in the physical world was not naturally possible, although it was supernaturally possible by God's absolute power. A few of the arguments against the vacuum were based on experiences that would have been recognized by many who heard or read about them. Albert of Saxony mentions an experience that may have been widely known. When anyone withdraws the air in a straw, he observes that the water follows the air upward without losing contact. Thus does the water rise against its natural inclination to remain in its natural place. Nature so abhors a vacuum that, in order to prevent its formation, it operates contrary to its own laws.

Another argument against the vacuum that would have been widely recognized as true to experience was proposed by John Buridan, who, in asking "whether it is possible that a vacuum exist,"[103] appealed to experimental induction by asserting that "everywhere we find some natural body, namely air, or water, or some other [body]."[104] By implication, we never find a vacuum, which therefore does not exist. In another example, Buridan also seeks to show the nonexistence of a vacuum, but he now resorts to an imaginary, hypothetical illustration. Buridan introduces the "experiment" with the statement that "we experience that we cannot separate one body from another unless another body intervenes."[105] To illustrate this claim, he offers the following imaginary experiment.

If all the holes of a bellows (*follis*) were perfectly stopped up so that no air could enter, we could never separate their surfaces. Not even twenty horses could do it if ten were to pull on one side and ten on the other; they would never separate the surfaces of the bellows unless something were forced or pierced through and another body could come between the surfaces.[106]

Buridan establishes the hypothetical nature of the argument at the outset, when he begins the first sentence with "if" (*si*). Thus, *if* we assume that the bellows has been perfectly evacuated of all air *and* that all the holes are then perfectly stopped up to prevent the entrance of any air whatever, the sides of the bellows could never again be separated. Nature would not permit the formation of a vacuum, not even if ten horses were harnessed to each

---

[103] John Buridan, *Questions on the Physics* (Paris, 1509), bk. 4, qu. 7, fols. 72v, col. 2–73v, col. 2.

[104] Ibid., fol. 73v, col. 1.

[105] "Et iterum nos experimur quod non possumus unum corpus ab alio separare quin interveniat aliud corpus." Ibid. I have slightly altered my translation of Buridan's remarks from the version in my *Source Book in Medieval Science*, 326.

[106] Ibid.

side of the bellows and pulled with all their might. What was merely an imaginary exercise for Buridan in the fourteenth century, became a reality in the seventeenth century, when Otto von Guericke (1602–1686) used two teams of eight horses to pull apart the two halves of a copper sphere from which the air had been evacuated by means of an air pump. The vacuum in the sphere was good enough to prevent the horses from pulling the two halves apart. Thus, where Buridan's horses labored to show the nonexistence of a vacuum in the bellows, Otto von Guericke's horses showed that a vacuum existed in the sphere and that air pressure bearing on its two halves was sufficiently powerful to keep the two halves together.[107]

All the examples of experiments I have given thus far were about problems relevant to the real physical world. Few of them, however, were of the kind that the author reporting them might have actually witnessed, or actually performed. Of those that were witnessed, or were probably witnessed, is the example presented by Albert of Saxony in which he shows that nature seeks to avoid a vacuum at all cost, and offers as evidence a straw in water. When one withdraws the air in the straw, the water ascends and keeps in contact with the air above, thus preventing the formation of a vacuum. This may have been an experience shared by numerous individuals in the Middle Ages. Other experiences, such as Buridan's rotating top and smith's mill were devices that many would have had occasion to observe in the course of their lives. Indeed, in his *Questions on On the Heavens*, we saw that Buridan offered the smith's wheel as evidence for the existence of impetus. Even after you cease to supply a motive power to a smith's wheel it does not cease its motion immediately, but continues to rotate, thus indirectly indicating that it has within it an impressed force. This would have been a phenomenon that many would have had occasion to witness during the Middle Ages.

Most experiences, however, were not the kind that one could readily observe. They were conceived and introduced to serve as examples or counterexamples, or were meant to satisfy the needs of theory. Nicole Oresme provides an illustration of the latter, when, in his *Questions on On Generation and Corruption*, book 1, question 1, he explains that "an alteration is when one thing is changed into another, as hotness into coldness, and similarly fire into air."[108] What kind of experience is the conversion of fire into air? How does Oresme know that fire is converted into air? It is not directly observable. It is unclear whether Oresme intended this assertion to count as an experience or observation. And yet why did medieval natural

---

[107] For an earlier, somewhat different emphasis on Buridan's experiment, see my *God and Reason in the Middle Ages*, 169, and my article "Medieval Natural Philosophy: Empiricism without Observation," in *The Dynamics of Aristotelian Natural Philosophy*, 156–159.

[108] For the Latin text, see Stefano Caroti, ed., *Nicole Oresme Quaestiones super De generatione et corruptione* (München: Verlag der Bayerischen Akademie der Wissenschaften, 1996), bk. 1, qu. 1, p. 4, lines 45–46.

philosophers believe that fire is transformed into air? Probably because all Aristotelian natural philosophers assumed that the four elements were transformable into each other. Just as, for example, water is convertible into air and vice versa, so is air convertible into fire and vice versa. The conversion of fire into air does not qualify as a proper observation, but is an "observation" that is required by theory, namely, the theory of the four elements and their mutual conversions. Most of the examples I have cited belong to these categories: they were either constructed for the purpose of serving as examples or counterexamples; or were assumed to be observable because they were required to satisfy the needs of some Aristotelian theory. Indeed most of the examples used in medieval questions in natural philosophy were of this kind. They involved no direct observation of a real phenomenon, and they were never performed, although there were often elements of experience embedded in them.

We must now consider "experiences" employed in counterfactual and hypothetical contexts, among which those attributed to the absolute power of God played a significant role.

The Role of Experience in Hypothetical Worlds
Whether a medieval natural philosopher or theologian imagined that, by His absolute power, God had performed some naturally impossible feat – for example, creating other worlds, or moving our entire spherical world with rectilinear motion, or annihilating matter within the world to create a vacuum – or whether he simply imagined a hypothetical or counterfactual situation, a major objective was to determine what actions and behavior patterns of material objects in our real world would be operative in the hypothetical domain. The numerous imaginary scenarios that were proposed in the Middle Ages were analyzed, almost of necessity, by pure reason. This was true for a discussion of other worlds, or the analysis of events in a vacuum imagined to exist in our world, or any other hypothetical situation that medieval ingenuity could imagine. In his analysis of Aristotle's rejection of a plurality of worlds, Nicole Oresme declared that "it is good to consider the truth of this matter without considering the authority of any human but only that of pure reason."[109] And yet there is an element of empiricism in medieval discussions of hypothetical situations, albeit a rather strange variety, as we shall see.

Earlier in this chapter, we saw that many laws of behavior in our world would be operative in the imaginary world along with numerous anomalies. The other worlds that God could create if He wished were assumed to be identical to our world, each with its own center and circumference; each containing the same four elements with their natural places related in the

---

[109] Oresme, *Le Livre du ciel et du monde*, ed. and trans., Albert D. Menut (Madison: University of Wisconsin Press, 1968), bk. 1, ch. 24, 167.

same way as in our world; and each with a celestial matter composed of the same incorruptible ether, and so on. Aristotle had argued that if other worlds existed, the earth of one world would seek to move to the center of another world.[110] This implies that any earth that seeks to move from its own center to the center of another world would have to rise up and depart its world in order to reach the center of another world. But the natural motion of earth is downward toward the center of its universe. It is impossible for it to move upward. Aristotle therefore concludes that it is impossible for other worlds to exist. By applying the laws of our world to other hypothetical worlds, Aristotle concludes that they cannot exist because they would violate those physical laws.

In their discussion of the possibility of other worlds, many medieval natural philosophers agreed with Aristotle that the earth of one world could not rise up and move toward the center of another world. This would violate the laws of physics. They solved the problem by assuming that no earth in any world would leave its place and seek the center of another world. The reason every earth would remain in its own world is that each identical world was assumed to be a self-contained cosmos, operating according to the laws of our world.[111]

In this hypothetical realm, both Aristotle and those scholastics who disagreed with him, appealed to the law of falling bodies, namely, that earth would fall naturally toward the center of the world and come to rest in that center. However, where Aristotle assumed that every earth would seek the center of another world and would move upward to do this, thereby violating one of the most fundamental laws of his physics, many medieval natural philosophers – although not all by any means – assumed that the earth of each world would remain immobile at the center of its own world and not violate any physical laws. To arrive at this conclusion, however, they had to assume that the existence of other worlds would be perfectly intelligible, as each would be a self-contained copy of our world. Thus, where Aristotle viewed the existence of other worlds as an absurdity, his medieval followers regarded those worlds as supernaturally possible and completely intelligible, because they were assumed to be replications of our world.

In all this, both Aristotle and medieval natural philosophers applied what they interpreted as experiences in our world to hypothetical worlds. They felt justified in doing so, because they made the momentous assumption that all other possible worlds were identical with ours. This enabled them to apply the ideas about the behavior of earthy bodies in our world to earthy bodies in other worlds. Nevertheless, they arrived at opposing conclusions about the behavior of the earths of other worlds. Although they could not

---

[110] For Aristotle's argument, see his *On the Heavens*, 1.8.276a.22–276b.22.

[111] Buridan offers an example to show that the earth of every world would remain at rest in the center of its own world. See Buridan, *Questions on De caelo*, bk. 1, 18, 86–87.

possibly empirically investigate any claims in this controversy, they may have felt that they had, in effect, done so, because their observations of physical phenomena in this world were applicable to the same physical phenomena in any other identical world. In his French commentary on Aristotle's *On the Heavens*, Oresme takes issue with Aristotle. He first describes Aristotle's position as follows: "all parts of the earth tend toward a single natural place, one in number; therefore the earth of the other world would tend toward the center of this world."[112] Oresme replies forcefully:

I answer that his argument has little appearance of truth.... For the truth is that in this world a part of the earth does not tend toward one center and another part toward another center, but all heavy bodies in this world tend to be united in one mass such that the center of the weight of this mass is at the center of this world, and all the parts constitute one body, numerically speaking. Therefore, they have one single place. And if some part of the earth in the other world were in this world, it would tend toward the center of this world and become united with the mass, and conversely. But it does not have to follow that the portions of earth or of the heavy bodies of the other world, if it existed, would tend to the center of this world because in their world they would form a single mass possessed of a single place and would be arranged in up and down order, as we have indicated, just like the mass of heavy bodies in this world. And these two bodies or masses would be of one kind, their natural places would be formally identical, and likewise the two worlds.[113]

Oresme regards the two worlds as formally identical, so they would behave in exactly the same way. If perchance the matter of one world were removed to the other world, it would become a part of that world and move with regard to the absolute places of that world. It would have no inclination to return to the world from which it came. Any event that occurs in our world would, under the same circumstances, also occur in any other formally identical world. This meant any empirical observation valid for an experiment in this world would be applicable to any other formally identical world.

The debate about other worlds gets much more remote and far-fetched as the medieval imagination came into play. "As a mental exercise," Oresme toyed with the idea that there could be worlds within worlds, "so that inside and beneath the circumference of this world there was another world similar but smaller."[114] Oresme adds that "although this is not in fact the case, nor is it at all likely, nevertheless it seems to me that it would not be possible to establish the contrary by logical argument."[115] The worlds within worlds are all concentric to one another, differing only in size. But that is irrelevant, because as Oresme declares, "*large* and *small* are relative, and not absolute terms used in comparisons." Oresme imagines that if between now and tomorrow, the world were made one hundred or one thousand times larger or smaller than it is now, with "all its parts being enlarged or diminished

---

[112] Oresme, *Le Livre du ciel et du monde*, bk. 1, ch. 24, 173.   [113] Ibid., 173–175.
[114] Ibid., 167.   [115] Ibid.

proportionally, everything would appear tomorrow exactly as now, just as though nothing had been changed." Oresme gives the startling argument that if we imagine a world lying within the concavity of our earth, our earth would still be regarded as at the center of the world "since the center of the world would be the middle or center of its weight," by which Oresme presumably means that the center of the world is the earth's center of gravity. Following a series of other arguments, Oresme asks whether worlds within worlds, or even worlds that are beyond our world but concentric to it, can exist. He replies that "the contrary cannot be proved by reason nor by evidence from experience, but I also submit that there is no proof from reason or experience or otherwise that such worlds do exist. Therefore, we should not guess nor make a statement that something is thus and so for no reason or cause whatsoever against all appearances; nor should we support an opinion whose contrary is probable; however, it is good to have considered whether such opinion is impossible."[116]

We saw earlier that Oresme believed that a real incorporeal void space existed beyond our world. He arrived at this because he was convinced that "if two worlds existed, one outside the other, there would have to be a vacuum between them for they would be spherical in shape."[117] But Oresme does not rest content to leave this void a mere hypothetical entity. He is convinced that "the human mind consents naturally, as it were, to the idea that beyond the heavens and outside the world, which is not infinite, there exists some space whatever it may be, and we cannot easily conceive the contrary." To buttress his claim, and what he regards as "a reasonable opinion," Oresme presents a thought experiment. He imagines that "if the farthest heaven on the outer limits of our world were other than spherical in shape and possessed some high elevation on its outer surface, similar to an angle or a hump and if it were moved circularly, as it is, this hump would have to pass through space which would be empty – a void – when the hump moved out of it." Oresme uses this thought experiment to conclude that "outside the heavens, then, is an empty incorporeal space, just as the extent of this time called eternity is of a different sort than temporal duration." He characterizes this space as "infinite and indivisible," and then identifies it with the "immensity of God and God Himself." Although our corporeal senses cannot inform us that this space exists beyond the heavens, we can rely on "reason and truth," to "inform us that it exists." Thus did Oresme join numerous other scholastic theologian-natural philosophers in accepting the existence of an infinite extracosmic void space that he identified with God's infinite immensity. Where experience was lacking, Oresme, and many others, substituted reason to arrive at their conclusions.

We have already seen that a number of significant conclusions were derived about bodies falling in a hypothetical vacuum. These vacua were usually

---

[116] The quotations after the last note are all from ibid., 169, 171.　　　[117] Ibid., 177.

imagined, or they were the results of God's annihilation of matter within part, or all, of the sublunar region. The most significant of these was the straightforward claim that motion in a vacuum would be intelligible, contrary to Aristotle's total rejection of such a notion. To counter Aristotle, scholastic natural philosophers devised concepts such as mixed bodies, internal resistance, and impressed force (or impetus) to make motion in a vacuum finite and successive, instead of instantaneous, as Aristotle would have it. They also devised a concept of homogenous bodies by means of which they could reject Aristotle's claim that all bodies fall with speeds proportional to their weights. Homogenous bodies of unequal weight would all fall with the same speed in a vacuum. Albert of Saxony even devised examples in which the same body would fall slower in a plenum than in a void.[118]

All of these anti-Aristotelian conclusions about motion in a vacuum were empirically unverifiable. However, the natural philosophers who formulated the conclusions that made finite, successive motion in a vacuum intelligible and plausible used Aristotle's ideas about the motion of bodies in a plenum. Whatever empirical content went into Aristotle's ideas became part of the legacy of medieval natural philosophers, who then altered those conditions as required to render motion in a vacuum intelligible. Of course, whether the empirical content of motion in a plenum applies to motion in a vacuum cannot be verified. Medieval natural philosophers had no way of knowing whether their conclusions about motion in a vacuum were really true. But they were convinced that if a vacuum did exist, the "laws" of motion they derived would be operative.

In his *Questions on the Physics*, book 4, question 11, Albert of Saxony presents an interesting and important discussion about motion in a void and its relations to our experiences in the real world. Following a few examples in which he drew consequences that were impossible in Aristotle's physics (see above, this chapter), Albert formulates yet another impossible consequence in Aristotle's physics when he argues that "the natural motion of some heavy body can be quicker in the beginning than in the end." Aristotle, and almost all his followers, assumed that in a natural downward motion, a body accelerates and thereby constantly increases its speed. Albert, however, argues that there might be instances in which a body would be slower at the end of its motion than it was at the beginning. "For example," he argues:

if a mixed [or compound] body of four elements should have one degree of fire, one of air, one of water, and four of earth and if everything were annihilated within the sides of the sky except this mixed body, and if the mixed body were placed where the fire was, then this mixed body would descend more quickly through the vacuum of fire (*vacuum ignis*) than through the vacuum of air (*vacuum aeris*), and so on, as can easily be deduced from this case.

[118] All of these were discussed earlier in this chapter.

Realizing that this was a radical departure from Aristotle's opinions, Albert goes on to pose a question that he knew would be on the minds of his reading and listening audience.

But you [now] say, What should be said, therefore, about the common assertion that natural motion is quicker in the end than at the beginning? One can say that is universally true of the motion of heavy and light bodies but not of the motion of heavy and light mixed [or compound] bodies.[119]

In arriving at the conclusion that heavy and light mixed bodies would move in a vacuum with a greater speed at the beginning of their motions than at the end, Albert assumes that the vacuum through which such bodies move is not an undifferentiated empty expanse. Indeed, it is divided into "a vacuum of fire" and a "vacuum of air." By this, Albert means to signify that although the place where fire was naturally located in the regular material world is now void and empty of fire, that entire place continues to function as if the fire was still present. The same for air and its now void natural place. This is a remarkable assumption. Albert would have no reason to believe that the vacuum of fire differs one whit from the vacuum of air. And yet, he assumes that the vacuum of fire has the same basic properties as the natural place of fire in the regular material world; and assumes the same for the vacuum of air, thus making the vacua of fire and air different kinds of void spaces. Thus, he applied the "experiences" of the physical world to the void, dividing it in the same manner as the sublunar plenum of the material world, as Aristotle described it.

As if to subvert his attributions of natural places to parts of the void, Albert concedes, in the next question (book 4, question 12), that "we have never experienced the existence of a vacuum, and so we do not readily know what would happen if a vacuum did exist. Nevertheless," he continues, "we must inquire what might happen if it existed, for we see that natural beings undergo extraordinarily violent actions to prevent a vacuum." As evidence of this, Albert offers the commonly cited example of a straw in water, mentioned earlier in this chapter, declaring that when one draws on the straw and withdraws the air, "the water follows by ascending, striving to remain contiguous with the air lest a vacuum be formed."[120] Albert reports that it is commonly said that "Nature abhors a vacuum infinitely." So intense is this abhorrence that Albert was convinced that "before the sky would allow a vacuum to remain, it would descend and fill it."[121]

---

[119] Ibid., for the previous quotations.

[120] Albert of Saxony, *Questions on the Physics*, bk. 4, qu. 12 in Grant, *Source Book in Medieval Science*, 339.

[121] Ibid. For a discussion of the medieval dictum that "nature abhors a vacuum," see Grant, *Much Ado about Nothing*, ch. 4 ("Nature's Abhorrence of a Vacuum"), 67–100.

Despite the fact that Albert and virtually all medieval natural philosophers thought the existence of a vacuum was not naturally possible, Albert believed that if a vacuum did exist, bodies could move in it with finite velocities and even changes of quality (alteration) could occur. "For if a vacuum existed," he declares,

men could, nonetheless, walk around on the earth – although birds could not fly. It is true that men could not walk around for long, since they could not live long without breathing. Similarly, if some heated water were assumed to exist in a vacuum, it would change itself to its prior state of coldness. It seems, therefore, that local motion as well as motion of alteration could be made in a vacuum, for as stated in the other question, mixed [or compound] bodies could be moved in a vacuum because they have internal resistance.[122]

Although he admits that we have no experience with vacua, Albert boldly declares the kinds of activities that could go on there if they did exist. Men could walk around on the earth that still lay at the center of the world, albeit surrounded by a vacuum all the way to the concave surface of the moon. Hot water would become cold, although this hardly differs from what would happen in our material world. And, as he had shown, local motion also could occur, as well as motions of alteration, that is, changes of qualities. And, of course, as we saw, he did not hesitate to divide the otherwise undifferentiated void into different natural places, one for fire and one for air, a move for which there was no justification whatever. Albert of Saxony followed the logic of his arguments. The conditions he establishes in his examples are contrived to yield the conclusions he sought to demonstrate.

Empiricism without Observation

In all of these hypothetical examples, whatever empiricism may have been included was overwhelmingly an empiricism without observation. The same may be said of most of the experiences that were used in nonhypothetical situations, that is, in attempts to explain physical phenomena in the real world. There was very little direct observation in the literature on Aristotle's natural books or in the tractates on specific themes in natural philosophy. Despite the fact that medieval Aristotelians emphasized empiricism and followed Aristotle in rooting knowledge in sense perception, there was a only a modicum of direct observation relevant to claims made about the physical world. It was unavoidable that that there should have been many empirical elements in medieval questions and in the numerous problems that were considered. But those who reported these empirical elements, and made use of them to support or refute an argument, rarely ever observed them. They

---

[122] Ibid. Albert refers here to his preceding question, bk. 4, qu. 11, which I have discussed earlier.

either obtained the observation from another author, or devised it from their imaginations. In an illuminating passage, A. C. Crombie explains that:

From the beginning of the 14th century to the beginning of the 16th there was a tendency for the best minds to become increasingly interested in problems of pure logic divorced from experimental practice, just as in another field they became more interested in making purely theoretical, though also necessary, criticisms of Aristotle's physics without bothering to make observations.[123]

Medieval natural philosophers did not introduce observational elements into their discussions to learn more about the operations of the physical world. Experiences, whether embedded in real or hypothetical world situations, were usually meant to uphold this or that apriori view about those worlds. It is not that they did not recognize that sense perception was the basis of Aristotelian epistemology. But they were more apt to recall Aristotle's admonition that although the senses "give the most authoritative knowledge of particulars . . . they do not tell us the 'why' of anything – e.g. why fire is hot; they only say that it is hot."[124] Medieval natural philosophers were in agreement with Aristotle on this highly significant point. It explains why empiricism was, and remained, the servant of the analytic and apriori during the late Middle Ages. Only the analytic and a priori could provide the "why" of things to explain and interpret the empirical world.[125] John Murdoch has perceptively argued that although it is true that:

empiricist *epistemology* was dominant in the fourteenth century . . . this did not mean that natural philosophy then proceeded by a dramatic increase in attention being paid to experience and observation (let alone anything like experiment) or was suddenly overwrought with concern about testing or matching its results with nature. On the contrary, its procedures were increasingly *secundum imaginationem* (to use an increasingly frequently occurring phrase) and when some "natural confirmation" of a result is brought forth, more often than not it too was an "imaginative construct."[126]

The most powerful tool medieval natural philosophers possessed was not empiricism as manifested by observation per se, but rather experience as adapted for use in thought experiments (*secundum imaginationem*). Most

[123] A. C. Crombie, *Medieval and Early Modern Science* (Garden City, NY: Doubleday, 1959), vol. 2, 22–23.

[124] Aristotle, *Metaphysics*, 1.1.981b.10–11.

[125] From this point to the end of this chapter, I follow the analysis with which I concluded my article, "Medieval Natural Philosophy: Empiricism without Observation," in Leijenhorst, Luthy, and Thijssen, eds. *The Dynamics of Aristotelian Natural Philosophy from Antiquity to the Seventeenth Century*, 167–168.

[126] John E. Murdoch, "The Analytic Character of Late Medieval Learning: Natural Philosophy Without Nature," in Lawrence D. Roberts, ed., *Approaches to Nature in the Middle Ages* (Binghamton, NY: Center for Medieval & Early Renaissance Studies, 1982), 174.

of the experiences cited in this article are really thought experiments designed to refute or uphold a theory. But the "experiences" were not actually performed – in most cases, they had not even been experienced by the author – although they were usually examined and analyzed with great seriousness. They only had to appear plausibly appropriate and relevant to be accepted and then utilized as part of an overall argument for or against some real or imagined position.

It was one thing to write about induction and observation, and to uphold their importance, as did Roger Bacon, Albertus Magnus, John Buridan, and others, but it was quite another to come to the realization that it was essential to make observations in the real world on a routine basis, and to design experiments to learn things about the world that could not be derived from raw observation and experience. And to make all this a routine and regular feature of natural inquiry. This stage of development was not reached in the Middle Ages. It had to await the seventeenth century, the century of Newton. But if scholastic natural philosophers developed empiricism without observation, and focused attention on hypothetical, rather than real and direct observations, they did, at least, recognize that experience and observation were important ingredients in doing science and natural philosophy.

Because they failed to realize the importance of regular and direct observation and the need for devising experiments to yield nature's patterns of behavior, medieval natural philosophers did the next best thing. They sought to uphold the laws of Aristotle's world as well as they could. Where they found it at variance with reason and observation, they changed those laws and perceptions. But they did this in the way Aristotle had taught them, and also by means of a new tool that they had devised for themselves. That is, they used observation and sense perception, guided by reason, to support the positions they believed true, but they relied most heavily on their imaginations, which were guided by reason in the form of analytic techniques and logical analysis. It was in this manner that they concocted thought experiments for the real world, as well as for the world Aristotle had regarded as naturally impossible, the world of imaginary void space. By these methods, they arrived at some rather startling theories and conclusions, such as the mean speed theorem, impetus theory, the possibility of finite motion in a vacuum, and claims for the existence of extracosmic void space. They achieved all this with a "natural philosophy without nature," to use John Murdoch's perceptive and felicitous phrase, and, perhaps not surprisingly, by employing an "empiricism without observation."

In light of all this, one is inexorably driven to ask: did medieval scholastic natural philosophers believe that their responses to the multitude of questions they posed about the workings of nature provided them with truths about the structure and operations of the physical world? To this question, we must, I believe, respond in the affirmative, since we have no evidence to the contrary. To reply in the negative is to assume that they knowingly and willingly labored to no purpose, an untenable assumption.

When we realize that the contributions described here, and others, were made without the sophisticated methodologies that would become a routine part of scientific inquiry in the seventeenth century, we should recognize that medieval natural philosophers deserve a much greater measure of respect than has hitherto been accorded them.

WAS ARISTOTELIAN NATURAL PHILOSOPHY SCIENCE?

In Chapter 10, I shall describe dramatic changes that occurred in natural philosophy in the seventeenth century, when that ancient discipline was transformed into something far beyond anything Aristotle had contemplated. Here I shall consider whether or not we should regard Aristotle's natural philosophy as science.

We saw that Aristotle regarded natural philosophy as the third of three theoretical sciences, namely, metaphysics, mathematics, and natural philosophy, in that order. For Aristotle and his medieval followers, natural philosophy was primarily about bodies undergoing motion and change. Although there was almost universal agreement on this point, we also saw that there was considerable disagreement on the scope of natural philosophy. Some would include medicine, or music, or perspective; others thought that metaphysics and natural philosophy were interrelated despite the fact that metaphysics was concerned with immaterial substances. But all were agreed that natural philosophy, or natural science, as it also was called, was primarily concerned with mobile bodies. As John Buridan declared in his *Questions on Aristotle's Physics*, "this term 'mobile being' (*ens mobile*) is the proper subject that should be assigned in natural science, because it is the most common term among the things considered in natural science and does not transcend the limits of natural science."[127]

We also saw that as a discipline that is concerned with mobile being, natural philosophy considers inanimate and animate bodies and thus embraces the subjects Aristotle treated in his natural books, ranging from inanimate celestial and terrestrial bodies in his treatises on cosmology and physics, to animate plants and animals in his biological works and in his treatise on the soul. Was Aristotle writing scientific treatises when he composed his natural books? In effect, was he doing science in his books on natural philosophy? And by extension, were medieval natural philosophers doing science, or

---

[127] "Secunda conclusio est quod iste terminus ens mobile est subiectum proprium in scientia naturali assignandum quia est terminus communissimus inter considerata in scientia naturali et non transcendens limites scientie naturalis." *Johannis Buridani subtilissime questiones super octo Phisicorum libros Aristotelis*, bk. 1, qu. 3, fol. 4r, col. 1.

writing scientific treatises, when they wrote questions and commentaries on Aristotle's natural philosophy? Or when they wrote thematic works on topics in natural philosophy?

Aristotle provided a preliminary response to these questions when, as we observed earlier, he distinguished mathematics from natural philosophy, or physics, and then excluded the "middle [or exact] sciences" from natural philosophy because they were subject areas in which mathematics was applied to natural phenomena, as in astronomy and optics. When we leave well-established disciplines like optics and astronomy, how do we classify the numerous instances in which mathematics was applied to specific topics about motion that turn up in thematic works on natural philosophy, as, for example, the works on ratios by Thomas Bradwardine and Nicole Oresme mentioned earlier in this chapter. Those two treatises involved the application of mathematics to problems of motion. Should such treatises be excluded from natural philosophy? Where do they belong?

But what of other sciences, those in which little, or no, mathematics was involved, as in biology, geology, meteorology, and chemistry? Moreover, there were numerous questions in physics that were nonmathematical, as, for example, questions on impetus theory and possible motion in a vacuum, and so on. There can be no doubt that Aristotle clearly intended those of his treatises in which the subject matters that we would assign to one or another of these sciences to belong to natural philosophy. Of course, Aristotle could not assign these subject discussions to the sciences I have named, because they did not exist in his day. Indeed, they only emerged as distinct sciences between the seventeenth and nineteenth centuries. Thus, even if we eliminate the exact sciences from the domain of natural philosophy, and even if we take into account the fact that the scientific disciplines we have mentioned did not materialize as distinct sciences until after the seventeenth century, the subject areas of those nonmathematical sciences fall into the domain of natural philosophy. This is evident from an inspection of medieval questions on Aristotle's natural books.

In a question "whether the whole earth is habitable," John Buridan found occasion to discuss earthquakes and mountain formation.[128] Although geology did not become a distinct scientific discipline until the eighteenth or nineteenth century, Buridan's brief treatment of earthquakes and mountain formation belong to the history of geology. Anyone writing such a history would be obligated to include Buridan's account. Scholastic natural philosophers regularly asked questions that may appropriately be assigned to the subject area of some particular modern science, as the following list indicates.

---

[128] See Buridan, *Questions on On the Heavens*, bk. 2, qu. 7. I have translated the question in Grant, *A Source Book in Medieval Science*, 621–624. For the relevant discussion, see 623–624. Also see my *God and Reason in the Middle Ages*, 148–149.

PHYSICS

"whether local motion can produce heat."[129]

"whether the natural places of heavy and light bodies are the causes of their motions."[130]

"whether in its proper region [or place], air is heavy or light; or neither heavy nor light."[131]

"whether the existence of a vacuum is possible."[132]

"whether a resisting medium is required in every motion of heavy and light bodies."[133]

"whether if a vacuum did exist, a heavy body could move in it."[134]

"whether, in local motion, velocity is measured according to distance traversed."[135]

"whether a ratio (*proportio*) of velocities in motions varies as a ratio of ratios of the motive powers to the resistances."[136]

"we inquire what it is that moves a projected body upwards after separation from what has projected it."[137]

Questions on Visual Rays, Rainbows, and Halos

"whether every visual ray is refracted in meeting a denser or rarer medium."[138]

"whether every visual ray is reflected when it meets a denser medium."[139]

"whether a halo appears because of the refraction of rays in the vapor interposed between the eye and a luminous body around which it appears."[140]

"whether the colors appearing in the rainbow are where they seem to be and are true colors."[141]

"On the supposition that a rainbow can occur by reflection of rays, we inquire whether such reflection occurs in a cloud or whether it occurs in tiny dewdrops or raindrops."[142]

---

[129] Buridan, *Questions on On the Heavens*, bk. 2, qu. 16; see Grant, *Source Book*, 205.

[130] Ibid., bk. 4, qu. 2. Grant, ibid.   [131] Ibid., bk. 4, qu. 7. Grant, ibid.

[132] Albert of Saxony, *Questions on the Eight Books of Aristotle's Physics*, bk. 4, qu. 8. Grant, *Source Book in Medieval Science*, 201.

[133] Albert of Saxony, ibid., bk. 4, qu. 10. Grant, ibid.

[134] Albert of Saxony, ibid., bk. 4, qu. 11. Grant, ibid.

[135] Albert of Saxony, ibid., bk. 6, qu. 4. Grant, ibid., 202.

[136] Albert of Saxony, ibid., bk. 7, qu. 6. Grant, ibid., 203.

[137] Albert of Saxony, ibid., bk. 8, qu. 13. Grant, ibid.

[138] Themon Judaeus, *Questions on the Four Books of Aristotle's Meteors*, bk. 3, qu. 1. In bk. 3 of his treatise, Themon has twenty-severn questions, all of which are on visual theory, rainbows, and halos. Only a few of these are cited here. For all of the questions, see Grant, *A Source Book in Medieval Science*, 208–209.

[139] Themon, ibid., bk. 3, qu. 3.   [140] Themon, ibid., bk. 3, qu. 5.

[141] Themon, ibid., bk. 3, qu. 11.   [142] Themon, ibid., bk. 3, qu. 14.

"whether every rainbow must be three-colored."[143]

"whether at the time of a rainbow's appearance, it is necessary that the center of the sun, the center of the horizon, the center of the rainbow, and the poles or pole of the rainbow, be in a straight line."[144]

GEOLOGY

"whether the whole earth is habitable."[145]

"whether the waters of springs and rivers are generated in the concavities of the earth."[146]

"whether a motion of the earth is possible."[147]

"whether the tranquility of the air is a sign of the earth's motion to come."[148]

CHEMISTRY

"whether elements remain [or persist] formally in a compound [or mixed] body."[149]

"whether a compound is possible."[150]

"whether a compound (*mixtio*) is natural. By "compound" we understand what Aristotle understands in the first book of this [*On Generation and Corruption*], namely a "compound" properly so called is that whose every part is [also] said to be a compound [or mixed] body."[151]

METEOROLOGY

"whether the middle region of air is the place where rain is generated."[152]

"whether hail occurs more in spring and autumn."[153]

"whether thunder is caused by fire extinguished in a cloud."[154]

---

[143] Themon, ibid., bk. 3, qu. 15.   [144] Themon, ibid., bk. 3, qu. 20.

[145] Buridan, *Questions on On the Heavens*, bk. 2, qu. 7; see Grant, *A Source Book in Medieval Science*, 204.

[146] Themon Judaeus, *Questions on the Four Books of Aristotle's Meteors*, bk. 1, qu. 19; see Grant, ibid., 207–208.

[147] Themon Judaeus, ibid., bk. 2, qu. 7. See Grant, ibid., 208. This question is primarily concerned with earthquakes and tremors.

[148] Themon Judaeus, ibid., bk. 2, qu. 8. Grant, ibid. This question is also about earthquakes and tremors.

[149] Albert of Saxony, *Questions on the Two Books of Aristotle's On Generation and Corruption*, bk. 1, qu. 19. See Grant, ibid., 206.

[150] Albert of Saxony, ibid., bk. 1, qu. 20. Grant, ibid.

[151] Albert of Saxony, bk.1, qu. 21. Grant, ibid.

[152] Themon Judaeus, *Questions on the Four Books of Aristotle's Meteors*, bk. 1, qu. 15. See Grant, *A Source Book in Medieval Science*, 207.

[153] Themon Judaeus, ibid., bk. 1, qu.17. Grant, ibid.

[154] Themon Judaeus, ibid., bk. 2, qu. 9. See Grant, ibid., 208.

"whether a typhoon and a hurricane are made from a hot and dry exhalation."[155]

"whether lightning is fire descending from a cloud."[156]

Other examples could be cited, but these should suffice to indicate the important truth that natural philosophy encompassed questions relevant to a variety of sciences. There were bits and pieces of different sciences scattered through the questions on Aristotle's natural books. During the seventeenth to nineteenth centuries, those bits and pieces materialized as the distinct sciences already mentioned. And if we recognize that the middle, or exact, sciences of optics, astronomy, and mechanics also were once part of natural philosophy before they became independent by the time of Aristotle, or earlier, it seems highly appropriate to regard medieval natural philosophy as the "mother of all sciences." Although natural philosophy was not a particular science, it included discussions and analyses of questions relevant to the sciences we have mentioned, and is therefore a legitimate, and important, part of the history of science.

[155] Themon Judaeus, ibid., bk. 2, qu. 10. Grant, ibid.
[156] Themon Judaeus, ibid., bk. 2, qu. 11. Grant, ibid.

9

The Relations between Natural Philosophy and Theology

Although the early centuries of Christianity played a significant role in shaping the relations between natural philosophy and theology in the Middle Ages, developments in the twelfth and thirteenth centuries were more directly relevant in establishing the interplay between these two enormously important disciplines, an interplay that may appropriately be interpreted as surrogate for the relationship between science and religion. Nevertheless, to better appreciate and understand the long-term connections and antipathies between these two disciplines, it is desirable to describe briefly the manner in which they began their relations in the early centuries of the rise and development of Christianity.[1]

When Christianity first emerged within the Roman Empire, pagan culture and literature were already centuries old. The fact that Christianity, in contrast to Islam some centuries later, was disseminated rather slowly – it was not until 392 AD that Christianity became the state religion, almost four centuries after it first emerged – enabled Christians to adjust to pagan philosophy and literature and to contemplate what role, if any, it might play in their religion. Those who were instrumental in shaping the attitudes of the early Church toward pagan philosophy, and especially natural philosophy are known collectively as the Church Fathers. They came to represent two major approaches toward pagan natural philosophy. The first approach was quite hostile, regarding pagan science and philosophy as of little use to Christianity and even potentially harmful. Certain Church Fathers, such as Tatian, Eusebius, Theodoret, and Saint Basil viewed pagan Greek science as a source of misinformation and confusion. Saint Basil wrote "the wise men of the Greeks wrote many works about nature, but not one account among them remained unaltered and firmly established, for the later account always overthrew the preceding one. As a consequence there is no need for us to

---

[1] I draw here on my two earlier accounts in *The Foundations of Modern Science in the Middle Ages* (Cambridge: Cambridge University Press, 1996), ch. 1 ("The Roman Empire and the first six centuries of Christianity"), 1–17, and *God and Reason in the Middle Ages* (Cambridge: Cambridge University Press, 2001), ch. 2, 31–45.

refute their words; they avail mutually for their own undoing."[2] Saint Basil and others were distrustful of Greek philosophy, but Tertullian (ca. 150–225) gave expression to the most extreme form of hostility to Greek learning, when he advocated its complete exclusion from Christian thought, declaring that:

What indeed has Athens to do with Jerusalem? What concord is there between the academy and the Church? What between heretics and Christians? ... Away with all attempts to produce a mottled Christianity of Stoic, Platonic, and dialectic composition! We want no curious disputation after possessing Christ Jesus, no inquisition after enjoying the gospel! With our faith we desire no further belief.[3]

Fortunately, for Western Civilization, this negative, hostile attitude toward Greek science and philosophy did not triumph, although it had adherents well into the thirteenth century. There were always a few churchmen who thought that Sacred Scripture was capable of interpretation without the aid of pagan philosophy. But those who viewed Greek thought in a very different light were destined to prevail. A number of Christians were convinced that there was much of value in Greek philosophy and that it was capable of providing insights into Christianity itself. Biblical passages were invoked to reinforce this attitude: just as the Lord had instructed Moses to plunder the wealth of the Egyptians (Exodus 3:22, 11:2, and 12:35), so Christians might use pagan thought to better comprehend Sacred Scripture. Moreover, just as David slew Goliath with the latter's own sword (1Samuel 17:51), so should Christians use the words and ideas of the pagan philosophers to overcome those arguments that were contrary, or offensive, to Christians.

From such ideas, Christians developed the concept that philosophy and science are "handmaids to theology," an idea that had been developed earlier by Philo Judaeus (ca. 25 BC–AD 50), a Hellenized Jew who lived in Alexandria, Egypt. Philo had urged that Greek philosophy be used to understand revealed theology. Numerous Greek Church fathers – among whom were Clement of Alexandria, Gregory Nazianzen, Saint Basil, John of Damascus (John Damascene) – embraced this attitude toward pagan learning. In addressing his fellow Christians, Clement, for example, declared "We shall not err in alleging that all things necessary and profitable for life came to us from God, and that philosophy more especially was given to the Greeks, as a covenant peculiar to them, being, as it were, a stepping-stone to the philosophy according to Christ."[4] John of Damascus expressed the handmaiden idea in similar

---

[2] *Saint Basil Exegetic Homilies*, trans. Sister Agnes Clare Way, vol. 46 of *The Fathers of the Church: A New Translation* (Washington, DC: Catholic University of America Press, 1963), 5.

[3] *On Prescription against Heretics*, chapter 7, trans. Peter Holmes in *The Ante-Nicene Fathers*, ed. Alexander Roberts and James Donaldson, 10 vols. (New York: Charles Scribner's Sons, 1896–1903), vol. 3, 246.

[4] From *Miscellanies*, VI, 8 as translated in *The Ante-Nicene Fathers: Translations of the Writings of the Fathers Down to AD 325*. Vol. 2: *Fathers of the Second Century: Hermas, Tatian,*

fashion when, in the preface of his *Fount of Knowledge (Fons scientiae)*, he informs his readers that "I shall set forth the best contributions of the philosophers of the Greeks, because whatever there is of good has been given to men from above by God, since 'every best gift and every perfect gift is from above, coming down from the Father of lights.'" A few lines below, John explains that "In imitation of the method of the bee, I shall make my composition from those things which are conformable with the truth and from our enemies themselves gather the fruit of salvation," thus emphasizing the Biblical theme of despoiling the Egyptians.[5]

The handmaiden concept was embraced by Saint Augustine (354–430), the greatest of the Latin Church fathers. In *On Christian Doctrine*, he gave forceful expression to the subordinate role of philosophy. "If those," he declared,

who are called philosophers happen to have said anything that is true, and agreeable to our faith, the Platonists above all, not only should we not be afraid of them, but we should even claim back for our own use what they have said, as from its unjust possessors. It is like the Egyptians, who not only had idols and heavy burdens, which the people of Israel abominated and fled from, but also vessels and ornaments of gold and silver, and fine raiment, which the people secretly appropriated for their own, and indeed better, use as they went forth from Egypt; and this not on their own initiative, but on God's instructions, with the Egyptians unwittingly lending them things they were not themselves making good use of.[6]

Augustine strongly urged Christians not to seek secular knowledge for its own sake but to take only what is useful for a better understanding of scripture.

The handmaiden theory of secular knowledge also tended to emphasize the role of authorities, from the divine Scriptures themselves to the church fathers who had interpreted Scripture. The handmaiden tradition remained strong in Western Europe up to the eleventh and twelfth centuries, during the period when natural philosophy was relatively weak. By then, however, occasional figures appear who challenge the traditional reliance on authoritative texts and who rely on their own reason to understand the physical world, and urge others to do the same. In Chapter 5, we described the views of some

---

Athenagoras, Theophilus, and Clement of Alexandria (Entire). American edition, chronologically arranged, with notes, prefaces, and elucidations by A. Cleveland Coxe, D. D. (Grand Rapids, MI: Wm. B. Eerdmans Publishing Co., 1983), 495, col. 2; also quoted by Etienne Gilson, *History of Christian Philosophy*, 567, n. 8.

[5] Frederic H. Chase, tr., *Saint John of Damascus, Writings*, in *The Fathers of the Church, A New translation*, vol. 37 (New York: Fathers of the Church, Inc., 1958), p. 5 for both quotations.

[6] Augustine, *On Christian Doctrine (De doctrina Christiana)* in *The Works of Saint Augustine, A Translation for the 21st Century*, ed. John E. Rotelle, O. S. A.: Part 1, Vol. 11: *Teaching Christianity (De doctrina Christiana)*, introduction, translation, and notes by Edmund Hill, O. P. (Hyde Park, NY: New City Press, 1996), bk. 2, 159–160, sect. 60. Hill, the translator, changed the customary title *On Christian Doctrine* to *Teaching Christianity*.

of these unusual scholars: Adelard of Bath, William of Conches, and John of Salisbury. Others, such as Berengar of Tours, Anselm of Laon, and Peter Abelard, sought to apply reason to theology and thereby challenged religious authorities, provoking a reaction.

Thus when Aristotle's natural philosophy entered Western Europe by way of translations from Greek and Arabic into Latin in the twelfth century, it entered a society that was already beginning to question the role of religious authorities and authority in general. Although there was still considerable resistance to the new antiauthoritarian approaches to nature, a surprisingly rationalistic attitude had taken root in the course of the twelfth century in Western Europe. The introduction of Aristotle's natural philosophy, and the accompanying Arabic commentary literature on his works, as well as other Greek and Arabic treatises in astronomy and medicine, provided an enormous stimulus to the development of a rationalistic natural philosophy. The new literature was warmly welcomed by a Europe starved for knowledge.

As the translations spread through Europe, they generated an enormous interest in natural philosophy, metaphysics, and logic. This interest found an institutional home in the new universities of Paris, Oxford, and Bologna that came into existence by 1200. Although most scholars and students enthusiastically received Aristotle's natural philosophy, some theologians and Church authorities viewed it with suspicion. Just as certain aspects of Aristotle's natural philosophy posed serious problems for Muslims in the civilization of Islam, so also did it confront Christians with similar dilemmas in the Latin West. For Christians, Aristotle's most objectionable and offensive beliefs about the world were the following:

1. His insistence that the world is eternal, without beginning or end, a judgment in direct opposition with the Christian belief that the world had a beginning and would have an end.
2. That everything that comes to be has come from preexisting matter, from which it follows that something cannot come from nothing and therefore God could not have created the world from nothing, as Christians believed.
3. Only the rational part of the human soul is immortal, with the rest of it perishing with the body, a view that conflicted with the profound Christian belief that the undivided soul was eternal.
4. All accidents must inhere in a substance; they cannot exist independently. Christians found this objectionable because it conflicted with the doctrine of the Eucharist, or Mass, wherein after God transforms the bread and wine of the Mass into the body and blood of Christ, the accidents of the bread and wine continue to exist without inhering in any substance or substances.

Christian concern for the opinions of Aristotle was not confined to his positive beliefs but also was directed against hypothetical phenomena that

Aristotle deemed naturally impossible. Aristotle regarded the existence of vacuum as utterly impossible, inside or outside of our universe. Motion in such an utterly empty environment, devoid of any kind of external resistance, would, in Aristotle's judgment, be instantaneous and therefore impossible; or, it would not move at all. For a number of reasons, Aristotle was firmly convinced that there is only one world in existence and that it is impossible for other worlds to exist. Because Aristotle regarded these hypothetical phenomena as naturally impossible, some theologians interpreted Aristotle's attitude to mean that, even if He wished to do so, God could not create a vacuum, or create other worlds. In effect, they concluded that Aristotle had set limits to God's absolute power to do anything He pleases, short of a logical contradiction.

For all these reasons, Aristotle's natural philosophy was suspect, although even those who feared his influence found much that was useful and important in his treatises. Theologians also may have viewed Aristotle as a threat because the total body of his work, ranging over many topics and subjects, was wholly secular in character and outlook, and could therefore be viewed as a potential rival to a Christian interpretation of the world. Because Paris was the theological center of Christendom, it is not surprising that the uneasiness about Aristotle's philosophy was largely manifested in Paris, the home of the University of Paris. In 1210, Church authorities in Paris tried initially to ban all of Aristotle's works, a ban repeated in 1215. On April 13, 1231, Pope Gregory IX issued a papal bull in which Aristotle's works were no longer banned, but were now to be purged of errors.[7] On April 23, the pope appointed a three-man commission to eliminate errors from the works of Aristotle. No report of this committee has ever been found. Except for his books on ethics and logic, Aristotle's works on natural philosophy remained under a ban and were not publicly taught until the mid-thirteenth century. In 1255, a list of texts used in lecture courses at the University of Paris reveals that by then all of Aristotle's works were taught at Paris. Indeed, Roger Bacon lectured at the University of Paris on Aristotle's *Physics* in the mid-1240s. The first phase of Christendom's efforts to deal with Aristotle's thought came to an end. Aristotle would not be banned again anywhere in Europe.

But the struggle was not over. A number of conservative theologians still feared Aristotle's potential influence, among which St. Bonaventure (John Fidanza) was the most eminent. A new tactic was utilized to control Aristotle's impact: condemn those of his ideas that seemed most dangerous and threatening to the faith. In 1270, at the urging of a group of traditional-minded theologians, the bishop of Paris, Etienne Tempier, banned thirteen articles drawn from the writings of Aristotle and from his great Islamic commentator, Averroes (Ibn Rushd). This action was apparently ineffective,

---

[7] The bull is known as *Parens scientiarum* and is often regarded as the Magna Carta of the University of Paris.

but reports of controversies in Paris over Aristotle's ideas reached Pope John XXI, who, in 1277, requested the bishop of Paris, still Etienne Tempier, to investigate the turmoil at the University of Paris. On March 7, 1277, three weeks after the Pope had ordered him to begin an investigation, bishop Tempier, following the advice of his theological advisers, issued a condemnation of 219 articles.[8] The penalty for defending any one of the condemned articles was excommunication.

The condemned articles were drawn up in haste. They were a mélange of individual propositions, some, perhaps even many, of which were not actually taught at the University of Paris, but were disseminated orally among students and faculty. Many of the articles were irrelevant to natural philosophy. The major theme of the Condemnation was undoubtedly the eternity of the world, which was targeted in approximately twenty-seven separate articles. Among the propositions denouncing the eternity of the world in one form or another were the following:[9]

9. That there was no first man, nor will there be a last; on the contrary, there always was and always will be the generation of man from man.

87. That the world is eternal as to all the species contained in it; and that time is eternal, as are motion, matter, agent, and recipient; and because the world is [derived] from the infinite power of God, it is impossible that there be novelty in an effect without novelty in the cause.

93. That celestial bodies have eternity of substance but not eternity of motion.

94. That there are two eternal principles, namely the body of the sky and its soul.

98. That the world is eternal because that which has a nature by [means of] which it could exist through the whole future [surely] has a nature by [means of] which it could have existed through the whole past.

107. That the elements are eternal. However, they have been made [or created] anew in the relationship which they now have.

Earlier, I had occasion to mention articles 34 ("That the first cause could not make several worlds"), which denied that God could make other worlds,

---

[8] Oxford University was the scene of another condemnation on March 18, 1277, when Robert Kilwardby, the archbishop of Canterbury, condemned thirty errors in grammar, logic, and natural philosophy. He had the full support of the Oxford faculty. See Leland Edward Wilshire, "The Condemnations of 1277 and the Intellectual Climate of the Medieval University," in Nancy Van Deusen, ed., *The Intellectual Climate of the Early University: Essays in Honor of Otto Grundler*, Studies in Medieval Culture, XXXIX, Medieval Institute Publications, Western Michigan University (Kalamazoo, Mich., 1997), 154–155; the article extends over pages 151–193.

[9] The articles cited here are drawn from my translation in Grant, *A Source Book in Medieval Science*, 48–50.

and 49 ("That God could not move the heavens [that is, the world] with rectilinear motion; and the reason is that a vacuum would remain"), which as we saw denied that God could move the entire, finite, spherical cosmos with a rectilinear motion, because a vacuum would be left behind in the space that the cosmos formerly occupied, which is naturally impossible because Aristotle had declared it so. These articles were obviously condemned because they placed restrictions on God's absolute power to do anything short of a logical contradiction. Among other articles that placed restrictions on God's power were articles 35 ("That without a proper agent, as a father and a man, a man could not be made by God [alone]") and 48 ("That God cannot be the cause of a new act [or thing], nor can he produce something anew"). Other articles denied that God could make an accident exist without a subject (articles 140, 141).

Perhaps the most important condemned article that sought to restrict God's absolute power to perform a naturally impossible act is article 147, which declares: "That the absolutely impossible cannot be done by God or another agent. – An error, if impossible is understood according to nature." That is, it is an error to argue that God cannot do what is regarded as naturally impossible, which was probably meant to encompass the kinds of actions that Aristotle had regarded as naturally impossible. But if it had to be conceded that God could perform any naturally impossible act short of a logical contradiction, almost all would have denied that God could perform a logically impossible action. In his thirteenth-century theological commentary on the *Sentences of Peter Lombard*, Richard of Middleton explains:

that God cannot make two contradictories exist simultaneously, not because of any deficiency in his power, but because it does not make any sense to [His] power in any way. And if you should ask why this does not make possible sense, it must be said that with respect to this [problem] no other argument can be given except that such is the nature, or the disposition, of affirmation and negation, just as if we sought why every whole comprehends a part, no other argument would be forthcoming than that such is the nature of whole and part.[10]

Certain of the condemned articles seem to reveal a tension that had developed between the faculties of theology and arts, that is, between the theologians and the arts masters. For example, article 150 condemns the idea "That on any question, a man ought not to be satisfied with certitude based upon authority." The theologians often argued from authority, but natural philosophers in the arts faculty usually tried to follow the logic of an argument independently of authority. Other articles are far more explicit

---

[10] From my translation in *God and Reason in the Middle Ages*, 227 (the Latin text is given on page 227, n. 55). The passage appears in Richard of Middleton's *Commentary on the "Sentences of Peter Lombard*," 4 vols. (Brescia, 1591; facsimile reprint Frankfurt: Minerva, 1963), vol. 1, bk. 1, dist. 42, qu. 4, 374, col. 2.

in revealing the hostility that developed between the theologians and arts masters, as can be seen in the following three articles:

152.   That theological discussions are based on fables.
153.   That nothing is known better because of knowing theology.
154.   That the only wise men of the world are philosophers.

If these propositions actually circulated in the faculty of arts, we can readily see why the theologians would have used their considerable influence to have them condemned. If such sentiments circulated orally among the arts masters, it is a virtual certainty they would not have dared include them in any of their treatises or lectures. The condemnation was not only a controversy between theologians and arts masters, but it also set conservative theologians against theologians who were more congenial to Aristotle, the most important of these being St. Thomas Aquinas. Indeed a number of condemned articles were deliberately included because they were views held by Aquinas. After Thomas was canonized in 1323, the bishop of Paris proclaimed all articles null and void that were specifically directed against St. Thomas. Over the next few centuries, the condemned articles of 1277 were occasionally mentioned.

The Condemnation of 1277 had an impact on natural philosophy in the late thirteenth and fourteenth centuries. Specific articles from it were cited by numerous scholastic natural philosophers from the fourteenth to the seventeenth century.[11] It was a source of counterfactual arguments about our world and other possible worlds, all of which involved phenomena that were regarded as naturally impossible in Aristotle's interpretation of nature and its laws. It produced interesting and important speculative arguments in which Aristotle's ideas were frequently tested, and occasionally abandoned or subverted. God's absolute power to do anything short of a logical contradiction was often the vehicle for subtle and imaginative problems in the form of counterfactual arguments, although scholastic natural philosophers were fully capable of conjuring up hypothetical physical situations from their own fertile imaginations. Although the Condemnation of 1277 indicates tensions that had developed between conservative theologians and arts masters at the University of Paris, the condemned articles did not seriously affect the development of natural philosophy. On the contrary, not only was it not a significant impediment to natural philosophy, but one might even argue that it actually served to stimulate the development of natural philosophy by encouraging natural philosophers to ponder whether God might not have made the world very differently than Aristotle envisioned it. Moreover, some of the problems and solutions that were derived from certain condemned articles concerned with God's absolute power continued to exercise

---

[11] For a number of examples, see my article "The Condemnation of 1277, God's Absolute Power, and Physical Thought in the Late Middle Ages," *Viator* 10 (1979), 211–244.

an influence in the sixteenth and seventeenth centuries on both scholastic and nonscholastic authors.[12]

Although Aristotle's works were under somewhat of a cloud for much of the thirteenth century, most theologians and natural philosophers from the late thirteenth century onward enthusiastically embraced them. The medieval universities could not turn away from the greatest and most significant source of learning they had. Whatever the relations between natural philosophy and theology, or science and religion, in the late Middle Ages, they were not seriously affected by the Condemnation of 1277. We must now inquire about those relations. How were natural philosophy and theology viewed in a disciplinary sense by medieval scholastics? Was natural philosophy heavily influenced by theological demands and considerations? Was there an attempt to Christianize natural philosophy? And, contrarily, did natural philosophy have an impact on theology?

THE DISCIPLINARY RELATIONS BETWEEN NATURAL PHILOSOPHY AND THEOLOGY

Up to the thirteenth century, philosophy as a whole, and natural philosophy in particular, were viewed as the handmaidens of theology. As we saw, natural philosophy was regarded as a tool or instrument for explicating theology and the articles of faith. It was not to be studied for its own sake or as an end itself. With the translation of Aristotle's logic and natural philosophy into Latin and the absorption of those treatises into the university system of Western Europe, the handmaiden attitude underwent a significant change, one that was greatly aided by the transformation of the discipline of theology into a science. This began in the thirteenth century with Alexander of Hales, who may have been the first to begin his *Commentary on the Sentences* with a prologue that considered the question: "whether theology is a science." It was Thomas Aquinas who, in his *Summa of Theology*, presented the most influential arguments for classifying theology as a science. Indeed, he regarded it as the higher of two kinds of sciences. Some sciences "proceed from principles known by the natural light of the intellect;" and then there are sciences that "proceed from principles known by the light of a higher science." In the first category, for example, "the science of optics proceeds from principles established by geometry and music from principles established by arithmetic." By contrast, "sacred science," or theology, "is a science because it proceeds from principles made known by the light of a higher science, namely the science of God and the blessed. Hence, just as music accepts on authority the principles taught by the arithmetician, so sacred science accepts

[12] See ibid., 242–244.

the principles revealed by God."[13] By analogy, theology is a science like the sciences distinguished by Aristotle and natural philosophers. In the course of the thirteenth century, "the scales had been definitively tipped in favor of a rational conception of theology, as faith seeking understanding, as an investigation of the data of revelation with the help of the sources of reason."[14] Rational theology became a characteristic feature of commentaries on the *Sentences* of Peter Lombard.

Classifying theology as a science had momentous consequences. It made theology a separate discipline that used rational arguments to arrive at its conclusions. As theology became an independent scientific discipline, philosophy, and natural philosophy, also became autonomous disciplines that used rigorous reasoning from fundamental scientific principles.[15] Thus did natural philosophy and theology become separate disciplines. Natural philosophy could now be studied for its own sake. It was no longer the handmaiden of theology, although it would play that role for many theologians who used natural philosophy to analyze the faith. What was of great significance, however, was the fact that theologians and natural philosophers all recognized that natural philosophy was a powerful instrument for the study and analysis of both the physical world and the faith. As part of natural philosophy, and by means of the new autonomy of natural philosophy, the sciences began the long road to their own independence. Although they became autonomous disciplines in the thirteenth century, natural philosophy and theology, and natural philosophers and theologians, interacted in numerous important ways that affected their histories.

DID GOD AND THEOLOGY PLAY AN INTEGRAL ROLE IN MEDIEVAL NATURAL PHILOSOPHY?

It is important to recognize at the outset that in the large body of commentary literature on the various works of Aristotle's natural philosophy, the authors of those treatises – usually masters who taught natural philosophy, or theologians who had been trained in natural philosophy – firmly believed that God had created the world from nothing and that He was the ultimate cause of all natural effects. Because all natural philosophers believed this, proclaiming it in a treatise on natural philosophy was unilluminating and largely formulaic, although it was occasionally mentioned in questions and commentaries. One certainly did not have to offer evidence in substantiation of this claim. Does the fact that medieval natural philosophers unanimously

---

[13] From *Introduction to Saint Thomas Aquinas*, edited with an Introduction by Anton C. Pegis (New York: The Modern Library, 1948), 5–6.
[14] J. M. M. H. Thijssen, *Censure and Heresy at the University of Paris 1200–1400* (Philadelphia: University of Pennsylvania Press, 1998), 113.
[15] See Monika Asztalos, "The Faculty of Theology," in *A History of the University in Europe*, Vol. 1: *Universities in the Middle Ages*, editor Hilde de Ridder-Symoens (Cambridge: Cambridge University Press, 1992), 423–424.

believed that God created the world from nothing and was the ultimate cause of all natural effects mean that their objective in doing natural philosophy was essentially theological or religious? Does this signify that their natural philosophy was about God and the faith? Based on our knowledge of medieval natural philosophy, the response to these queries must be in the negative. In what follows, I shall attempt to justify my response.

Certain features of the medieval university proved detrimental to the intermingling of theology and natural philosophy. In medieval universities, theology and natural philosophy were taught in two independent faculties, the former in the faculty of theology, the latter in the faculty of arts. Teachers in the arts faculty were not trained in theology and rarely introduced theological matters into their lectures and writings. As if to reinforce this tendency, the faculty of arts at the University of Paris, beginning in 1272, required all masters to swear an oath that they would not introduce theological matters into their disputations. In proclaiming the statute, the masters decreed and ordained "that no master or bachelor of our faculty should presume to determine or even to dispute any purely theological question, as concerning the Trinity and incarnation and similar matters, since this would be transgressing the limits assigned him, for the Philosopher says that it is utterly improper for a non-geometer to dispute with a geometer."[16] Here we have a clear indication that natural philosophy is not a vehicle for discussing theology or matters of faith. Natural philosophers would be transgressing the bounds of their discipline if they introduced theology into their natural philosophy. As the Philosopher, namely Aristotle, declared, it would be wholly inappropriate to do that because "it is utterly improper for a non-geometer to dispute with a geometer." In sum, the disciplines of theology and natural philosophy have nothing to do with one another and should be kept apart. The oath further stipulated that if a question touched both faith and philosophy, it was to be resolved in favor of the faith. Any master who failed to do so would be branded a heretic. The Paris oath was required of all masters who taught in the arts faculty and, apparently, remained in effect until the end of the fourteenth, or the very beginning of the fifteenth century.[17]

If medieval natural philosophy was actually about God and the faith, we might appropriately ask why Popes, and other church officials, often objected to the introduction of natural philosophy into theology?[18] Indeed, if natural philosophy was about God and faith, why, in the course of the

---

[16] The oath is translated in Lynn Thorndike, *University Records and Life in the Middle Ages* (New York: Columbia University Press, 1944), 85–86.

[17] In his commentary on Aristotle's *Physics* composed between 1506 and 1511, Ludovicus Coronel asserts that "I . . . do not recall that when I was promoted to the degree in arts I took, or knew of any of my fellows taking, such an oath, but alas that laudable custom of the university along with others had become obsolete." Thorndike, *University Records*, 87–88. See also Monika Asztalos, "The Faculty of Theology," in H. de Ridder-Symoens, ed., *A History of the University in Europe*, Vol. 1: *Universities in the Middle Ages*, 424.

[18] See Asztalos, ibid., 421–422.

thirteenth century, did Church authorities at first ban it, then try to expurgate it, and then condemn a number of Aristotle's fundamental concepts as dangerous to the faith? And why, in the fourteenth century, did they attempt to minimize the intrusion of natural philosophy into theology?[19] If natural philosophy was really about God and the faith, the hostile reaction to it by some Church authorities seems rather bizarre and self-defeating.

In truth, with perhaps a few exceptions, the overwhelming number of those who taught and wrote about natural philosophy in the late Middle Ages regarded it as a distinct discipline, independent of theology. But even if they sought to intrude theology and matters of faith into their natural philosophy, they would have faced a formidable challenge. The subject matter of natural philosophy militates against its theologization. The object of Aristotelian natural philosophy is to provide natural explanations for natural phenomena. By its very nature, then, it is difficult to inject theology and matters of faith into natural philosophy. Doing so to any extent would convert natural philosophy into supernatural philosophy.

Before leaving the role of God in medieval natural philosophy, let me bring to your attention a thesis about the role of God in natural philosophy that is radically opposed to my own. Andrew Cunningham has argued in a number of places that natural philosophy, whether medieval or early modern, was always about God. He declares that "no-one ever undertook the practice of natural philosophy without having God in mind, and knowing that the study of God and God's creation – in a way different from that pursued by theology – was the point of the whole exercise." Indeed, God was so central to natural philosophy "that natural philosophy was not just 'about God' and his creation at those moments when natural philosophers were explicitly talking or writing about God in their natural philosophical works or activities. It was, by contrast, 'about God' and His creation the whole time."[20] My response to Cunningham's arguments prompted an *Open Forum* exchange in *Early Science and Medicine*, where the two of us reacted to each others claims.[21] I shall not attempt to summarize those arguments here. Instead,

[19] See Grant, *God and Reason in the Middle Ages*, 280–282 ("The Reaction to Analytic Theology").

[20] See Andrew Cunningham, "How the Principia Got Its Name; Or, Taking Natural Philosophy Seriously," *History of Science* 29 (1991), 388; see also 381 and Andrew Cunningham, "Getting the Game Right: Some Plain Words on the Identity and Invention of Science," *Studies in History and Philosophy of Science* 19 (1988), 383–384.

[21] The exchange began with my article "God, Science, and Natural Philosophy in the Late Middle Ages," in Lodi Nauta and Arjo Vanderjagt, eds., *Between Demonstration and Imagination: Essay in the History of Science and Philosophy Presented to John D. North* (Leiden: Brill, 1999), 243–267. This provoked the exchange with Dr. Cunningham titled "Open Forum: the Nature of 'Natural Philosophy'," in *Early Science and Medicine: A Journal for the Study of Science, Technology and Medicine in the Pre-modern Period*, vol. 5, no. 3 (2000), 258–300. This consists of "An Introduction to the Exchange between Edward Grant and Andrew Cunningham," by the editors of the *Journal*, p. 258, followed by

I shall focus on one aspect of our debate that seems to me to negate his position.

If it is true, as Cunningham would have it, that natural philosophy is about God even when God is not mentioned, then, significant consequences must follow from this truth. Those who believe that natural philosophy is always about God would surely interpret natural philosophy and its impact quite differently than those who failed to recognize that profound truth. And yet Cunningham mentions not a single important consequence that flows from these radically different approaches to the comprehension of natural philosophy. I suspect this is because there are no substantive differences. One's interpretation of any text in natural philosophy where the effect of some natural cause is under consideration would be fundamentally the same no matter what position one took on this issue. If we are studying a treatise on natural philosophy in which God is not mentioned, and we also assume that natural philosophy is always about God even when God is not mentioned, how will this change our interpretation and understanding of the text? If it does not affect one's interpretation of the text in a manner that would produce an understanding of it that differs markedly from the understanding of it by someone who did not believe natural philosophy was about God even when God is not mentioned, then it obviously does not matter which assumption one makes. Unfortunately, Cunningham fails to raise this question and seems unaware of the problem.

We can take this a step further. As long as a treatise in natural philosophy can be about God even when God is not mentioned or implied, then we might, with equal justification, say that it is about angels, or government, or society, or anything else that comes to mind, even though no mention is made of these things.

HOW A FEW SIGNIFICANT NATURAL PHILOSOPHERS VIEWED THE RELATIONS BETWEEN NATURAL PHILOSOPHY AND THEOLOGY

What was the opinion of the natural philosophers themselves? Did they think natural philosophy was about God and the faith? Very few expressed themselves on this question. Among those who did were Albertus Magnus and Thomas Aquinas in the thirteenth century, and John Buridan in the fourteenth. The opinions of Albertus and Thomas are especially important, because they were both theologians when they wrote their commentaries on Aristotle's natural philosophy. In the opening passage of his commentary on

Andrew Cunningham, "The Identity of Natural Philosophy. A Response to Edward Grant," 259–278, which is followed by my reply to Cunningham: "God and Natural Philosophy: the Late Middle Ages and Sir Isaac Newton," 279–298. The exchange concludes with "A Last Word" by Dr. Cunningham, 299–300.

Aristotle's *Physics*, Albertus informs his readers that his Dominican brothers had requested that he "compose a book on physics for them of such a sort that in it they would have a complete science of nature and that from it they might be able to understand in a competent way the books of Aristotle."[22] Albert explains that he will expound on what is called physics, that is, natural philosophy, in the manner of the Peripatetics, or Aristotelians. He announces that he will not introduce anything from his own knowledge, presumably knowledge of theology, "for if, perchance, we should have any opinion of our own, this would be proffered by us (God willing) in theological works rather than in those on physics."[23] Thus did Albertus inform his Dominican colleagues that he intended to treat Aristotle's natural philosophy in the customary manner of Peripatetics, that is, naturally. Should theological issues arise, he would treat them in theological treatises. Albertus leaves no doubt that he believes that theology should not be intruded into natural philosophy.

Albertus wrote before the Condemnation of 1277 had made any significant impact and indicates that he was not interested in what God could do by His absolute power, but wanted only to consider what was possible in nature. On the widely discussed issue of a plurality of worlds, Albertus concedes, in his *Commentary on De caelo*, that God could make more worlds if He wished, but explains that he does not wish to discuss that possibility. Rather, he assumes with Aristotle that the existence of other worlds is impossible "and that it is necessary that there be one [world] only. Here we understand about [i.e., we are concerned about] the impossible and necessary – that is, [we are concerned about] the world with regard to its essential and proximate causes. And there is a great difference between what God can do by means of his absolute power and what can be done in nature [or by nature]."[24] It is obvious that Albertus is not interested in what God can do supernaturally, but what nature can do by its customary operation. That is what he wishes to convey to his fellow Dominicans.

For Thomas Aquinas, I include only one quotation, but it is telling. Near the end of his life, in 1271, Thomas explained why he avoided mixing faith with natural philosophy. In considering a question on the rational soul in man he asserts, "I don't see what one's interpretation of the text of Aristotle has to do with the teaching of the faith."[25] Vernon Bourke argues that Aquinas did not believe he was "required to make Aristotle speak like a Christian"

---

[22] Translated in E. Synan, "Introduction: Albertus Magnus and the Sciences," in J. A. Weisheipl O. P., ed., *Albertus Magnus and the Sciences: Commemorative Essays 1980* (Toronto, 1980), 9. I am drawing on my discussion in Grant, "God, Science, and Natural Philosophy in the Late Middle Ages," 252–254.

[23] Synan, ibid., 10.

[24] My translation taken from Grant, "God, Science, and Natural Philosophy," 253.

[25] The translation is by Vernon Bourke in *St. Thomas Aquinas Commentary on Aristotle's "Physics,"* translated by R. J. Blackwell, R. Spath, and W. E. Thirlkel; Introduction by V. J. Bourke (New Haven, CT: Yale University Press, 1963), xxiv. The statement does not occur in

and he undoubtedly "thought that a scholarly commentary on Aristotle was a job by itself, not to be confused with apologetics or theology."[26]

We see that two of the greatest theologians in the Middle Ages thought one should try to avoid intermingling natural philosophy with theology. Most theologians, when they wrote on natural philosophy, usually followed the practice of Albertus Magnus and St. Thomas Aquinas.

As a master of arts who was not a theologian or trained in theology, John Buridan sometimes sought to mollify the theologians and to assure them that he was a faithful Christian who did not subscribe to those of Aristotle's opinions that were patently contrary to the Christian faith. In a question that inquired "whether a resting, or unmoved, heaven should be assumed beyond the heavens that are moved,"[27] Buridan was actually asking whether an empyrean heaven exists. The empyrean heaven was a purely theological construct that many theologians assumed was the dwelling place of God, the angels, and the blessed.[28] After presenting arguments for both sides, Buridan explains that "you may choose any side you please. But, because of the arguments of the theologians, I choose the first part [that is, the existence of an immobile, empyrean heaven]."[29] Because Aristotle rejected the existence of any immobile heaven, he would obviously have opposed the existence of an empyrean heaven. Knowing this, Buridan feels it necessary to defend the faith, declaring: "And one can reply to Aristotle's argument that he assumes many things against Catholic truth because he wished to assume nothing that could not be deduced from the senses and experience. Thus it is not necessary to believe Aristotle in many things, namely where he clashes with Sacred Scripture."[30] Earlier, in the same treatise, however, Buridan managed to preserve Aristotle's position in a question similar to, but much broader than, the one about the empyrean heaven. In his *Questions on De caelo*, book I, question 20,[31] Buridan asks "whether something exists beyond the heaven or world, namely beyond the outermost heaven; and Aristotle assumes this as obvious. But you ought to have recourse to the theologians [in order to learn] what must be said about this according to the truth of faith or constancy."[32]

Thomas's *Commentary on the Physics* but in a response to some questions by a Dominican colleague. See Grant, "God, Science, and Natural Philosophy," 256–257.

[26] The two quotations were translated by Vernon J. Bourke in *St. Thomas Aquinas Commentary on Aristotle's "Physics,"* xxiii and xxiv. Also in Grant, "God, Science, and Natural Philosophy," 257.

[27] Ernest A. Moody, ed., *Iohannis Buridani Quaestiones Super Libris Quattuor De caelo et mundo* (Cambridge, MA: The Mediaeval Academy of America, 1942), bk. 2, qu. 6, 149–153.

[28] For a discussion of the empyrean heaven, see my book, *Planets, Stars and Orbs: The Medieval Cosmos, 1200–1687*, ch. 15 ("The Immobile Orb of the Cosmos: The Empyrean Heaven"), 371–389.

[29] My translation from Grant, *God and Reason in the Middle Ages*, 188.   [30] Ibid.

[31] Moody, ed., *Iohannis Buridani Quaestiones Super Libris Quattuor De caelo et mundo*, bk. 1, qu. 20, 91–95.

[32] My translation from my *Source Book in Medieval Science*, 51, n. 4. The Latin text is in Moody, ed., ibid., 93.

Buridan then saves Aristotle by declaring: "Let it be assumed that there is an empyrean heaven beyond all the heavens that are in motion. Then we could say that this [empyrean] heaven belongs to this [that is, our] world, enclosing the rest of the world; therefore, we can revert to what Aristotle said, [namely] that there is no body beyond the last heaven, because it is not assumed that there is any body beyond this empyrean heaven." Thus, if we assume that the empyrean heaven is the outermost sphere of our world, and therefore the outermost part of our world, we still arrive at Aristotle's position that there is no body beyond our world.

In addition to the two examples just cited, in which Buridan clearly tried to observe the oath of 1272 and avoid offending the theologians, he reacted similarly in his *Questions on the Physics*, book 8, question 12, in which he inquires "whether a projectile, after it leaves the hand of the projector, is moved by air, or by what is it moved?"[33] After suggesting that when God created the world, He might have impressed a force, or impetus, into the celestial spheres, which thereafter kept them in motion, Buridan recognized that his suggestion had theological implications about the creation and that he was probably treading into the theological domain. He therefore informs the theologians "This, I do not say assertively, but [tentatively] so that I might seek from the theological masters what they might teach me in these matters as to how these things take place."[34]

From an earlier question in his *Questions on the Physics* (book 4, question 8), Buridan seems to reveal why he was sensitive to the potential criticism of theologians. In this question, Buridan asks, "whether it is possible that a vacuum exist by means of any power."[35] Buridan replies that a vacuum can indeed exist by virtue of God's power. After all,

God could annihilate everything under the lunar orb with the magnitude and figure of the lunar orb preserved. Then the concave orb of the moon, which is now a plenum in the lower world, would be a vacuum, just as a pitcher would be a vacuum if God annihilated the wine in it while preserving the pitcher and where no other body enters or is made in the pitcher.

What Buridan describes here was apparently offensive to some theologians, because Buridan now tells us that:

some of my lords and masters in theology have reproached me on this, [saying] that sometimes in my physical questions I intermix some theological matters which do not pertain to the artists [that is, Masters of Arts]. But with [all] humility I respond that

---

[33] John Buridan, *Acutissimi philosophie reverendi Magistri Johannis Buridani subtilissime questiones super octo phisicorum libros Aristotelis* (Paris, 1509), bk. 8, qu. 12, fols. 120r, col. 2–121r, col. 2.

[34] Buridan, ibid., fol. 121r, col. 1. With a slight emendation, I have used my translation in Grant, *Source Book in Medieval Science*, 51, n. 4.

[35] I again rely on my translation of the relevant parts of this question in Grant, *Source Book*, 50–51.

I very much wish not to be restricted [with respect] to this, namely that all masters beginning in the arts swear that they will dispute no purely theological question, nor [dispute] on the incarnation; and they swear further that if it should happen that they dispute or determine some question which touches faith and philosophy, they will determine it in favor of the faith and they will destroy the arguments (*rationes*) as it will be seen that they must be destroyed. Now it is evident that if any question touches faith and theology, this is one of them, namely whether it is possible that a vacuum exist. And so, if I wish to dispute it, it is necessary that I say about it what appears to me must be said according to theology, or to perjure myself and avoid the arguments on the opposite side insofar as this will seem possible for me. But I could not resolve these arguments [on the opposite side] unless I produce them. Therefore, I am compelled to do these things. I say, therefore, that "vacuum" can be imagined in two ways...

Buridan's remarks are puzzling. It is not obvious why Buridan and the theologians thought that "if any question touches faith and theology, this is one of them, namely whether it is possible that a vacuum exist." Because all agreed with Aristotle that a vacuum was not naturally possible, a vacuum was only possible if God produced it supernaturally. Apparently this made it a theological question. Perhaps God also made a vacuum in which to create our world. As we saw earlier, vacuum also was involved in articles 34 and 49 of the Condemnation of 1277. If God made other worlds (article 34), many believed that void spaces would lie between them and if God moved the world with a rectilinear motion (article 49), a void would be left where the world formerly rested. For all these reasons, questions about the possible existence of a vacuum may have been regarded as a question that involved the faith and theology.

The oath of 1272 was obviously in full effect when Buridan wrote his treatises on natural philosophy. It is likely that natural philosophers at Paris normally avoided the injection of theology, or theological issues, into their discussions. But, as with Buridan, they did not know what might be regarded as theological. The vacuum was a topic that Aristotle discussed at some length and its existence or nonexistence was central to natural philosophy. Certain articles in the Condemnation of 1277, however, seem also to have made it a theological problem. But all any natural philosopher had to do was say that God could make a vacuum if He wished and then move on to discuss any aspect of such a vacuum. After Buridan, natural philosophers discussed the vacuum and what might occur in it without theologians looking over their shoulders.

Indeed, they freely discussed any action stemming from God's absolute power. They had only to concede that God could do it if it did not involve a logical contradiction. To my knowledge there was only one exception to this rule. Many held that God could not create an actual infinite, because if He did, He would be unable to create anything larger and therefore His absolute power would be limited. Buridan discussed this issue in his *Questions on the Physics* (book 3, question 15), asking, "whether there is some infinite magnitude."[36] In this question, Buridan explains that "it is not necessary to believe that God could create an actually infinite magnitude, because when it has been created he could not create anything that is greater, since it is repugnant [or absurd] that there should be something greater than an actual infinite."[37] After asserting this argument, Buridan, once again, painfully aware that he was in the domain of the theologians and fearful of their criticism, declares that "with regard to all of the things that I say in this question, I yield the determination of them to the lord theologians, and I wish to acquiesce in their determination."[38]

From Buridan's experience, we may infer that the theologians wanted the arts masters to refrain from considering theological issues in their natural philosophy. We may plausibly assume that natural philosophers also sought to avoid theological issues, but certain problems in natural philosophy compelled them to speak about what God might or might not do and to cope with Aristotle's interpretations that were contrary to some article of the Christian faith. Although he was compelled to concede God's absolute power to do this or that naturally impossible action, Buridan often emphasized that God's power to do these things ought not to imply that He had done so, or would do so. Thus, although Buridan was prepared to concede that "we hold on faith that just as God made this world, so could He also make another, or others,"[39] he preferred to believe that if God wished to create more creatures of the kind that inhabit our world, He would simply double the size of our world, or make it one hundred times greater than its present size."[40] Similarly, God could create a finite or infinite space beyond the limits of our world, but we have no warrant to assume that He did, as the ordinary sources of evidence, namely, sense experience, natural reason, and the authority of Sacred Scripture, fail to indicate the existence of such a space beyond our world.[41] For the most part, Buridan was not attracted to the physics and cosmology of "what God might have done."

---

[36] On fols. 57r, col. 2–58r, col. 2 in the Latin edition cited earlier.

[37] I cite my translation from Grant, *God and Reason in the Middle Ages*, 233. For a discussion of this issue, see ibid., 228–234 ("God and the Infinite").

[38] Translation from *God and Reason in the Middle Ages*, 233.

[39] My translation from Edward Grant, "Scientific Thought in Fourteenth-Century Paris: Jean Buridan and Nicole Oresme," in Madeleine Pelner Cosman and Bruce Chandler, eds., *Machaut's World: Science and Art in the Fourteenth Century*, Vol. 314 of *Annals of the New York Academy of Sciences*, 108. Buridan's remarks appear in his *Questions on De caelo*, bk. 1, qu. 18, edited by Ernest A. Moody, 84.

[40] This response comes not from Buridan's *Questions on De caelo*, but from his *Questions on the Physics*, bk. 3, qu. 15. For the Latin text, see my article cited in the previous note, p. 119, n. 19.

[41] For Buridan's discussion of this issue, see his *Questions on De caelo*, bk. 1, qu. 17, p. 79 of Moody's edition. See also my article, "Scientific Thought in Fourteenth-Century Paris: Jean Buridan and Nicole Oresme," 108.

Like most of his arts master colleagues, however, Buridan was not eager to ensnare himself in theological problems. His primary concern, as it was for all natural philosophers doing natural philosophy, was to explain natural phenomena by means of natural causes. He and his colleagues sought to defend Aristotelian science as the best means of understanding the natural processes of the physical world. Natural philosophers were interested in the "common course of nature" (*communis cursus nature*), all the while recognizing that God could intervene supernaturally in cause-effect relationships. He could make fire cold, or He could make a log burn without using fire, and so on. Natural philosophers could not explain such divine interventions in the workings of nature. Despite such uncertainties, Buridan and his colleagues believed that truth about nature was attainable.

Buridan exemplifies this approach in a question asking "whether the grasp of truth is possible for us."[42] In the course of his response, Buridan almost certainly represents the great majority of natural philosophers, when he replied that truth is indeed attainable provided "the common course of nature (*communis cursus nature*) obtains in natural things, and in this way it is evident to us that fire is warm and that the heaven moves, although the contrary is possible by God's power."[43] Although God could alter the course of natural events at any time, Buridan holds firmly to the conviction that "in natural philosophy, we ought to accept actions and dependencies as if they always proceed in a natural way."[44] Neither the occurrence of miracles nor anomalous chance events affect the validity of natural science.

We have now seen how some of the most significant medieval natural philosophers viewed the relations between natural philosophy, on the one hand, and theology and faith, on the other. I shall now attempt to convey the role played by theology and faith in the questions and commentaries on Aristotle's natural books.

THE RELATIONSHIP AS REFLECTED IN THE QUESTIONS AND COMMENTARIES ON THE WORKS OF ARISTOTLE

The best way to determine the role of theology and faith in treatises on natural philosophy is to examine the substantive content of the treatises. The examples and data I present here are drawn from Aristotelian commentaries by Albertus Magnus, Thomas Aquinas, and Roger Bacon in the thirteenth

---

[42] *Questions on the Metaphysics*, bk. 2, qu. 1 in the reprint edition *Johannes Buridanus, Kommentar zur Aristotelischen Metaphysik* (Paris, 1588; reprinted Frankfurt: Minerva, 1964), fol. 8r, col. 1. I draw here on my *Foundations of Modern Science in the Middle Ages*, 145 and 211, n. 7.

[43] Ibid., fols. 8v, col. 2–9r, col. 1. Also, *Foundations*, 145, n. 8.

[44] My translation from Buridan's *Questions on De caelo*, bk. 2, qu. 9 on p. 164 of Moody's edition.

century; and by John Buridan, Nicole Oresme, Themon Judaeus, and Albert of Saxony in the fourteenth century.[45]

As I have already indicated, when Albertus Magnus and Thomas Aquinas were explaining Aristotle's texts in natural philosophy, they sought to avoid the introduction of theological analyses, feeling that it was inappropriate to do so. As theologians, both Albertus and Thomas were free to inject God into their deliberations wherever they pleased. But they chose to do so very sparingly. In the 261 chapters that comprise the eight books of his *Commentary on the Physics*, Albertus Magnus found occasion to mention God (*deus* and its variants, such as "First Mover," or "First Cause," and so on) in only twenty-four chapters, or in only 9 percent of his chapters; and in the 111 chapters of his *Commentary on De caelo*, Albertus mentions God in only nine chapters, or in approximately 8 percent of the total. Thomas Aquinas behaved in a similar manner. In his *Commentary on the Physics*, which is divided by his modern editors into 2,550 paragraphs, Thomas mentions God in only twenty-one paragraphs; mentions "Prime Mover" and its variants in forty-three paragraphs; mentions the expression "First Cause" in ten paragraphs; and mentions matters of faith in eight paragraphs. Thus Thomas found occasion to mention God and matters of faith in a total of eighty paragraphs, which is approximately 3 percent of the 2,550 paragraphs.

A striking example of how Albertus and Thomas consciously sought to avoid injecting theology into natural philosophy is apparent from the fact that neither mentions the empyrean heaven in their commentaries on *De caelo*. In another work – his *Commentary on the Sentences of Peter Lombard* – we learn why Thomas would not include the empyrean heaven in a work on natural philosophy. "The empyrean heaven," Thomas explains, "cannot be investigated by reason because we know about the heavens either by sight or by motion. The empyrean heaven, however, is subject to neither motion nor sight ... but is held by authority." The empyrean heaven is a theological construct and not a subject for reason. Consequently, it would have been inappropriate to include it in a treatise on natural philosophy.

By contrast, Roger Bacon, who was not a theologian, urged one and all to apply theology to natural philosophy and natural philosophy to theology. In his *Questions on the Eight Books of Aristotle's Physics*, Bacon included 461 brief questions but mentions God and the supernatural in only twenty-three questions. The religious or theological content in these twenty-three questions is minimal. In a treatise titled *On the Heavens* (*De celestibus*), which is really a discussion of Aristotle's book *On the Heavens* (*De caelo*), Bacon mentions the faith on only 2 of the 147 pages of the printed text. Treatises on cosmology were ideal vehicles for injecting theological matters,

---

[45] The data and citations were first presented in my article cited earlier, "God, Science, and Natural Philosophy in the Late Middle Ages," in Nauta and Vanderjagt, *Between Demonstration and Imagination*, 243–267.

but Bacon chose not to avail himself of this opportunity, despite his firm conviction that theology and natural philosophy should be intermingled. Like Albertus and Thomas Aquinas, however, he also makes no mention of the empyrean heaven, which, as we saw, was solely a theological concept without cosmological significance.

Most mentions of God and the faith in natural philosophy fall into certain categories. These categories provided occasions for injecting God or the faith into the discussion. The first is one in which Aristotle mentions something about God, or gods, or had occasion to express himself on the divine. In his *Commentary on Aristotle's On the Heavens*, book 1, Thomas Aquinas illustrates this when he explains that the place "up" (*sursum*) is the place of all divine things and that all men attribute to God the place that is called "up."[46] This is Thomas's summation of Aristotle's statement that "we recognize habitually a special right to the name 'heaven' in the extremity or upper region, which we take to be the seat of all that is divine."[47] A second category of theological comment occurs in those places where opinions contrary to the faith, often Aristotle's arguments, are discussed. In those instances, it was incumbent on the author to support the faith – either by rebutting such errors, or by showing how they might be compatible with the faith. Buridan, for example, asserts that

Aristotle says many things that cannot be properly saved.... For he holds indeed that nothing corruptible, or having potency for not being, can always exist in the future; and this is in fact false and against the faith because all things except God are corruptible and at some time they are not able to be because they could be annihilated by God.[48]

A third category in which God or the faith was often injected into a discussion in natural philosophy was by way of analogy, example, or comparison that was intended to illuminate something about the natural world. Thus, in a question about "whether the sun and moon ought to be moved with fewer motions than the other planets,"[49] Buridan found occasion to assert that "just as all order in the world arises from God, so in a city [does order arise] from a prince."[50] Nicole Oresme offered this analogy: "Some power makes this or that operation anew without changing itself, just as is obvious

[46] Thomas Aquinas, *Commentary on De caelo*, bk. 1, lectio 20, para. 199 in *S. Thomae Aquinatis in Aristotelis Libros De caelo et mundo, De generatione et corruptione, Meteorologicorum Expositio*, ed. Raymundi M. Spiazzi (Turin: Marietti, 1952), 98.

[47] Aristotle, *On the Heavens*, 1.9.278b.13–15, translated by J. L. Stocks in *The Complete Works of Aristotle, The Revised Oxford Translation* (Princeton, NJ: Princeton University Press, 1984).

[48] Buridan, *Questions on De caelo*, bk. 1, qu. 26, p. 127 of Moody edition. My translation from "God, Science, and Natural Philosophy in the Late Middle Ages," 259.

[49] See Buridan, *Questions on De caelo*, bk. 2, qu. 21, in Moody ed., 223–225.

[50] For the Latin text, ibid., 224.

with God who continuously produces new effects without any change in Himself,"[51] and Themon Judaeus makes this comparison: "a pure element is understood [to be] simple, but not simple absolutely, as is God, or an intelligence."[52]

A fourth category of situations in which God or the faith were likely to be inserted into a discussion in natural philosophy[53] concerns God's absolute power. Many examples could be cited. Most natural philosophers found numerous occasions to invoke God's absolute power, as did Albert of Saxony, who assumed that "God could create another body around this world; and around that body [He could] create another body; and so on to infinity. Nevertheless, these bodies are not mutually continuous."[54] In another question, Albert uses God's absolute power in a number of different contexts. Thus, he assumes that God annihilates all celestial bodies except the moon, which rotates from east to west. In what sense, Albert inquires, can the moon be said to be in motion under such circumstances. In the same vein and in the same question, Albert assumes that God fuses all the celestial spheres and all the bodies below the moon into one solid, continuous whole, which He then sets into rotation from east to west, or in any way He pleases. Once again, Albert asks how should we understand a motion that does not relate to anything outside of itself.[55] John Buridan and Nicole Oresme both assumed that God moved the whole world with a rectilinear motion.[56] Other examples by scholastic natural philosophers already have been mentioned, namely, God creating a vacuum by annihilating all or part of the matter between the concave surface of the lunar sphere and the earth; and God creating other worlds.

I have now described the four major categories into which theological discussions in medieval commentaries and questions may justifiably be divided. To determine the role theological citations played in the overall context of the treatises in which they were inserted, I examined all of the questions

---

[51] Nicole Oresme, Questions on De anima, bk. 3, qu. 2, in Nicholas Oresme's "Questiones super libros Aristotelis De anima": A Critical Edition with Introduction and Commentary by Peter Marshall (Ph.D. diss., Cornell University, 1980), 517.

[52] Themon Judaeus, Questions on the Meteorology, bk. 4, qu. 5, fol. 213v. The examples from Oresme and Themon appear in my article, "God, Science, and Natural Philosophy in the Late Middle Ages," 260.

[53] There is a fifth category that I shall omit from consideration here. It involves mentions of God and faith that are not classifiable in any of the four categories that I have included.

[54] Albert of Saxony, Questions on the Physics, bk. 3, qu. 11, fol. 39r, col. 2.

[55] For these examples, see my article "God, Science, and Natural Philosophy in the Late Middle Ages," 261.

[56] Buridan did so in his Questions on De caelo, bk. 1, qu. 16, 75–76 of Moody's edition and Oresme in his Questions on De anima, bk. 2, qu. 15; see Nicholas Oresme's "Questiones super libros Aristotelis De anima," by Peter Marshall (Ph.D diss., Cornell University, 1980), 386. Also see Grant, "God, Science, and Natural Philosophy in the Late Middle Ages," 261–262.

in the following five treatises in natural philosophy:[57] John Buridan, *Questions on De caelo (On the Heavens)* (fifty nine questions); Albert of Saxony, *Questions on the Physics* (107 questions); Nicole Oresme, *Questions on De anima (On the Soul)* (forty-four questions); Albert of Saxony, *Questions on Generation and Corruption* (thirty-five questions); and Themon Judaeus, *Questions on the Meteorology* (sixty-five questions).

Of the 310 questions included in these five treatises, a total of 217 are wholly free of any hint of theological discussion. From the content of the questions, one could not determine whether the author is Christian, Muslim, Jewish, agnostic, or even atheist. The remaining ninety-three questions mention God and the faith. However, fifty-three of the ninety-three questions mention God in a cursory manner. Of the remaining forty questions, only ten have relatively detailed discussions about God or the faith. In terms of the four categories identified earlier, the first (in which Aristotle, or some Greek or Islamic commentator, mentions God or gods) appears only twice in the 310 questions; the second category (which involves ideas contrary to the faith) turns up twelve times; the third category (where God or faith are used by way of analogy, example, or comparison) appears thirty-four times; and the fourth category (involving references to God's absolute power) also occurs thirty-four times. A fifth category encompasses sixteen mentions of God and faith that do not fit any of the first four categories and have not been included. All told there are eighty-two instances of the first four categories and a total of ninety-eight if we add the sixteen from the fifth category. It is important to add, however, that even where God and faith are mentioned in a question, the part of the question devoted to God and faith usually represents a small fraction of the total question, probably less than 5 percent.

It is obvious that the dominant concern of these fourteenth century natural philosophers were the third and fourth categories. They found occasions when mention of God or the faith served to exemplify some point they were making in natural philosophy. Instances in which they invoked God's absolute power to perform this or that naturally impossible act usually served to offer an opportunity to test the application of Aristotelian physics and cosmology under hypothetical circumstances that Aristotle regarded as impossible. The concept of God's absolute power encouraged the creation of many counterfactual examples with imaginary conditions.

In both of these categories – the use of God and faith as analogies or examples and their use in devising counterfactuals based on God's absolute power – genuine religious content was lacking. Thus, we may generalize that the penetration of substantive religious material into natural philosophy was minimal during the late Middle Ages. For the most part, medieval natural philosophers focused their attention on the study of natural phenomena in a rational and secular manner. Once again, John Buridan provides an example

---

[57] Grant, "God, Science, and Natural Philosophy in the Late Middle Ages," 243–267.

that illuminates the way natural philosophers approached their subject. In considering a question as to whether every generable thing will be generated, Buridan immediately acknowledges that a natural philosopher can treat this problem naturally: "as if the opinion of Aristotle were true concerning the eternity of the world, and that something cannot be made from nothing," *or* he could treat it supernaturally, in which event God could prevent a generable thing from generating naturally by simply annihilating it. "But now," Buridan declares, "with Aristotle, we speak in a natural mode, with miracles excluded."[58]

Because fewer than one-third of the 310 questions used in my sample had theologically relevant content, and most of these questions usually included less than five percent that was relevant to God, the faith or church doctrine, we may rightly conclude that God and faith played a minimal role in medieval natural philosophy. The explanation for this is simply that discussions about God, faith, and Church doctrine were irrelevant to the objective of medieval natural philosophers, which was to provide natural explanations for natural phenomena. It seems an inevitable conclusion that natural philosophy did not need theology to accomplish its mission. But did theology need natural philosophy? We must now address this vital question.

DID NATURAL PHILOSOPHY INFLUENCE MEDIEVAL THEOLOGY?

When we ask if natural philosophy influenced, or penetrated, medieval theology, we are really asking whether natural philosophy influenced the theologians who wrote commentaries on the *Sentences* of Peter Lombard. For, as we saw in Chapter 5, sometime between 1155 and 1158, Peter Lombard wrote a theological treatise in four books, titled *Sentences*. Peter's book became the basic textbook in theology until the seventeenth century, a period of approximately five hundred years, roughly the same length of time during which Aristotle's natural philosophy was dominant in Western Europe. In the four books, Peter devoted the first book to God, the second book to the creation, the third to the Incarnation, and the fourth to the sacraments. In the first half of the thirteenth century, Alexander of Hales (ca. 1186–1245) who taught theology at the University of Paris, adopted Peter Lombard's *Sentences* as a textbook in theology and, for convenience, divided the work into a large number of distinctions, each of which consisted of a number of questions that had much the same form as the questions that were later used to elucidate Aristotle's natural philosophy. Soon thereafter, all students who matriculated for a degree in theology had to comment on Peter Lombard's

---

[58] See Buridan, *Questions on De caelo*, bk. 1, qu. 25; in Moody edition, 123. I have drawn on my article "God, Science, and Natural Philosophy in the Late Middle Ages," 263–264.

four books of *Sentences*. Our knowledge of medieval theology is derived from the large number of extant commentaries on the *Sentences*.[59]

University theologians – especially at the University of Paris – were destined to transform theology into a thoroughly analytic discipline that made heavy use of logic and natural philosophy. This is not surprising when we realize that virtually all who became theological masters had previously studied logic and natural philosophy in the faculty of arts. They found the urge to apply these disciplines to problems of theology irresistible. In response, therefore, to the question that serves as the heading for this section – *Did natural philosophy influence medieval theology?* – we can straightaway declare that the influence of natural philosophy on theology was enormous. From time to time, the Church became alarmed at the overwhelming emphasis on natural philosophy in theology and Popes tried to deemphasize its use. In 1366, at the University of Paris, those who taught Peter Lombard's *Sentences* were admonished to avoid as much as possible the intrusion of logic and natural philosophy into their commentaries. But all such efforts proved futile. This is evident from a statement by John Major, a theologian at the University of Paris in the sixteenth century, who informs us that "for some two centuries now, theologians have not feared to work into their writings questions which are purely physical, metaphysical, and sometimes purely mathematical."[60] It is noteworthy that at no time did the authorities resort to a ban on natural philosophy and logic in *Sentences* commentaries, as they had done with Aristotle's works in the thirteenth century. They conceded that it was permissible to use natural philosophy and logic if they were deemed essential to the resolution of a question. Theologians found it necessary to use those disciplines in the resolution of most questions.

There were two major categories of questions relevant to natural philosophy in any commentary on the *Sentences*. One included questions that were essentially theological, but in which natural philosophy played a more or less important role. The other category involved questions that were essentially about natural philosophy rather than theology. There were, of course, questions that were purely theological and devoid of natural philosophy, as a few illustrations from Thomas Aquinas's second book of his commentary on the *Sentences* reveal: "whether created angels are blessed"; "whether in angels there could be sin"; "whether good angels could sin"; and "whether angels guard men."[61] These are irrelevant to our discussion.

---

[59] For a more thorough discussion of Peter Lombard's *Sentences*, see my *God and Reason in the Middle Ages*, 209–212, 217–219.

[60] Translated by Walter Ong, *Ramus, Method, and the Decay of Dialogue: From the Art of Discourse to the Art of Reason* (Cambridge, MA: Harvard University Press, 1958), 144.

[61] Thomas Aquinas, *Scriptum super libros Sententiarum Magistri Petri Lombardi Episcopi Parisiensis*, new ed., 4 vols. (Paris: P. Lethielleux, 1929–1947), vol. 2, 132–134, 143–145, 179–182, 270–272. See also Grant, *God and Reason in the Middle Ages*, 255.

The most extensive use of natural philosophy occurred in commentaries on the second book of the *Sentences*, which was devoted to creation and angels. Here we find numerous theological questions that were about creation or angels, but in which natural philosophy was extensively used. The following questions, drawn from a variety of authors, are concerned with different aspects of the creation account:

> "Whether light was created on the first day."[62]
> "Whether light made on the first day is corporeal or spiritual."[63]
> "Whether waters are above the heavens."[64]
> "Whether the crystalline heaven is moved."[65]
> "Whether the firmament has the nature of fire."[66]
> "Whether the firmament has the nature of inferior bodies."[67]
> "Whether the empyrean heaven is a body."[68]
> "Whether the empyrean heaven has stars."[69]
> "Whether the empyrean heaven exerts an influence on inferior things [that is, on things in the terrestrial region]."[70]

These questions were about basic elements in the creation account. Only trained, professional theologians were deemed appropriate to discuss them. Although they contained much natural philosophy, they were not questions that natural philosophers would have considered in their questions on Aristotle's natural books.

The empyrean heaven is an interesting exception. Aristotle had argued that nothing lies beyond the last moving celestial sphere. Natural philosophers considered it in their domain to inquire about the various possibilities beyond the outermost sphere of our world. It was in this spirit that Buridan asked "whether beyond the heavens that are moved there should be assumed a heaven that is resting or unmoved"; and Albert of Saxony similarly inquired

---

[62] In Peter Aureoli, *Commentariorum in primum [-quartum] librum Sententiarum, pars prima [-quarta]*, 2 vols. (Rome: Aloysius Zannetti, 1596–1605), vol. 2, bk. 2, dist. 3, qu. 1, art. 1, 1809, col. 1–185, col. 2.

[63] Saint Bonaventure, *Commentaria in quattuor libros Sententiarum Magistri Petri Lombardi: In secundum librum Sententiarum*, bk. 2, dist. 3, qu. 1, art. 1 in *Opera Omnia* (Ad Claras aquas [Quaracchi]: Collegium S. Bonaventurae, 1882–1901), vol. 2 (1885): 311–313.

[64] Thomas Aquinas, *Commentary on the Sentences*, bk. 2, dist. 14, qu. 1, art. 1 in his *Scriptum super libros Sententiarum*, vol. 2, 356–349.

[65] Richard of Middleton, *Super quatuor libros Sententiarum Petri Lombardi questiones*, bk. 2, dist. 14, art. 1, qu. 2, vol. 2, 168, col. 1–169, col. 1.

[66] Ibid., qu. 3, vol. 2, 169, col. 1–170, col. 2.

[67] Thomas Aquinas, *Commentary on the Sentences*, bk. 2, dist. 14, art. 2, vol. 2, 349–351.

[68] Alexander of Hales, *Summa theologica*, inq. 3, tr. 2, qu. 2, tit. 1, memb. 1, ch. 1, art. 2, in *Summa theologica*, Tomus II: Prima pars secundi libri (Florence: Collegium S. Bonaventurae, 1928), 328–329.

[69] Richard of Middleton, *Sentences*, bk. 2, dist. 2, art. 3, vol. 2, 44, cols. 1–2.

[70] Richard of Middleton, ibid., bk. 2, dist. 2, art. 3, qu. 3, vol. 2, 44, col. 2–45, col. 2.

"whether every heaven is mobile, or whether we must assume some heaven that is at rest"; and Themon Judaeus "whether something should be assumed [to exist] beyond the ninth sphere." We observe that none of these natural philosophers used the term "empyrean" in the enunciation of their questions, which were framed wholly in Aristotelian terms. Most natural philosophers would have conceded the existence of an empyrean heaven as a matter of faith, as did Buridan, although it is obvious he had his doubts. Indeed, after presenting the arguments for and against an empyrean heaven, Buridan says, as we saw earlier in this chapter: "you may choose any side you please. But, because of the arguments of the theologians, I choose the first part [that is, the existence of an immobile, empyrean heaven]."[71] Judging from other arguments and discussions, Buridan was really ambivalent about the existence of an empyrean heaven, but was undoubtedly fearful of arousing the theologians against him.[72] By contrast, in the course of the discussion in his relevant question, Albert of Saxony emphatically rejected the existence of an immobile heaven, and did so without any theological repercussions.[73]

The treatment of angels in commentaries on the *Sentences* was regarded as exclusively the province of theologians. No natural philosopher would have included a question on angels in explicating any of Aristotle's natural books. Many questions about angels, cited earlier in this chapter, are exclusively theological and devoid of natural philosophy. But just as many, it seems, involved problems in natural philosophy. In his *Summa theologiae*, Thomas Aquinas shows how deeply natural philosophy penetrated theology. In question 52, which is titled "On the Relationship of Angels to Places," Thomas includes the following three articles, which, as is obvious, are really questions:

Article 1: does an angel exist in a place?
Article 2: can an angel be in several places at once?
Article 3: can several angels be in the same place at once?

In question 53, "On the Local Motion of Angels," Thomas includes these three questions:

Article 1: can an angel move from place to place?
Article 2: does an angel, moving locally, pass through an intermediate place?
Article 3: whether an angel's motion occurs in time or in an instant.[74]

Here Thomas is dealing with problems of place and motion, which were major themes in natural philosophy. By analogy with how physical bodies are in places, and how they move from place to place, Thomas, like most other

---

[71] My translation from Grant, *God and Reason in the Middle Ages*, 188.
[72] For a discussion of this, see Grant, *Planets, Stars, and Orbs*, 375–376.   [73] See ibid., 376.
[74] I draw here on my discussion in *God and Reason in the Middle Ages*, 256–257.

theologians, wants to know how immaterial substances like angels do the same things. To investigate such matters, the theologians had to use natural philosophy. For the most part, when angels performed actions that physical bodies did, angels did them differently. Concepts from natural philosophy were imported into theology to deal with problems of immaterial substances. From these questions, we can see that theologians applied concepts of place, time, and instant, as well as concepts of successive motion, to explain how angels occupied places, how they moved from place to place, and whether they moved continuously and successively or instantaneously.

In the fourteenth century, some theologians used angels as a springboard for the formulation of lengthy expositions on a variety of philosophical problems. In some discussions, the angels fade into the background and are barely mentioned. One of the most interesting theologians in this group was Gregory of Rimini (ca. 1300–1358). In his *Commentary on the Sentences*, Gregory asked "whether angels were created before time [began], or after time [began]."[75] At the outset, Gregory explains his intent: "In this question, it is first necessary to see whether time is something created; and if so, what it is. Then we will see about what has been sought."[76] In a lengthy question that extends over pages 235 to 277, Gregory does not mention angels again until page 275, virtually at the end of the question. Instead of discoursing on angels, Gregory chose to present a detailed treatment of Aristotelian philosophical themes, mostly about time. When he finally turns his attention to angels, he does so to conclude that "no time was created before angels," and that "however, time is taken, angels were created before any time was time."

In the next question (question 2), Gregory asks "whether an angel exists in a divisible or indivisible place." He divides the question into two articles, the first of which is then divided into three conclusions, which he discusses on pages 278 to 331. He describes the three conclusions as follows:

The first is that no magnitude is composed of indivisibles, from which it follows that any magnitude is composed of magnitudes. The second [conclusion is] that any magnitude is composed of an infinity of magnitudes. The third [conclusion is] that in no magnitude is there something indivisible that is intrinsic to it.[77]

This first article has nothing to do with angels. Indeed, the word "angel" (*angelus*) occurs only once, on page 331, the last page of the article. The subjects on which Gregory focuses his exclusive attention are mathematics, physics, and logic. He applies these disciplines to indivisibles, a topic that immediately involves him with the nature of instants and the mathematical

75 Gregory of Rimini, *Lectura super primum et secundum Sententiarum*, 7 vols. (Berlin: Walter de Gruyter, 1979–1987), vol. 4, bk. 2, dist. 2, qu. 1, 235–277. In what follows, I draw on Grant, *God and Reason in the Middle Ages*, 259–264.
76 Gregory of Rimini, ibid., 236.   77 Ibid., bk. 2, dist. 2, qu. 2, art. 1, vol. 4, 278.

continuum. In the course of his fifty-three page discourse on the mathematical themes mentioned in the three conclusions, Gregory cites Euclid's *Elements* numerous times and includes fourteen geometrical diagrams. All of this is carried on within the usual scholastic format, where Gregory raises doubts about his own conclusions that he then answers.

In the second article, on pages 331–339, Gregory inquires whether an angel exists in a place. During the discussion, Gregory mentions an article condemned in 1277, article 204, which decreed it an error to assume that a separated substance, such as an angel, could not be in a place, or move from one place to another place, without operating there. Gregory's position was in accordance with the condemned article: "I say therefore that an angel is in a place not only by [its] operation, but also by its substance," and that it can be in a place even if does not operate in that place.

In his continuing discussion about angels, Gregory reveals the powerful role of natural philosophy in a typical theological treatise. In the first of three questions in book 2, distinction 6 of his *Commentary on the Sentences*, Gregory poses three questions about angels. The first of these asks "whether an angel could be moved locally by itself."[78] Gregory divides this question into four articles, of which the third is concerned with whether bodies can move themselves locally. Of the four articles, the third is seventeen pages long, much longer than the other three articles, and does not even mention angels.[79] In the third article, Gregory says that he is concerned only with the natural motions of simple – that is, elemental – bodies. In the course of the seventeen pages devoted to this article, Gregory presents six conclusions by means of which he considers common problems of motion drawn from Aristotle and Averroes. He shows that simple bodies are not moved directly by the heavens (first conclusion); that they are not moved actively by the places toward which they tend (second conclusion); that they are not moved by the media in which they happen to be (third conclusion); that they are not moved by the things that generated them (fourth conclusion); but that they are moved by some mover that lies within themselves, and not by something external (fifth conclusion); and, finally (sixth conclusion), that these simple bodies move themselves per se and not accidentally. Thus did Gregory import many ideas about natural motion into a theological treatise.

But why did Gregory think it necessary to have a lengthy discussion about the natural motion of elemental bodies? What has the simple motion of elemental bodies to do with the motion of angels from place to place? Gregory provides an answer to this question in the fourth article, where he explains that "since an angel could be moved locally, as is obvious from the first article, it is not impossible that it could move itself locally, as is obvious

---

[78] Ibid., bk. 2, dist. 6, qu. 1, art. 3, vol. 5, 12–21.
[79] The first article is one page long; the second article is nine pages long; and the fourth article is slightly longer than two pages.

from the third [article, since] it is possible that God could confer on it the ability to move itself."[80] The lengthy discussion of the local motion of bodies was intended to suggest that if God conferred the power of self-motion on elemental bodies, one might plausibly conjecture that He also could confer it on the more perfect angels.[81]

Gregory's various analyses of possible angelic activities were drawn from discussions about physical bodies in questions in Aristotelian natural philosophy. In another question, we can clearly see how ideas about motion that were applied to physical bodies were altered when applied to angels. In asking "whether an angel could be moved from place to place in an instant," Gregory arrives at two conclusions. In the first, which is the only one I shall consider, he proclaims that an angel can move from one place to another in an instant, even though it passes through the midpoint of the distance that separates the two places. He proves the conclusion by analogy with the motion of bodies in a vacuum. Aristotle had argued that bodies would move in a vacuum instantaneously, because there is no resistance in a vacuum. Without resistance bodies would move in an instant. The absurdity of instantaneous motion was one of the consequences that prompted Aristotle to reject the possibility of a vacuum. Gregory applies Aristotle's argument about bodies moving in a resistanceless vacuum to angels moving from one place to another. He assumes that no body or medium offers resistance to an angel. Therefore, an angel could move from one place to another in an instant and also move through the middle point that separates the two distances.

Many instances could be presented to show the impact of natural philosophy on what were ostensibly theological issues. Medieval theologians, however, took the final step and devoted some of their questions wholly to natural philosophy. Indeed, the questions were imported from natural philosophy. Theologians recognized the need to use natural philosophy to explicate the creation and other occurrences of natural phenomena in the Bible. They upheld St. Augustine's attitude as expressed in his commentary on *Genesis*.

In matters that are obscure and far beyond our vision, even in such as we may find treated in Holy Scripture, different interpretations are sometimes possible without prejudice to the faith we have received. In such a case, we should not rush headlong and so firmly take our stand on one side that, if further progress in the search of truth justly undermines this position, we too fall with it.[82]

---

[80] Gregory of Rimini, ibid., bk. 2, dist. 6, qu. 1, art. 4, vol. 5, 29.   [81] Ibid.

[82] St. Augustine, *The Literal Meaning of Genesis: De genesi ad litteram*, bk. 1, ch. 18, para. 37, ed. and tr., John Hammond Taylor in Johannes Quasten, Walter J. Burghardt, and Thomas Comerford Lawler, eds., *Ancient Christian Writers: The Works of the Fathers in Translation* (New York: Newman, 1982), vol. 41, 41.

This was an important sanction for the importation of natural philosophy into theology, even though Augustine insisted that Scripture should be taken literally whenever feasible. Thomas Aquinas added his powerful support when he declared that:

Augustine teaches that two points must be kept in mind when resolving such questions. First, the truth of Scripture must be held inviolable. Secondly, when there are different ways of explaining a Scriptural text, no particular explanation should be held so rigidly that, if convincing arguments show it to be false, anyone dare to insist that it is still the definitive sense of the text. Otherwise unbelievers will scorn Sacred Scripture, and the way to faith will be closed to them.[83]

A good illustration of how natural philosophy affected biblical interpretations is apparent from a question that was usually included in commentaries on the *Sentences*, namely, "whether the heaven [or firmament] has a spherical shape." Following Aristotle and the astronomers, theologians routinely accepted the claim that the heavens are spherical or orbicular. But in Psalm 103, the heaven is said to be stretched like a skin, or a tent, or an arched roof.[84] St. Bonaventure declares: "Scripture, condescending to poor, simple people, frequently speaks in a common way." Hence it describes the heaven as it appears to our senses and speaks of it as "a skin (*pellis*), or a stretched, arched roof (*camerae extensum*)."[85] Responding to this question, however, Bonaventure informs us that theologians "say, both according to reason and according to the senses, that the heaven has an orbicular shape."[86] In his *Sentences*, Richard of Middleton gave much the same response to this question, as did Bonaventure.[87] When Durandus de Sancto Porciano asked the same question in the fourteenth century, he omitted the biblical objections, very likely because they were no longer regarded as proper counterarguments.

The sense that it was acceptable to interpret important aspects of the creation account by the application of natural philosophy undoubtedly facilitated the massive intrusion of natural philosophy into theology. As we already saw, many questions were theological but were explicated by natural philosophy. The tendency to treat theological questions with natural philosophy was extended to the point where some theologians simply replaced many theological questions with questions in natural philosophy. The most blatant example of this trend appears in the *Sentence Commentary* of Peter John Olivi (1248–1298). Although only the second book of Olivi's *Sentence Commentary* exists, it is very long, consisting of 118 questions extending

---

[83] St. Thomas Aquinas, *Summa theologiae*, vol. 10: Cosmogony, pt. 1, q. 68, art. 1, 71–73.

[84] In the Vulgate, the Latin reads: "extendens caelum sicut pellem."

[85] Bonaventure, *Sentences*, bk. 2, dist. 14, pt. 1, art. 2, qu. 1, 342, col. 2.

[86] Bonaventure, ibid., 342, col. 1.

[87] Richard of Middleton, *Sentences*, bk. 2, dist. 14, art. 1, qu. 4, 170, col. 2–171, col. 1. Also see Grant, *God and Reason in the Middle Ages*, 267.

over three lengthy volumes in the modern edition.[88] If we liberally interpret Olivi's sense of what constitutes a theological question, there are only
forty-five theological questions and seventy-three nontheological questions,
a striking disparity when one realizes that commentaries on the *Sentences*
constituted the basic theological treatise of the Middle Ages.[89] God and
faith are mentioned occasionally in the nontheological questions, but they
are minimal and of no relevance to the questions themselves. For example,
in question 26, in which Olivi asks "whether the first impressions of all
agents are made by them in an instant," a question of eighteen pages, it is
not until the final two pages that Olivi mentions God twice, and Christ once.
Theological considerations are even more minimal in question 23, in which
Olivi asks "whether every agent is always present to its patient, or to its
first effect."[90] Not until the penultimate sentence of the question does Olivi
mention the "divine power" (*de divina virtute*).

There are questions where even this minimal concession to theology is
absent. Thus, in questions 18 to 21, extending over some twenty-five pages
(vol. 1, pp. 363–388), Olivi found no occasion to mention, God, faith, or anything else relevant to theology. The same may be said of questions 24 (vol.
1, pp. 434–438), 29 (vol. 1, pp. 499–504), 87 (vol. 3, pp. 198–203), and
88 (vol. 3, pp. 203–204). Evidence that Peter John Olivi did not regard his
work as a theological treatise is apparent from the fact that in his 118 questions he ignores all the usual questions about the heavens and creation that
were customarily included in the second book of a *Sentence Commentary*.
There is no discussion of Genesis, the heavens, the firmament, the crystalline
sphere, or the waters above and below the firmament, and so on. Instead,
Olivi included one question about the possible existence of other worlds,
which was not a commonly asked question in the second book of a *Sentence
Commentary*.

There can be little doubt that *Sentence Commentaries* were becoming
more and more like straightforward questions treatises on natural philosophy. Although many of the questions in a *Sentence Commentary* were seemingly theological, their substantive content was not theological, but physical,
that is, rooted wholly in natural philosophy. Another sign that theology was
becoming more natural philosophy than theology is the fact that medieval
theologians did not rest content with merely applying natural philosophy to
ostensibly theological problems. They imported a large number of questions

---

[88] Bernard Jansen, S. I., ed., *Fr. Petrus Iohannis Olivi, O. F. M. Quaestiones in Secundum
Librum Sententiarum* (Ad Claras Aquas [Quaracchi]: Ex Typographia Collegii S. Bonaventurae, 1922–1926).

[89] If we followed Bernard Jansen's categorization, only seventeen questions would be classified
as properly theological, leaving 101 as nontheological.

[90] Bernard Jansen, S. I., ed., *Fr. Petrus Iohannis Olivi, O. F. M. Quaestiones in Secundum
Librum Sententiarum*, bk. 2, qu. 23, vol. 1, 422–433.

that were in no way theological but belonged exclusively to the domain of natural philosophy. Indeed, some of these questions had a dual life: they appear in both theological treatises and works on Aristotelian natural philosophy, as the following questions exemplify:

"Whether the heaven is composed of matter and form."[91]
"On the number of the spheres, whether there are eight or nine, or more, or less."[92]
"Whether the heaven is spherical in shape."[93]
"Whether the heavens are animated."[94]
"Whether the whole heaven from the convexity of the supreme [or outermost] sphere to the concavity of the lunar orb is continuous or whether the orbs are distinct from each other."[95]
"Whether celestial motion is natural."[96]
"Whether the stars are self-moved or are moved only by the motions of their orbs."[97]

If we extended our range beyond cosmology to other subject areas, such as matter, motion, sense perception, and other themes, we could find other questions that were imported from natural philosophy into theology. There can be no doubt that medieval theologians were heavily into logic and natural

[91] This question appears in the *Sentence Commentaries* of Peter Aureoli and John Major; and in the *Questions on De caelo* by John of Jandun and Johannes de Magistri. For the precise citations, see my *Planets, Stars, and Orbs*, 694, qu. 79. For the full titles of these treatises, see the Bibliography in *Planets, Stars, and Orbs*.

[92] Thomas Aquinas and John Major discuss this question in their *Sentence* commentaries, whereas Albert of Saxony and Themon Judaeus consider it in questions on *De caelo*.

[93] St. Bonaventure and Durando de Sancto Porciano consider this question in the *Sentence Commentaries*, whereas Albert of Saxony, Johannes de Magistris, Johannes Versor, and John Major included it in their *Questions on De caelo*. To these, we may add Michael, Scot, Themon Judaeus, and Pierre d'Ailly, who included this question in their commentaries and questions on John of Sacrobosco's *Treatise on the Sphere*. See Grant, *Planets, Stars, and Orbs*, 703, qu. 126.

[94] Richard of Middleton and Peter Aureoli in their *Sentences*; and John of Jandun and Johannes de Magistris in their *Questions on De caelo*; Benedictus Hesse in his *Questions on the Physics*. See Grant, *Planets, Stars, and Orbs*, 703–704, qu. 128.

[95] Bonaventure and Richard of Middleton in their *Sentence Commentaries*; and Albert of Saxony and Paul of Venice in their *Questions on De caelo*. Others who included this question in commentaries on the *Sphere* of Sacrobosco and in questions on Aristotle's *Meteorology* are cited in Grant, *Planets, Stars, and Orbs*, 704–705, qu. 132.

[96] Durandus de Sancto Porciano in his *Sentences*; Johannes de Magistris in his *Questions on de Caelo*. For others, see Grant, *Planets, Stars, and Orbs*, 709–710, qu. 173.

[97] Bonaventure, Richard of Middleton, and Durandus de Sancto Porciano considered this question in their *Sentence Commentaries*; Roger Bacon, John of Jandun, John of Buridan, Albert of Saxony, Nicole Oresme, and a few others considered this question in their *Questions on De caelo*. See Grant *Planets, Stars, and Orbs*, 715, qu. 211.

philosophy. "One can point to numerous *Sentence Commentaries*," observes Edith Sylla,

in which natural science is used extensively, and there are some *Sentence Commentaries* which in fact seem to be works on logic and natural science in disguise – in response to each theological question raised, the author immediately launches into a logical-mathematical-physical disquisition and then returns only briefly at the end to the theological question at hand.[98]

So deeply did natural philosophy permeate theology that occasionally parts of *Sentence* commentaries were extracted and circulated as separate treatises on natural philosophy. John Murdoch explains that:

genuine parts of fourteenth century theological tracts...successfully masqueraded as straightforward tracts in natural philosophy. Thus, Gerard of Odo's examination of the problem of the composition of continua was detached from the *Sentence Commentary* to which it belongs and circulated separately. So totally without theological relevance (it was shorn of its introduction) it appears exactly as if it could be the initial *question* of Book VI of a commentary on Aristotle's *Physics*.[99]

Why did medieval theologians transform so many questions in theology into questions on natural philosophy, many of the latter emphasizing logico-mathematical techniques that had been developed in medieval natural philosophy? Beginning in the late eleventh century, theologians had begun to rationalize and systematize their discipline, a process that came to fruition in the twelfth century with Peter Abelard's *Sic et Non* (*Yes and No*) and Peter Lombard's Four *Books of Sentences*. But there was little natural philosophy in these treatises. All this was dramatically changed by the educational procedures at medieval universities, where the study of natural philosophy, as taught in the faculty of arts, was made an essential prerequisite of study in the higher faculties of medicine, law, and theology. As a consequence, virtually all theologians were well trained in logic and natural philosophy. Many of them had been arts masters before matriculating in the faculty of theology. They may even have taught natural philosophy to arts students before they entered the theology faculty to begin a lengthy course of study that ranged

98 Edith D. Sylla, "Autonomous and Handmaiden Science: St. Thomas Aquinas and William of Ockham on the Physics of the Eucharist." In *The Cultural Context of Medieval Learning*, edited with an Introduction by John E. Murdoch and Edith Dudley Sylla (Dordrecht, Holland: D. Reidel, 1975), 352. For a brief description of Oxford theologians with the same attitude, see W. A. Courtenay, "Theology and Theologians from Ockham to Wyclif," in J. I. Catto and Ralph Evans, eds., *The History of the University of Oxford*, vol. 2 (Oxford: Clarendon Press, 1992), 7.

99 John E. Murdoch, "From Social into Intellectual Factors: An Aspect of the Unitary Character of Late Medieval Learning," in John E. Murdoch and Edith Dudley Sylla, eds., *The Cultural Context of Medieval Learning* (Dordrecht: Holland: D. Reidel, 1975), 276.

from twelve to sixteen years.[100] For most of the late Middle Ages, a theological candidate was required to have reached the minimum age of thirty-five before he could acquire the degree. Thus, they spent much less time in the arts faculty than they did in the faculty of theology. Theologians generally reached philosophical maturity while in the theology faculty, where they finally had a good opportunity to develop their philosophical ideas. Consequently, they tended to present their mature thoughts about natural philosophy in their *Sentence Commentaries*. The strong desire to philosophize in their *Sentence Commentaries* perhaps explains why they included questions that could only be answered by the introduction of logico-mathematical techniques that had been developed in natural philosophy. Without these techniques it would not have been possible to cope with questions such as: "whether God could make the future not to be";[101] "whether an angel is in a divisible or indivisible place";[102] "whether God could cause a past thing [or event] to have never occurred";[103] "whether an angel could sin or be meritorious in the first instant of his existence";[104] "whether God could know something that He does not know";[105] "whether [an angel] could be moved from place to place without passing through the middle [point]."[106] Logic and natural philosophy were applied to the deepest mysteries of the Christian faith: the Trinity and Eucharist.

The background in natural philosophy, including logico-mathematical techniques, that theologians had learned as students and teachers in the arts faculty enabled them to transform theology into a rationalistic, analytic discipline during the thirteenth and fourteenth centuries. Theology was often more natural philosophy than it was theology. All that I have said in this chapter about the relations between natural philosophy and theology points to the unavoidable and overwhelming conclusion that while natural philosophy was virtually independent of theology, theology was utterly dependent on natural philosophy.

---

[100] See Monika Asztalos, "The Faculty of Theology," in Hilde de Ridder-Symoens, ed., *A History of the University in Europe*, vol. 1: *Universities in the Middle Ages*, 419.

[101] Hugolin of Orvieto, *Sentences*, bk. 1, dist. 40, qu. 3, art. 3, vol. 2, 341.

[102] Gregory of Rimini, *Sentences*, bk. 2, dist. 2, qu. 2, vol. 4, 277.

[103] Richard of Middleton, *Sentences*, bk. 1, dist. 42, art. 1, qu. 5, vol. 1, 375, col. 2.

[104] Gregory of Rimini, *Sentences*, bk. 2, dist. 3–5, qu. 1, art. 2, vol. 4, 345. The article is discussed on pages 369–379.

[105] Thomas Aquinas, *Sentences*, bk. 1, dist. 39, qu. 1, art. 2, vol. 1, 922.

[106] Robert Holkot, *Sentences*, bk. 2, qu. 4, art. 5. The book is unpaginated, but the fifth article occurs on the page where AA appears in the right margin.

10

The Transformation of Medieval Natural Philosophy from the Early Modern Period to the End of the Nineteenth Century

From its high point in the in the fourteenth century, medieval natural philosophy underwent significant changes in the course of the sixteenth and seventeenth centuries. But just as important as the changes that occurred directly to medieval natural philosophy are the changes that significantly altered almost everything around it. By the seventeenth century, Western Europe had undergone a great transformation from what it had been like in the fourteenth century. Beginning with Gutenberg's invention of the printing press around 1450, followed by Columbus's voyage to America in 1492, and, in the seventeenth century, the inventions of the microscope and telescope, the world in which Aristotle's natural philosophy was developed and nurtured had largely vanished. No doubt other factors of change might be cited, but one that also must be mentioned is the Protestant Reformation, which directly challenged the Catholic Church and therefore the culture within which Aristotle's natural philosophy had flourished. Aristotle's dominance in natural philosophy during the late Middle Ages is partially, if not largely, explicable by the fact that until the first half of the fifteenth century, Aristotelian natural philosophy had no rivals. From the mid-fifteenth century on, this began to change dramatically, as Greek works previously ignored or unknown were translated into Latin and vernacular languages and began to have an impact. Soon rival philosophies emerged among which were Platonism, Atomism, Stoicism, Neoplatonism, Hermeticism, and Copernicanism.

THE FATE OF MEDIEVAL NATURAL PHILOSOPHY DURING THE SIXTEENTH AND SEVENTEENTH CENTURIES

With the occurrence of so many significant changes in the sixteenth and seventeenth centuries, the attitude of many scholars and natural philosophers toward Aristotelian scholastic philosophy changed dramatically, even though the universities in Catholic Europe continued to teach medieval versions of Aristotelian natural philosophy well into the seventeenth century. In that century, scholastic natural philosophers cited Thomas Aquinas and Albert of Saxony, whose works were available in printed editions, just as readily as they did contemporary authorities, such as the Coimbra Jesuit commentaries on the works of Aristotle. But conceptions of the physical world had been drastically altered. In the course of the seventeenth century, scholastics found themselves trying to accommodate cosmological changes that had been instituted by Copernicus, Galileo, Tycho Brahe, and Kepler. Not only had Columbus and subsequent explorers shown an earth very different from Aristotle's conception of it, but the just mentioned astronomers and cosmologists had challenged the very structure of Aristotle's cosmos. Copernicus had proclaimed his heliocentric system in 1543, which was condemned by the Catholic Church in 1616; Tycho Brahe showed that the comet of 1577 was moving in the celestial region beyond the Moon, thereby shattering Aristotle's universally accepted idea that comets are sublunar phenomena. From the motion of comets in the celestial region, Tycho concluded that the heavens are composed of fluid matter, rather than hard celestial spheres, as was usually assumed in the late Middle Ages. If hard orbs, or even solid, fluid orbs, existed in the heavens, the comet would either smash them apart, or be unable to move through them. The celestial nature of comets was dramatic proof that the celestial region was not incorruptible and unchangeable, as Aristotle and his followers believed.

Aristotle's world was further subverted when Galileo turned his telescope skyward in 1610 and saw that the moon was mountainous and not a perfect sphere as all planets were assumed to be; he also observed that the sky had stars never before seen, and that Jupiter had four satellites. This was a world that Aristotle had never seen or envisioned. The invention of the telescope was instrumental in destroying the Aristotelian cosmos. Aristotle's authority was seriously eroded.

Scholastic natural philosophers, who were Aristotle's supporters, were divided in their response to the new realities. Some remained staunch defenders of the traditional cosmology and rejected the new discoveries and theories. Others, however, sought to adjust their interpretations and take cognizance of some of the new ideas that had emerged. They abandoned the doctrine of celestial incorruptibility and accepted change in the heavenly region.[1] Some assumed a celestial fluid in the heavens that was sometimes said to be fire, or comprised of more than one of the traditional terrestrial elements. Numerous Jesuit natural philosophers adopted Tycho Brahe's geoheliocentric system in which the earth remained at the center of the cosmos – thus conforming to the dictates of the Church that the earth must be located at the center of the world – but all of the planets were then assumed to move around the Sun as their center of motion as the Sun orbited the earth. Despite some effort to adjust to the new cosmology that was taking shape in the seventeenth century, scholastic natural philosophers were often condemned by a new breed of natural philosopher that rejected scholastic Aristotelianism

---

[1] I rely here on my discussion in Grant, *Planets, Stars, and Orbs: The Medieval Cosmos, 1200–1687*, 675–679 ("Conclusion").

because they were convinced that it could not reveal the manifest and hidden causes of nature.

Francis Bacon (1561–1626), for example, rejected Aristotle's syllogistic logic and its use in natural philosophy. In his *New Organon* (*Novum Organum*), Bacon denounced the way Aristotle had made natural philosophy subservient to logic. In the *Advancement of Learning* (1605), Bacon denounced the schoolmen for their superficial and narrow discussions based on a few authors. He regarded their efforts as useless.[2] Thomas Hobbes (1588–1679) condemned Aristotle's natural philosophy, which he calls "*physics*, that is, the knowledge of subordinate and secondary causes of natural events." Hobbes regarded Aristotle's natural philosophy, or physics, as nothing "but empty words."[3] As an illustration, Hobbes offers a few examples, of which the first is this:

If you desire to know why some bodies sink naturally downwards toward the earth, and others go naturally from it; the Schools will tell you out of Aristotle, that the bodies that sink downwards, are *heavy*; and that this heaviness is it that causes them to descend. But if you ask what they mean by *heaviness*, they will define it to be an endeavour to go to the centre of the earth: so that the cause why things sink downward, is an endeavour to be below: which is as much to say, that bodies descend, or ascend, because they do.[4]

From such examples, Hobbes concluded "if such *metaphysics*, and *physics* as this, be not vain philosophy, there was never any; nor needed St. Paul to give us warning to avoid it."[5] Although John Locke (1632–1704) admired Aristotle, he regarded Aristotle's scholastic followers with contempt, as nothing more than disputatious wordmongers. His sentiments were made explicit in *An Essay Concerning Human Understanding*, where, among other criticisms, he characterized the *Peripatetick* philosophy as having taught the world nothing more than the "Art of Wrangling," that is, the art of contentious disputation.[6]

[2] For Bacon's condemnation of the scholastics, see my *God and Reason in the Middle Ages*, 315–317.
[3] The quotations and sentiments are from Thomas Hobbes, *Leviathan. Or the Matter, Form, and Power of a Commonwealth Ecclesiastical and Civil*, pt. 1, ch. 5, in *The English Works of Thomas Hobbes of Malmesbury*, Now First Collected by Sir William Molesworth, 11 vols. (London: John Bohn, 1839–1845); reprint, Darmstadt: Scientia Verlag, Aalen, 1966), Vol. 3, 33.
[4] Hobbes, ibid., 674–675.
[5] Ibid., 680. For more on Hobbes's denunciations of scholastic thought, see Grant, *God and Reason in the Middle Ages*, 317–321.
[6] See John Locke, *An Essay Concerning Human Understanding*, edited with an Introduction by Peter H. Nidditch (Oxford: Clarendon Press, 1975), bk. IV, ch. VII, sec. II, 601. For more on Locke's criticisms of the scholastics, see Grant, *God and Reason in the Middle Ages*, 321–323.

Many other seventeenth-century critics of scholastic natural philosophy could be cited, but the most significant of all was undoubtedly Galileo Galilei (1564–1642). Galileo tirelessly emphasized the slavish devotion of Aristotle's medieval followers to the words and ideas of their intellectual master. In his *Letters on Sunspots*, he complained that Aristotelians had denied the reality of his telescopic discoveries, calling them illusions.[7] Galileo insisted that we "abase our own status too much... when we attempt to learn from Aristotle that which he neither knew nor could find out, rather than consult our own senses and reason. For she [that is, nature] in order to aid our understanding of her great works, has given us two thousand more years of observations, and sight twenty times as acute as that which she gave Aristotle."[8] In the *Dialogue Concerning the Two Chief World Systems*, Galileo shows respect for Aristotle, but only scorn for his followers. He has Salviati, who represents Galileo's point of view, report that a certain doctor claimed that Aristotle invented the telescope and the invention was taken from him. To this claim, Salviati declares to Simplicio, Aristotle's defender:

Tell me, are you so credulous as not to understand that if Aristotle had been present and heard this doctor who wanted to make him inventor of the telescope, he would have been much angrier with him than with those who laughed at this doctor and his interpretations? Is it possible for you to doubt that if Aristotle should see the new discoveries in the sky he would change his opinions and correct his books and embrace the most sensible doctrines, casting away from himself those people so weak-minded as to be induced to go on abjectly maintaining everything he had ever said?[9]

Elsewhere in the *Dialogue*, Galileo has Sagredo, who plays the role of neutral observer, recount a story in which an anatomist had shown "that the great trunk of nerves, leaving the brain and passing through the nape, extended on down the spine and then branched out through the whole body, and that only a single strand as fine as a thread arrived at the heart."[10] Aristotle had argued that the nerves originate in the heart, not in the brain. Sagredo informs us further that a peripatetic philosopher was present when the anatomist showed these discoveries about the nerves. The peripatetic philosopher was asked if he was convinced "that the nerves originated in the brain and not in the heart." After considering for awhile, he replied: "You have made me see this matter so plainly and palpably that if Aristotle's texts were not contrary to it, stating clearly that the nerves originate in the heart, I should be forced to admit it to be true."[11]

---

[7] *Letters on Sunspots*, in Stillman Drake, *Discoveries and Opinions of Galileo*, 140–141.
[8] Galileo, ibid., 142–143. The bracketed words are mine.   [9] Galileo, ibid., 110–111.
[10] Galileo, ibid., 107–108.   [11] Ibid.

Christia Mercer has observed that from humanist works by Petrarch in the fourteenth century and Gianfrancesco Pico della Mirandola (1469–1533) in the late fifteenth and sixteenth centuries,

anti-Aristotelians acquired a set stock of complaints. The standard criticisms were that the Peripatetics are more committed to Aristotle than to the pursuit of truth and, hence, are removed from the proper source of knowledge; talk about many things but understand little; do not even agree among themselves; and use obscure terminology which they neither properly define nor fully understand....[12]

The complaints against Aristotelian natural philosophers grew more intense and severe as the Aristotelian natural philosophy faded away. By the beginning of the eighteenth century, few scholars were familiar with it but the legacy of criticism and complaints about Aristotelian natural philosophy provided them with powerful rhetoric to denounce Aristotelians, long after it played any significant role in intellectual life.

THE NEW NATURAL PHILOSOPHY OF THE SEVENTEENTH CENTURY

Europe was in process of dramatic changes in the course of the sixteenth and seventeenth centuries. Perhaps the most striking evidence of this is the widespread use of the word "new" in titles of books. Two of the best known were Galileo's *Two New Sciences* and Johannes Kepler's *New Astronomy*.[13] The sense of "newness" that prevailed in the seventeenth century was undoubtedly a consequence of the belief that much of the knowledge being made public was new, and that it represented significant departures from Aristotelian natural philosophy and the traditional wisdom of the ancients. No one was as brilliant a spokesman for the new age of learning and science as was Sir Francis Bacon (1561–1626). His contributions to seventeenth-century science were based almost wholly on his vision of science – especially as he expressed it in his *New Organon* and *New Atlantis* – rather than on any substantive scientific achievements of his own.[14] Bacon called on the human race to gain mastery over the universe. In his *New Organon*, he identified three kinds of ambition for mankind. The first involved the acquisition of power in one's own country, while the second was for humans to aid their

[12] Christia Mercer, "The Vitality and Importance of Early Modern Aristotelianism," in Tom Sorrell, ed., *The Rise of Modern Philosophy: The Tension between the New and Traditional Philosophies from Machiavelli to Leibniz* (Oxford: Clarendon Press, 1993), 35.

[13] See Steven Shapin, *The Scientific Revolution* (Chicago: University of Chicago Press, 1996), 65. For a much more extensive list, which also cites a few medieval usages of the term "new" in book titles, see Lynn Thorndike, "Newness and Novelty in Seventeenth-Century Science and Medicine," in Philip P. Wiener and Aaron Noland, eds., *Roots of Scientific Thought: A Cultural Perspective* (New York: Basic Books Publishers, 1957), 443–457.

[14] See Mary Hesse, "Bacon, Francis," in *Dictionary of Scientific Biography*, vol. 1 (1970), 376.

country to extend its "dominion among men." These were, however, not the qualities of the human race that Bacon favored. It was the third kind that he urged upon his readers. For "if a man endeavor to establish and extend the power and dominion of the human race itself over the universe, his ambition (if ambition it can be called) is without doubt both a more wholesome thing and more noble than the other two. Now the empire of man over things depends wholly on the arts and sciences. For we cannot command nature except by obeying her."[15] Bacon's vision for the human race was to use the arts and sciences to gain power and control over nature. To achieve this, which is "the true and lawful goal of the sciences," it is essential "that human life be endowed with new discoveries and powers."[16] In the process of gaining mastery over nature, not only would useful knowledge be created, but continuous progress would result.

To show his readers how the human race might achieve mastery over nature, Bacon wrote his *New Atlantis*, which was published posthumously and became, perhaps, his most influential treatise. Bacon imagines the island of Bensalem in which there exists a foundation, college, or society, devoted to science and which is called "Solomon's House." Bacon goes into considerable detail in describing four basic functions of "Solomon's House," using the Father, or head, of Solomon's House as his spokesman. The first function expresses the purpose of the Foundation, which is to advance natural philosophy, the latter being "the knowledge of Causes and secret motions of things; and the enlarging of the bounds of Human Empire, to the effecting of all things possible."[17] The second function is "the Preparations and Instruments we have for our works," which is a lengthy and detailed description of the laboratories and instruments that Bacon envisioned as essential for scientific experimentation and discovery.[18] The third function involves the "Employments and Offices" of the fellows of Solomon's House. Different numbers of fellows are assigned to different tasks. Some collect experiments from books; some try new experiments which they think worth performing; three of the fellows are assigned to examine the experiments performed by their colleagues "and cast about how to draw out of them things of use and practice for man's life, and knowledge as well as for works as for plain demonstration of causes, means of predictions, and the easy and clear discovery of the

---

[15] See *Francis Bacon: Selected Philosophical Works*, edited, with Introduction, by Rose-Mary Sargent (Indianapolis: Hackett Publishing Co., 1999), *New Organon*, bk. 1, 147.

[16] Ibid., *New Organon*, book one, Aphorism 81, 117. Bacon was not optimistic that this goal would be achieved, as he indicates a few lines below the passage just cited: "But in general, so far are men from proposing to themselves to augment the mass of arts and sciences, that from the mass already at hand they neither take nor look for anything more than what they may turn to use in their lectures, or to gain, or to reputation, or to some similar advantage." Ibid.

[17] See *Francis Bacon: Selected Philosophical Works*, ed. Rose-Mary Sargent, 261.

[18] Ibid., 261–267.

properties and parts of bodies. These we call Dowry-men or Benefactors."[19] The fourth, and final, function is labeled "Ordinances and Rites." The Father of Solomon's House, who is describing the four functions, says of the fourth function that "we have two very long and fair galleries, in one of these we place patterns and samples of all manner of the more rare and excellent inventions, in the other we place the statues of all principal inventors."[20] This seems to be a museum for the display of inventions and inventors. Here we find "the statue of your Columbus, who discovered the West Indies, also the inventor of ships, your monk [Roger Bacon] that was the inventor of ordnance and of gunpowder, the inventor of music, the inventor of letters, the inventor of printing, the inventor of observations of astronomy, the inventor of works in metal, the inventor of glass. . . . " For worthy inventions made by their own members, "we erect a statue to the inventor and give him a liberal and honorable reward."[21] For the well-being of the people, the members of the House of Solomon "also make predictions of diseases, plagues, swarms of hurtful creatures, scarcity, tempests, earthquakes, great floods, comets, temperature of the year, and diverse other things," and, continues the Father of Solomon's House, "we give counsel thereupon what the people shall do for the prevention and remedy of them."[22]

Francis Bacon's ideas about the ways to study science and to apply the results of scientific research to the well-being of the general populace are truly extraordinary. And, yet, "although Bacon was a contemporary of William Gilbert, Johannes Kepler, and William Harvey, he was curiously isolated from the scientific developments with which they were associated."[23] Bacon rejected the Copernican heliocentric system and failed to recognize the importance of mathematics, even criticizing Kepler's theory of perspective because it did not go beyond geometry. But Bacon's vision of a science that was supported by the state and which sought to advance itself by collective research activities in order to benefit the human race has no parallels in the history of science prior to his own efforts. Bacon's ideas and vision had a great impact on the role that science came to play in Western society for the rest of the seventeenth century and beyond.

The Transformation of Natural Philosophy

The basic idea that natural philosophy was essentially a series of comments and questions on the texts of Aristotle was largely abandoned. Although Aristotle would continue to have his defenders and the commentary tradition lived on, scholars and investigators in the early modern period had a quite different view of the function of natural philosophy. Unlike their scholastic

[19] Ibid., 267.　　[20] Ibid., 268.　　[21] Ibid.　　[22] Ibid.
[23] Mary Hesse, "Bacon, Francis," *Dictionary of Scientific Biography*, vol. 1, 373.

predecessors, many early modern natural philosophers were convinced that they had to approach nature directly. They were determined to avoid the linguistic and logical emphasis of the scholastics and the characteristic lip service to empiricism.

Beginning with Copernicus's proclamation of the heliocentric system as the true system of the world in his *On the Revolutions of the Heavenly Spheres* in 1543, the Aristotelian cosmos suffered a sequence of blows from which it could not recover. Tycho Brahe, the great Danish astronomer, provided the next momentous blow when his study of the comet of 1577 showed that it was moving in the celestial region beyond the moon, and not, as Aristotle had argued, in the realm below the lunar sphere. With this startling observation, Tycho further denied the existence of hard, solid celestial orbs that were always believed to carry the planets around in their orbits. If hard orbs existed in the celestial region, the comets would smash them apart. But no such events are observed, because, as Tycho insists, celestial matter is fluid. In his book of 1588 (*De mundi aetherei recentioribus phaenomenis*), in which Tycho describes his research on the comet, he explains that he "first showed and clearly established that by the motions of comets [the heaven] is fluid and that the celestial mechanism is not a hard and impervious body filled with various real orbs, as has been believed by many up to this point, but that it is very fluid and simple with the orbits of the planets free and without the efforts and revolutions of any real spheres."[24] Tycho did not, however, accept the truth of Copernicus's heliocentric system and fashioned a compromise in which the earth remained immobile at the center of the world, but all the planets circled around the sun, while the sun revolved around the stationary earth. In the seventeenth century, many astronomers and natural philosophers found Tycho's compromise system appealing. It was thus a rival system to the Aristotelian-Ptolemaic and Copernican systems, and, as we saw, was popular in the seventeenth century. To these new astronomical and cosmological interpretations, Johannes Kepler (1571–1630) added his revolutionary contributions, among which was his claim that the planetary orbits around the sun – he was a strong Copernican – were not circular, but elliptical with the sun at one focus. And as we saw at the beginning of this chapter, Galileo's telescopic discoveries of Jupiter's satellites and mountains on the moon played a momentous role, perhaps more important than any other of the contributions we have mentioned. For with his telescope, Galileo had given visual evidence of a cosmos that differed radically from Aristotle's. There were things in the heavens that had never been seen before, and there was reason to believe that other celestial bodies might be brought from a state of invisibility to one of dramatic visibility.

[24] See Edward Grant, "Celestial Orbs in the Latin Middle Ages," in *Isis* 78 (June 1987), 155. The Latin text appears on 155, n. 9.

Although all of these incredible theories and discoveries shattered the medieval Aristotelian cosmos, they did not yet replace it with a new cosmic structure. This was achieved by Sir Isaac Newton (1642–1727) in his monumental *Mathematical Principles of Natural Philosophy* (*Philosophiae Naturalis Principia Mathematica*), first published in 1687 (second edition 1713). In his *Mathematical Principles*, Newton proclaimed his universal theory of gravitation, which has been described as follows: "The universe is composed of particles of matter all of which attract each other with a force proportional to the products of their masses and inversely proportional to the square of the distance between them."[25] As the final and most dramatic departure from Aristotle, Newton proclaimed that our world is in an infinite space that is God Himself. In the General Scholium, by which he concludes his great treatise, Newton identifies God and infinite space:

He is eternal and infinite, omnipotent and omniscient; that is, his duration reaches from eternity to eternity; his presence from infinity to infinity; he governs all things, and knows all things that are or can be done. He is not eternity and infinity, but eternal and infinite; he is not duration or space, but he endures and is present. He endures forever, and is everywhere present; and by existing always and everywhere, he constitutes duration and space.[26]

With the enunciation of his universal theory of gravitation operating in an infinite space,[27] Newton rang down the curtain on Aristotle's cosmos and introduced a wholly new conception of our world and its physical operations, although the identification of God with infinite space had already been made in the Middle Ages (see Chapter 8).[28]

If all that occurred in science and natural philosophy during the sixteenth and seventeenth centuries is what I have briefly outlined here, it would have undoubtedly constituted a Scientific Revolution: the replacement of one world by another, the Aristotelian world by the Newtonian. But historians of science and social historians have showed us that much more did indeed happen. Natural philosophy was altered in significant ways. Although medieval scholastic natural philosophers frequently disagreed with Aristotle, he continued to be the unrivaled authority for scholastic natural philosophers into the seventeenth century. But as Copernicus, Galileo, Kepler, and others made contributions that subverted Aristotle's natural philosophy, his authority was

---

[25] Richard S. Westfall, *The Construction of Modern Science: Mechanisms and Mechanics* (New York: John Wiley & Sons, 1971), 155.

[26] *Sir Isaac Newton's Mathematical Principles of Natural Philosophy and His System of the World.* Trans. into English by Andrew Motte in 1729. The translations revised and supplied with an historical and explanatory appendix by Florian Cajori (Berkeley: University of California Press, 1947), 545.

[27] For a detailed description of Newton's ideas about infinite space, see Edward Grant, *Much Ado about Nothing: Theories of Space and Vacuum from the Middle Ages to the Scientific Revolution* (Cambridge: Cambridge University Press, 1981), 240–255.

[28] See Grant, ibid., ch. 6, 116–147.

seriously eroded. This process was significantly aided by the recently activated works of Plato and other Greek scientists and natural philosophers
who had been largely unknown in the Middle Ages. Aristotle was no longer
the unrivaled authority but was now one of many. In fact, although some
seventeenth-century natural philosophers still held Aristotle in high regard,
Galileo and others had taught them that Aristotle was wrong on many fundamental points. Medieval natural philosophers had previously shown many
weaknesses in Aristotle's natural philosophy. They rejected his explanation
of projectile motion and replaced it by the famous impetus theory. Many
denied his insistence on a moment of rest between the up and down motion
of a stone thrown into the air. They often showed that ideas thought impossible by Aristotle were in fact possible under certain conditions that could
be produced supernaturally. They did this by the assumption of counterfactual conditions and thought experiments. But despite these subversions of
Aristotle's concepts, medieval scholastics retained Aristotle's cosmos much
as they had received it.

The likes of Copernicus, Galileo, Kepler, and Newton made Aristotle's
cosmos untenable. Indeed, not only was their cosmos radically different
from Aristotle's description of it, but the voyages of discovery in the fifteenth and sixteenth centuries disclosed large areas of our earth about
which Aristotle had no knowledge whatever. The mission of seventeenth-
century natural philosophers was to investigate nature and to discover the
truths about it that had eluded Aristotle and his medieval followers. They
continued the strong medieval tradition of reasoned argument but came
to emphasize what had been largely ignored in the Middle Ages: regular
observation of nature's activities by the naked eye and by instruments, and
the use of experiments to coax nature to yield her secret operations by
artificial means. Aristotle and his medieval followers did not believe that
experiments that intruded into nature could produce informative results.
The very act of constructing an experiment would artificially distort nature
and prevent the experimenters from learning about the way nature acted
"naturally."

In the seventeenth century, many natural philosophers approached nature
in a radically different manner. They came to view nature as a machine that
could be taken apart to see how one part acts on another part. In brief, they
adopted what came to be called the "mechanical philosophy," an approach
that sought to determine what made things "tick." Indeed, "tick" is a key
word, because the machine that came to symbolize nature's actions was the
mechanical clock. "Of all the mechanical constructions whose characteristics
might serve as a model for the natural world," Steven Shapin explains, "it
was the clock more than any other that appealed to many early modern natural philosophers. Indeed," he continues, "to follow the clock metaphor for
nature through the culture of early modern Europe is to trace the main con-
tours of the mechanical philosophy, and therefore of much of what has been

traditionally construed as central to the Scientific Revolution."[29] Ironically, the clock analogy had its roots in the Middle Ages, when Nicole Oresme (ca. 1320–1382), the great French theologian-natural philosopher, was probably the first to invoke the clock metaphor. As Oresme expressed it, when God created the heavens, He "put into them motive qualities and powers just as He put weight and resistance against these motive powers in earthly things.... The powers against the resistances are moderated in such a way, so tempered, and so harmonized that the movements are made without violence; thus, violence excepted, the situation is much like that of a man making a clock and letting it run and continue its motion by itself."[30] But until the seventeenth century, no mechanical philosophy accompanied Oresme's clock analogy.

A natural world that was thought of as a machine that ran like a clock mechanism was one that invited the use of scientific instruments to examine its parts and their relations. In the course of the seventeenth century, various devices were invented or perfected for use as scientific instruments. Among the most important were the telescope, the microscope, the thermometer, the barometer, and the air pump. All were used as scientific instruments for the investigation of nature. During the Middle Ages, with the exception of a few astronomical instruments, instruments were not regarded as important for scientific research, largely because there was little that qualified as scientific research. In the mechanical philosophy of the seventeenth century, however, it came to be recognized that instruments were essential for the study of nature and its operations. Often, the instruments had already been invented with their potential for scientific research recognized subsequently. No instrument illustrates this better than the compound microscope (it includes two or more lenses), invented at the end of the sixteenth century in the Netherlands. Although a number of seventeenth-century figures used the microscope to make significant discoveries – one of the most important was the Dutchman, Antoni van Leeuwenhoek (1632–1723), who used a simple microscope – it was Robert Hooke, using a compound microscope, who published what he saw with it in his *Micrographia* (1665), the first treatise on microscopy. He described and drew numerous illustrations of microscopic phenomena, including seeds, stones, and insects. Hooke was able to manifest his genius in the new atmosphere of invention and science in the seventeenth century. "He added something to every important instrument developed in the seventeenth century."[31]

---

[29] Steven Shapin, *The Scientific Revolution*, 32.

[30] *Nicole Oresme: Le Livre du ciel et du monde.* Edited by Albert D. Menut and Alexander J. Denomy; translated with and Introduction by Albert D. Menut (Madison: University of Wisconsin Press, 1968), bk. 2, ch. 2, 289.

[31] See Richard S. Westfall, "Hooke, Robert," in Charles C. Gillispie, ed., *Dictionary of Scientific Biography*, vol. 6 (1972), 487. Westfall gives a lengthy list of instruments that Hooke improved and observes that he was later called "the Newton of mechanics."

Experiments

With an array of instruments and an intense desire to use those instruments to discover nature's many operational secrets, the seventeenth century made the phrase "experimental science" meaningful for the first time in the history of science. It came to be understood that, if possible, an experiment should be made to determine the veracity of any claims made about nature's operations. The list of seventeenth-century natural philosophers and craftsman who performed experiments is impressive. Among them we find Isaac Beeckman (1588–1637), who "determined experimentally the law of the velocity of the outflow of water" (usually attributed to Torricelli);[32] Francesco Redi (1626–1697/98), who performed numerous experiments on the effects of snakebites and also devised experiments to disprove spontaneous generation, a doctrine that had been accepted since the days of Aristotle;[33] Evangelista Torricelli (1608–1647), who performed barometric experiments using mercury in a tube to show that atmospheric pressure, not the force of a vacuum, was the cause of the rise of liquids in tubes;[34] Blaise Pascal (1623–1662), using a Torricellian barometer, showed that the levels of mercury varied with elevation and concluded: "All the effects ascribed to [the abhorrence of a vacuum] are due to the weight and pressure of the air, which is their only real cause";[35] Otto von Guericke's (1602–1686) most famous experiment was that of the Magdeburg hemispheres, "in which he placed together two copper hemispheres, milled so that the edges fit together snugly. He then evacuated the air from the resulting sphere and showed that a most heavy weight could not pull them apart."[36] (In 1657, he used two teams of horses to try – unsuccessfully – to pull them apart.) Numerous other experiments could be mentioned that were performed by famous seventeenth-century natural philosophers, such as Isaac Newton, Robert Boyle, Robert Hooke, and William Harvey, to name only the most famous.

The Dissemination of Scientific Knowledge

Johann Gutenberg's invention of printing from movable metallic type was of momentous significance for all aspects of Western society, but it was especially welcome in the sciences and natural philosophy. Since it appeared in

---

[32] R. Hooykaas, "Beeckman, Isaac," in *Dictionary of Scientific Biography*, vol. 1 (1970), 567.

[33] Luigi Belloni, "Redi, Francesco," in *Dictionary of Scientific Biography*, vol. 11 (1975), 341–342.

[34] Mario Gliozzi, "Torricelli, Evangelista," in *Dictionary of Scientific Biography*, vol. 13 (1976), 438–439.

[35] Cited by Steven Shapin, *The Scientific Revolution* (Chicago: University of Chicago Press, 1996), 41.

[36] Fritz Krafft, "Guericke (Gericke), Otto von," in *Dictionary of Scientific Biography*, vol. 5 (1972), 575.

the 1450s, printing had its greatest initial impact on the dissemination of traditional natural philosophy and science. The majority of printed books were scholastic treatises that were either commentaries or questions on one or another of Aristotle's works, or devoted to some traditional theme from the thirteenth or fourteenth century. Printed editions for classical scientific texts, such as Euclid's *Elements*, Ptolemy's *Almagest*, the medical treatises of Galen, Avicenna, and al-Rāzī (Rhazes), became quickly available in reasonably reliable editions throughout Europe.

But the printed book quickly became a revolutionary instrument for the rapid dissemination of the new science that materialized from the late fifteenth century onward. Indeed, the first momentous published scientific work was Nicholas Copernicus's *On the Revolutions of the Heavenly Spheres (De Revolutionibus orbium celestium)*, published in 1543 (an identical second edition was published in 1566) and proclaiming a sun-centered, rather than earth-centered, cosmos. *On the Revolutions* was widely read, as is obvious from the fact that at least six hundred copies of the first and second editions have survived. A number of the surviving copies have been annotated by their owners, who were themselves frequently eminent astronomers.[37] In the same year, 1543, Andreas Vesalius (1514–1564) published his landmark book on human anatomy, *On the Fabric of the Human Body (De humani corporis fabrica)*. In addition to the inclusion of numerous anatomical drawings in his treatise, "Vesalius made many contributions to the body of anatomical knowledge, by description of structures hitherto unknown, by detailed descriptions of structures known only in the most elementary terms, and by the correction of erroneous descriptions."[38] To these two landmark books, many others could be added, especially three by Galileo (1564–1642) in the seventeenth century. The first of these is the *Starry Messenger* (*Siderius Nuncius*) (Venice, 1610) in which Galileo describes his discoveries with the telescope, as he informs us on the title page:[39]

SIDEREAL MESSENGER
unfolding great and very wonderful sights
and displaying to the gaze of everyone,
but especially philosophers and astronomers,
the things that were observed by

---

[37] For an exciting description of the discovery and preservation of the numerous copies of Copernicus's book, see Owen Gingerich, *The Book Nobody Read: Chasing the Revolutions of Nicolaus Copernicus* (New York: Walker and Co., 2004). As Gingerich explains (p. 281), *The Book Nobody Read* is the story of making my *An Annotated Census of Copernicus' De Revolutionibus* (Nuremberg, 1543 and Basel, 1566) (Leiden: Brill, 2002).

[38] C. D. O'Malley, "Vesalius, Andreas," in *Dictionary of Scientific Biography*, vol. 14 (1976), 10–11.

[39] *Sidereus Nuncius or the Sidereal Messenger Galileo Galilei*, translated with introduction, conclusion, and notes by Albert Van Helden (Chicago: University of Chicago Press, 1989), 26.

GALILEO GALILEI
Florentine patrician
and public mathematician of the University of Padua
with the help of a spyglass lately devised by him,
about the face of the Moon, countless fixed stars,
the Milky Way, nebulous stars,
but especially about
four planets
flying around the star of Jupiter at unequal intervals
and periods with wonderful swiftness;
which, unknown by anyone until this day,
the first author detected recently
and decided to name
MEDICEAN STARS.

Galileo's two other major works are *Dialogue Concerning the Two Chief World Systems – Ptolemaic and Copernican* (1632), which supports the Copernican system, and *Dialogues Concerning Two New Sciences* (1638), which consists of four dialogues, the last two of which "are devoted to the treatment of uniform and accelerated motion and the discussion of parabolic trajectories. The first two deal with problems related to the constitution of matter; the nature of mathematics; the place of experiment and reason in science; the weight of air; the nature of sound; the speed of light; and other fragmentary comments on physics as a whole."[40] Finally, I mention the book that presented the first synthesis of mathematical physics: Isaac Newton's *Mathematical Principles of Natural Philosophy*. Newton's other famous publication is the *Opticks* in three books first published in English in 1704. Other editions followed, as well as a Latin translation. Newton included many experiments on various aspects of optics. In the opening words of the *Opticks*, he declares his intent: "My design in this Book is not to explain the Properties of Light by Hypotheses, but to propose and prove them by Reason and Experiments."[41] Numerous other significant books also appeared in the late sixteenth and seventeenth centuries. Because they were produced by the printing press, these books were rapidly disseminated throughout Europe. Scholars in different countries studied the same mathematical and scientific diagrams and illustrations and read the same texts of a given author. Science was thus standardized in a way that was unimaginable before the 1450s. Medieval scholars labored under the most difficult

---

[40] Stillman Drake, "Galilei, Galileo," in *Dictionary of Scientific Biography*, vol. 5 (1972), 245.

[41] *Opticks or a Treatise of the Reflections, Refractions, Inflections & Colours of Light, Sir Isaac Newton*. Based on the fourth edition, London 1730; with a Foreword by Albert Einstein; and Introduction by Sir Edmund Whittaker; a Preface by I. Bernard Cohen; and an analytical Table of Contents prepared by Duane H. D. Roller (New York: Dover Publications, 1952), 1.

conditions: they had to use hand-copied manuscripts frequently riddled with textual errors that often were quite different from one manuscript version to another. Diagrams and illustrations often were omitted or filled with errors made by copyists who had little idea of their meaning or purpose. Not only did the printing press make scientific texts uniform and of very readable quality, but they undoubtedly encouraged publication of texts that might otherwise have been extant in only a few variant manuscript copies that would be little known, or wholly unknown, and largely unread.

The Forms of Literature in the New Natural Philosophy

The printed books of natural philosophy in the seventeenth century were produced in very different formats than had been followed in the late Middle Ages. In the seventh and eighth chapters of this volume, I described the manner in which medieval natural philosophers composed their works. With the exception of certain thematic treatises, medieval natural philosophy was largely composed of a few types of commentaries on the natural books of Aristotle. They were either straightforward commentaries on one section after another of a given Aristotelian treatise, or they were sequences of questions on the substantive content of this or that treatise selected from Aristotle's works on natural philosophy (for an illustration of the questions format, see the sample question by John Buridan in Chapter 8). These handwritten and hand-copied commentaries and questions formed the basic texts in courses on natural philosophy in the arts faculties of medieval universities. Even with the advent of printing, many printed books were commentaries and questions on Aristotle's works. Among scholastic natural philosophers, this state of affairs continued into the seventeenth century.

The new natural philosophy that emerged in the sixteenth and seventeenth centuries differed radically from its scholastic predecessors and contemporaries. The idea of commenting on authoritative texts, such as Aristotle's, was abandoned. Although there were some revered texts in magic and occult endeavors, the days of Aristotle's all-encompassing intellectual authority in natural philosophy and metaphysics were, with the exception of a dwindling number of scholastic Aristotelians, at an end. And with the fall of Aristotle and his authoritative texts, the commentary and questions formats were abandoned. No new authority, or authorities, emerged to replace Aristotle and encourage a continuation of the commentary tradition. The format that came to prevail was what we may call the "tractate" or treatise, where an author pursued a theme or topic, or even a series of topics, and presented the subject matter in a form deemed most suitable for that particular work. During the late Middle Ages, a number of scholastic authors had written treatises on specific themes (see Chapter 8). One of them, Henry of Langenstein (1325–1397), composed more works in the treatise format than in the scholastic format.[42] By the seventeenth century, the growing number of non-scholastic authors – most of whom were no longer interested in commenting on the works of Aristotle but were much more interested in denouncing and rejecting his authority – abandoned the commentary tradition of doing natural philosophy and, instead, wrote treatises on a great variety of topics.

Scientific Societies

It was not, however, just a matter of a changed literary format that dramatically altered natural philosophy. It was the manner in which some of the most important products of that literature were publicized and recognized. A new scientific age dawned with the emergence of scientific societies and scientific journals, both of which were unheard of in the Middle Ages. Because the latter were the by-products of the former, I shall first briefly describe scientific societies.

The closest approximation to a scientific society in the late Middle Ages was the faculty of arts in almost any of the numerous universities of the period, especially the universities of Paris and Oxford. The mission of a medieval faculty of arts was to teach the three philosophies: natural, moral, and metaphysical.[43] Natural philosophy encompassed Aristotle's natural philosophy along with logic, as well as advancements beyond, and departures from, that natural philosophy. Each faculty of arts was bound together by the common interests of its members, who were almost wholly devoted to teaching. Faculties of arts in the Middle Ages were teaching organizations, not research institutions, although a number of significant contributions to science were made by these scholars. Indeed, theologians in the faculties of theology made a number of major contributions to science, perhaps more than did the members of the arts faculties. But these were incidental to their teaching duties. Research and experimentation were not integral aspects of medieval science and natural philosophy. This had to await the sixteenth and seventeenth centuries.

The scientific society was one of the major developments that emerged from the greater role science came to play in society, a role that Bacon had envisioned in his *New Atlantis*. The earliest organization that may qualify as a scientific society is the *Academy of the Lynxes (Accademia dei Lincei)*, an Italian society. Most of its members, some thirty-two, which included Galileo, worked independently but presented their findings at scheduled meetings. The *Academy of the Lynxes* lasted for the first thirty years of the

---

[42] See Nicholas H. Steneck, *Science and Creation in the Middle Ages: Henry of Langenstein (d. 1397) on Genesis* (Notre Dame, IN: University of Notre Dame Press, 1976), 16.

[43] See Gordon Leff, "The *Trivium* and the Three Philosophies," in H. de Ridder-Symoens, ed., *A History of the University in Europe*, Vol. 1: *Universities in the Middle Ages*, 308.

seventeenth century,[44] and was superseded by the Florentine *Accademia del Cimento (Academy of Experiment)* in 1657. A few years later, in 1660, the *Royal Society of London* was established, and in 1666, the *Académie Royale des Sciences* was founded in Paris. These new societies had as their major objective the production of new scientific knowledge rather than the preservation and elaboration of old knowledge, by which was meant Aristotelian scholastic natural philosophy. The new knowledge would flow from the collective effort of all the members of each society, in a manner similar to Francis Bacon's description in the New Atlantis.[45]

Scientific Journals

A momentous by-product of the new scientific societies was the scientific journal. The first recognizable scientific journal was the *Philosophical Transactions* established in 1665 by Henry Oldenburg, secretary of the *Royal Society of London*. In the same year, the *Académie Royale* began publication of the *Journal des sçavans*. Although the latter was actually begun a few months before the *Philosophical Transactions*, the latter is usually regarded as the first scientific journal. Italian and German scientific journals also were published in the seventeenth century. These journals, and others that came later, published the results of experiments and observations made by their members, who sometimes worked collaboratively. By printing and distributing many copies of each issue, scientific societies were able to quickly disseminate new scientific knowledge throughout Europe, thus contributing immeasurably to the advance of science.

Natural Magic

In widening the scope of the causes and movements that contributed to the Scientific Revolution – or the "transformation of science" in the sixteenth and seventeenth centuries, for those who find the expression "Scientific Revolution" inappropriate or misleading – historians of science have now included two significant features that were largely ignored in the early days of research into the Scientific Revolution: natural magic and religion, with particular emphasis on natural theology.

In the final segment of Chapter 7, I described the way medieval scholastic natural philosophers regarded natural magic, emphasizing the views of St. Thomas Aquinas. Put simply, Thomas insists that any powers that a body has derives from the elements that compose it. Thomas recognizes, however, that

---

[44] See A. R. Hall, *The Scientific Revolution, 1500–1800: The Formation of the Modern Scientific Attitude* (London: Longmans, Green and Co., 1954), 188.
[45] For a fine, brief discussion of scientific societies in the seventeenth century, see Shapin, *The Scientific Revolution*, 131–135.

some effects "cannot be caused by the powers of the elements: for example, the magnet attracts iron, and certain medicines purge particular humors, in definite parts of the body. Actions of this sort, therefore, must be traced to higher principles" (see Chapter 7, n. 106), by which he meant celestial bodies or separated substances, such as God, intelligences, angels, or even demons. Of these "higher principles," only the celestial bodies produced occult effects that are natural; all the others – God, intelligences, angels, demons – produced occult effects supernaturally. Occult qualities were not only uncaused by the elements in the body exhibiting occult powers but, unlike such qualities as colors, shapes, and odors, they also were imperceptible. Although we could observe the magnet attracting iron, we could not perceive the magnet's causal action. Scholastic natural philosophers did not regard it as in any sense an obligation or duty to determine what specific causes enabled a magnet to attract iron or what causative agent, or agents, might be operative in the purging actions of certain medicines. They were content to invoke the broad, general causes just mentioned. Consequently, natural magic played a relatively inconsequential role in medieval natural philosophy.

In the sixteenth and seventeenth centuries, natural magic came to play a significant role. What are we to understand by "natural magic"? It obviously concerns occult qualities. But what is an "occult quality"? Is it an effect whose cause cannot be attributed to the properties of one or more of the four elements that compose all bodies, as Thomas Aquinas believed? In time, many qualities, if not most, came to be viewed as "occult," because their causes were imperceptible – that is, "occult." Early modern science diverged from medieval natural philosophy by virtue of the fact that many natural philosophers of that period came to view science "as a hunt for nature's secrets," as William Eamon has expressed it. To compel nature to divulge its "secrets," natural philosophers had to employ instruments and experiments.[46]

This attitude toward nature was by no means universal, but it was an approach that was widely adopted. Its proponents sought to discover the hidden mechanisms that produced the external properties of things, a strategy that marked a significant departure from the medieval approach to nature. The best means of discovering the occult causes of all sorts of qualities was to devise experiments and, if possible, to use scientific instruments, such as the microscope or telescope. Natural philosophers such as Robert Boyle (1627–1691) sought to use the mechanical philosophy to explain occult causes. Because phenomena were caused by mechanical means, mechanical explanations arrived at on the basis of experiments were sought for occult phenomena. Indeed, Isaac Newton's universal theory of gravitation was regarded as

---

[46] William Eamon, "Magic and the Occult," in Gary B. Ferngren, ed., *The History of Science and Religion in the Western Tradition: An Encyclopedia* (New York: Garland Publishing, Inc., 2000), 538.

an occult phenomenon, which Newton himself could not explain.[47] Natural magic, however, included much more than the quest for scientific knowledge of occult causes and effects. It included astrology as a major component and much else that ranged far beyond science and natural philosophy. This is not surprising. Natural magic was embedded in a long tradition of magical thought that traced back to the Greek Hermetic literature of the second century BC. This was a collection of different treatises on a variety of subjects, including "alchemy, astrology, astronomy, physics, embryology, botany, medicine, and numerous topics of a purely magical, mystical, or religious character. . . . "[48] Although the Hermetic treatises were written by different authors at different times, they were linked by their common attribution to the god Hermes Trismegistus ("thrice great Hermes"). These treatises influenced Renaissance thought after Marsilio Ficino translated the first fourteen Hermetic works from Greek to Latin in 1471. Because it included astrology and other mystical elements, natural magic was much more than a search for occult natural causes. Scholars have given radically different and opposing interpretations of the role played by the occult qualities of natural magic in the Scientific Revolution. Ron Millen insists that "by the end of the seventeenth century, occult qualities not only had been incorporated into modern science; they had become the foundation"[49] and "that we take them for granted represents one of the triumphs of the scientific revolution."[50] By contrast, Brian P. Copenhaver declares that "the decline of natural magic as a normal and legitimate concern of Western natural philosophy . . . was one of the most important features of the Scientific Revolution."[51] As far as the fortunes of natural philosophy are concerned, most scholars would agree that that aspect of natural magic that encouraged investigators to seek the hidden causes of a great variety of natural effects and properties was a positive benefit and marked a significant departure from the late Middle Ages. Many other aspects of natural magic, especially astrology and other mystical pursuits, were, however, detrimental to the advance of science and, in that sense, we ought to agree with Copenhaver that "the decline of natural magic as a normal and legitimate concern of Western natural philosophy . . . was one of the most important features of the Scientific Revolution." And, finally, one should keep in mind the fact that although numerous scholars of the sixteenth and seventeenth centuries wrote favorably about the need to seek the causes of occult phenomena, relatively few of them made significant contributions

---

47 See ibid., 539.
48 Marshall Clagett, *Greek Science in Antiquity* (London: Abelard-Schuman Ltd, 1957), 120.
49 See Ron Millen, "The Manifestation of Occult Qualities in the Scientific Revolution," in Margaret J. Osler and Paul Lawrence Farber, eds. *Religion, Science, and Worldview: Essays in Honor of Richard S. Westfall* (Cambridge: Cambridge University Press, 1985), 215.
50 Ibid., 216.
51 Brian P. Copenhaver, "Natural Magic, Hermetism, and Occultism in Early Modern Science," in David C. Lindberg and Robert S. Westman, eds., *Reappraisals of the Scientific Revolution* (Cambridge: Cambridge University Press, 1990), 290.

to science, as, for example, Marsilio Fiicino, Giambattista Della Porta, and Giovanni Pico della Mirandola (1463–1494). One who did, Isaac Newton, wrote disapprovingly of occult causes in his *Opticks*, where in book 3, part 1, he accuses the Aristotelians of giving the

Name of occult Qualities, not to manifest Qualities, but to such Qualities only as they supposed to lie hid in Bodies, and to be the unknown Causes of manifest Effects: Such as would be the Causes of Gravity, and of magnetick and electrick Attractions, and of Fermentations, if we should suppose that these Forces or Actions arose from Qualities unknown to us, and uncapable of being discovered and made manifest. Such occult Qualities put a stop to the Improvement of natural Philosophy, and therefore of late Years have been rejected. To tell us that every Species of things is endow'd with an occult specifick Quality by which it acts and produces manifest Effects, is to tell us nothing.[52]

I have thus far indicated numerous departures of early modern natural philosophy and science from the approach to those disciplines in the late Middle Ages. Although a few of these departures had occurred in an embryonic form in the late Middle Ages – for example, the clock analogy, printing, and an occasional experiment – almost all of the dramatic changes cited earlier in this chapter were wholly new, or received their mature development during the sixteenth and seventeenth centuries – in the period between Copernicus and Newton, encompassing what is often described as the "Scientific Revolution." These dramatic changes included the dissemination of the new scientific literature by the printing press; the mechanical philosophy; experiments and discoveries made with the use of new scientific instruments; scientific societies; scientific journals; and the use of natural magic to identify occult causes.

Although there are other topics that affected changes in natural philosophy from the Middle Ages to the seventeenth century, the final theme with which I shall deal is the relationship between science and religion – or natural philosophy and theology – a theme too important to ignore. Despite some similarities, the relations between science and religion were considerably transformed as we move from the late Middle Ages to the early modern period. Indeed, the Christian religion itself had been dramatically altered with the advent of the Protestant Reformation in the sixteenth century. New attitudes emerged that had no genuine counterparts in the late Middle Ages. In short, the world of seventeenth-century Europe differed considerably from that of late medieval Europe.

Natural Philosophy and Religion

In the preceding chapter, I described the relations between the disciplines of natural philosophy and theology – or science and religion – during the late

---

[52] *Sir Isaac Newton "Opticks."* Based on the Fourth Edition, London, 1730, 401.

Middle Ages. I argued that although natural philosophy greatly influenced the doing of theology, the latter had relatively little impact on natural philosophy. The interrelations between science and religion – or, better, theology and natural philosophy – in any age can only be properly understood if we distinguish between what the authors of the period explicitly regarded as the true relationship between religion and natural philosophy, and what they did about it – that is, did what they actually wrote in their scientific treatises accurately reflect their ideas about the relations of religion and science or natural philosophy? Although we might try to find individuals who not only expressed their views and opinions about the relationship of science and religion but also wrote one or more substantive treatises on some theme or themes in natural philosophy, I shall instead describe the general attitude toward the interrelationship of science and religion in the seventeenth century, and then examine some treatises on natural philosophy to see if those attitudes and opinions are manifested in substantive works of the period.

The interrelations between science and religion during the Middle Ages were fairly straightforward in comparison to the situation that would obtain in the seventeenth century, when a variety of interpretations and attitudes appeared that posed problems but which seemed, for the most part, to co-exist. The appearance of a variety of opinions was due, in no small measure, to the fact that natural philosophers were no longer focused on the works of Aristotle. They were now free to use the literary sources of antiquity, where these were found useful, or to ignore them. In truth, they were more likely to ignore ancient authorities, because they were no longer regarded as "authoritative." Their intellectual horizons had expanded enormously by comparison to the late Middle Ages. But one old problem remained that would receive new interpretations in the early modern period: how should scientists and natural philosophers relate nature, as described in natural philosophy, and Scripture, as presented in the Bible? How to relate "God's Two Books": the Book of Nature and the Book of Scripture?[53] We saw earlier that, during the Middle Ages, natural philosophy was applied extensively to theological problems in theological commentaries (*Sentence Commentaries*), but, in sharp contrast, theology found little application in treatises on natural philosophy, which were largely commentaries on Aristotle's natural books. Medieval natural philosophers did not assert that their objective in studying the Book of Nature (an expression they did not use) was to better understand God's handiwork. Rather, they studied natural philosophy to understand a world that God created, but which operated according to Aristotelian principles and laws. Medieval theologians found that natural philosophy not only served to reveal the operations of an Aristotelian world, but they found it also was immeasurably helpful to resolve theological problems of all

---

[53] "God's Two Books" is drawn from the title of Kenneth J. Howell's book *God's Two Books: Copernican Cosmology and Biblical Interpretation in Early Modern Science* (Notre Dame, IN: University of Notre Dame Press, 2002).

kinds: from the articles of faith to innumerable problems about what God could and could not do. In the sixteenth and seventeenth centuries, all this changed, as the relations between science and religion were significantly altered.

In the seventeenth century, two antagonistic attitudes lived side-by-side with most scientists, or writers about science, supporting one or the other. There were those who wished to keep science – which was concerned with the Book of Nature – separate from the Book of Scripture. Kepler, Galileo, and Newton were in this group.[54] Numerous seventeenth-century scientists were interested only in the Book of Nature, and left the Book of Scripture to theologians and to those natural philosophers for whom religion played a major role. Scientists and natural philosophers were primarily interested in investigating nature by means of observation, experiments, instruments, and reason. Those who had mathematical ability were also interested in applying mathematics to the behavior of physical bodies in nature. Galileo spoke for this group in his *Il saggiatore (The Assayer)* (1623), in which he uttered a profound statement about what came to be called the "Book of Nature." "Philosophy is written in this grand book, the universe," Galileo explains,

which stands continually open to our gaze. But the book cannot be understood unless one first learns to comprehend the language and read the letters in which it is composed. It is written in the language of mathematics, and its characters are triangles, circles, and other geometric figures without which it is humanly impossible to understand a single word of it; without these, one wanders about in a dark labyrinth.[55]

For Galileo, the Book of Nature can only be read by the application of mathematics. What immediately catches our eye is the fact that he assigns no role to the Book of Scripture, or religion. Others in the seventeenth century explicitly linked religion and philosophy (that is, natural philosophy or science) as the essential instruments for dealing with the two books – the Book of God and the Book of Nature (God's handiwork) – but generally thought they should be studied separately. In the *Advancement of Learning*, at the conclusion of a section labeled "[Defence of Learning Against Divines]," Francis Bacon declared:

To conclude therefore, let no man upon a weak conceit of sobriety or an ill-applied moderation think or maintain, that a man can search too far, or be too well studied in the book of God's word, or in the book of God's works, divinity or philosophy; but rather let men endeavour an endless progress or proficience in both; only let men beware that they apply both to charity, and not to swelling; to use, and not to

---

[54] For a brief discussion, see Frank E. Manuel, *The Religion of Isaac Newton: The Fremantle Lectures 1973* (Oxford: Clarendon Press, 1974), 27–28.

[55] *Discoveries and Opinions of Galileo, Including: The Starry Messenger (1610); Letters on Sunspots (1613); Letter to the Grand Duchess Christina (1615); And Excerpts from The Assayer (1623)*, translated with an Introduction and Notes by Stillman Drake (New York: Doubleday Anchor Books, 1957), 237–238.

ostentation; and again, that they do not unwisely mingle or confound these learnings together.[56]

Bacon viewed both books as equally important, but advised his readers not to "mingle or confound these learnings together." He elaborated on this theme in the New Organon (book 1, aphorism 65). After denouncing the corrupting influence on philosophy of the ancient Greek Empirical school (in aphorism 64), Bacon goes on, in aphorism 65, to mention even greater dangers to philosophy when he explains that:

the corruption of philosophy by superstition and an admixture of theology is far more widely spread, and does the greatest harm, whether to entire systems or to their parts.... It shows itself likewise in parts of other philosophies, in the introduction of abstract forms and final causes and first causes, with the omission in most cases of causes intermediate and the like. Upon this point the greatest caution should be used. For nothing is so mischievous as the apotheosis of error; and it is a very plague of the understanding for vanity to become the object of veneration. Yet in this vanity some moderns have with extreme levity indulged so far as to attempt to found a system of natural philosophy on the first chapter of Genesis, on the book of Job, and other parts of the sacred writings; seeking for the dead among the living, which also makes the inhibition and repression of it the more important because from this unwholesome mixture of things human and divine there arises not only a fantastic philosophy but also a heretical religion. Very meet it is, therefore, that we be sober minded, and give to faith that only which is faith's.[57]

Although Bacon regarded the study of theology to be as important as the study of natural philosophy, he was emphatic in his conviction that the two should not be intermingled, because the result would be not only "a fantastic philosophy, but also a heretical religion." The sense that religion and science should be kept separate was widespread, especially in England. This is especially apparent in the Royal Society of London, where "no one ever presented a case for a scientific fact with a theological argument."[58] In his autobiography of 1697, John Wallis (1616–1703), the English mathematician and theologian, recounts meetings, held in London around 1645, by a group of scholarly individuals who were "inquisitive into Natural Philosophy and other parts of Humane Learning; and particularly of what hath been called the New Philosophy or Experimental Philosophy." Wallis explains that at these meetings "Our business was (precluding matters of theology and State Affairs) to discours and consider of Philosophical Enquiries, and such as related thereunto; as Physick, Anatomy, Geometry, Astronomy, Navigation, Staticks, Magneticks, Chymicks, Mechanicks, and Natural Experiments;

---

[56] Francis Bacon: The Advancement of Learning and New Atlantis, edited by Arthur Johnston (Oxford: Clarendon Press, 1974), 9–10.

[57] See Francis Bacon: Selected Philosophical Works, ed. Rose-Mary Sargent, 106.

[58] Manuel, The Religion of Isaac Newton, 30.

with the State of these Studies, as then cultivated, at home and abroad."[59] Thus, theology was excluded from these London meetings, even though some of the participants also were theologians. Indeed, it was not uncommon in seventeenth-century England for scholars to be both natural philosophers and theologians. Nevertheless, "they were able to follow the Baconian advice in all its parts; they studied both books diligently while making a show of keeping their inquiries separate, and seemed to don different caps for each of their occupations."[60] Although Isaac Newton was not a theologian by profession, his intense interest in theology is well known. Nevertheless, "when Newton was President of the [Royal] Society, the journal-books record, he banned anything remotely touching on religion, even apologetics."[61]

But the caution against intermingling the two did not mean that one should not see the hand of God in nature and in new scientific discoveries. It is in this sense that John Hedley Brooke has urged scholars to avoid the term "separation" with respect to science and religion in the scientific revolution and to recognize that although "a change in the status of natural philosophy was certainly achieved," it was accomplished "by differentiation from, and reintegration with, religious belief rather than by complete severance."[62]

The religious association or "reintegration," to use Brooke's term, was largely by way of Natural Theology, which was "the practice of establishing the existence and attributes of God from the evidence of nature."[63] The primary evidence for the existence of God was the "argument from design," which assumed that the cosmos functioned like a mechanical clock, and therefore could not have been produced by the mindless, undirected random motion of matter. The clocklike operations of the world strongly implied a divine clockmaker, who was none other than the creator God of Scripture. The intricate and harmonious design of our world led most seventeenth-century natural philosophers and scientists to assume its creation by a divine mind, the mind of God. Robert Boyle employed the clock metaphor to argue for design in the universe, when he declared:

When I see in a curious clock, how orderly every wheel and other parts perform its own motions, and with what seeming unanimity they conspire to show the hour, and accomplish the other designs of the artificer; I do not imagine, that any of the wheels, &c. or the engine it self is endowed with reason, but commend that of the workman, who framed it so artificially. So when I contemplate the actions of those several creatures, that make up the world, I do not conclude the inanimate pieces, at

---

[59] Cited by Christoph J. Scriba, "Wallis, John," in *Dictionary of Scientific Biography*, vol. 14 (1976), 151.

[60] Manuel, *The Religion of Isaac Newton*, 30.   [61] Ibid. I have added the bracketed word.

[62] John Hedley Brooke, *Science and Religion: Some Historical Perspectives* (Cambridge: Cambridge University Press, 1991), 58.

[63] Shapin, *The Scientific Revolution*, 142, n. 5.

least, that it is made up of, or the vast engine it self, to act with reason or design, but admire and praise the most wise Author, who by his admirable contrivance can so regularly produce effects, to which so great a number of successive and conspiring causes are required....[64]

Boyle, and others, extended the design argument to the human body. The "circular motion of the blood, and structure of the valves of the heart and veins... though now modern experiments have for the main... convinced us of them, we acknowledge them to be very expedient, and can admire God's wisdom in contriving them."[65]

For many seventeenth-century natural philosophers, religion seems to have served as a powerful incentive to investigate the workings of the divinely created universe. This is undoubtedly true for the scientist-theologians and for the likes of Robert Boyle and Isaac Newton, who, although not professional theologians, were strongly disposed toward religion. They never tired of emphasizing nature as God's creation and design. But, as Boyle reveals, they also defended the study of natural philosophy against theologians, or divines, who, "out of a holy jealousy (as they think) for religion, labour to deter men from addicting themselves to serious and thorough inquiries into nature, as from a study unsafe for a Christian, and likely to end in atheism, by making it possible for men... to give themselves such an account of all the wonders of nature, by the single knowledge of second causes, as may bring them to disbelieve the necessity of a first."[66] Thus, Francis Bacon, Boyle, and many others in the seventeenth century thought it important to keep religion and science apart, even as they sang the praises of God and His creation. Although we have no reason to believe that for many, if not most, praise of God and His creation in a natural philosophical work was genuine, we also must keep in mind the fact that some introduced God into their treatises to avoid charges of atheism, which usually did not mean a denial of God's existence, but, rather, largely ignoring, or downplaying, God's role and according too much power to material and mechanical causes in nature, as Isaac Newton is alleged to have done in the famous General Scholium to his *Mathematical Principles of Natural Philosophy* (see later). To illustrate the manner in which natural philosophers interrelated science and religion, or natural philosophy and theology, we can do no better than describe briefly the approaches and attitudes of the two greatest English scientists of the seventeenth century: Robert Boyle and Isaac Newton.

---

[64] From Robert Boyle, *Of the Usefulness of Experimental Natural Philosophy*, Part I: "Of its Usefulness in reference to the Mind of Man," cited in Marie Boas Hall, ed., *Robert Boyle on Natural Philosophy: An Essay with Selections from His Writings* (Bloomington: Indiana University Press, 1965), 147.

[65] Ibid., 144.   [66] Ibid., 141–142.

Robert Boyle (1627–1691)

As the modern fourteen-volume edition of his writings bears witness,[67] Robert Boyle was one of the most prolific scholars among those who combined a deep interest in science and religion in late-seventeenth-century England. He wrote much on natural philosophy, religion, and theology. In certain important ways, Boyle resembles a medieval theologian-natural philosopher. To assist in comprehending this claim, I cite a passage from Boyle's *The Excellency and Grounds of the Corpuscular or Mechanical Philosophy* (1674). In this passage, Boyle explains that:

> when I speak of the corpuscular or mechanical philosophy, I am far from meaning with the Epicureans, that atoms, meeting together by chance in an infinite vacuum, are able of themselves to produce the world, and all its phaenomena; nor with some modern philosophers, that, supposing God to have put into the whole mass of matter such an invariable quantity of motion, he needed do no more to make the world, the material parts being able by their own unguided motions to cast themselves into such a system (as we call by that name): But I plead only for such a philosophy, as reaches but to things purely corporeal, and distinguishing between the first original of things, and the subsequent course of nature, teaches, concerning the former, not only that God gave motion to matter, but that in the beginning he so guided the various motions of the parts of it, as to contrive them into the world he designed they should compose, (furnished with the seminal principles and structures, or models of living creatures) and established those rules of motion and that order amongst things corporeal, which we are wont to call the laws of nature. And having told this as to the former, it may be allowed as to the latter to teach, that the universe being once framed by God, and the laws of motion being settled and all upheld by his incessant concourse and general providence, the phaenomena of the world thus constituted are physically produced by the mechanical affections of the parts of matter, and what they operate upon one another according to mechanical laws.[68]

Just as his medieval scholastic predecessors would have done, Boyle rejects the kind of world in which atoms move about randomly in an infinite vacuum and form a world by "unguided motions." He advocates a world that God originally created and to which He not only gave original motion but also conferred on it "laws of nature." Once this was accomplished, and the laws of nature became operative, "the phaenomena of the world thus constituted are physically produced by the mechanical affections of the parts of matter, and what they operate upon one another according to mechanical laws." In this passage, Boyle has described the widely adopted mechanical philosophy, which assumed a created world that operates by causal necessity.

Boyle's natural philosophy is not about God. If we have any doubts about this, we should ponder Boyle's concluding section of *The Excellency and*

---

[67] See *The Works of Robert Boyle*, edited by Michael Hunter and Edward B. Davis, 14 vols. (London: Pickering & Chatto, 1999–2000).

[68] See Marie Boas Hall, *Robert Boyle on Natural Philosophy*, 189.

Grounds of the Corpuscular or Mechanical Philosophy. In the closing few pages of his treatise, which he titles "A Recapitulation" and in which he seeks to convince his readers of the validity of the mechanical philosophy, Boyle declares:

having first premised once for all, that presupposing the creation and general providence of God, I pretend to treat but of things corporeal, and do abstract in this paper from immaterial Beings (which otherwise I very willingly admit), and all agents and operations miraculous or supernatural.[69]

After pronouncing, as was customarily done, that God created the world, Boyle then declares quite clearly that he excludes from his consideration of the mechanical philosophy all matters that involve "miraculous or supernatural" agents as explanations of natural phenomena. Boyle has made it quite apparent that his natural philosophy is not about God. Nor did he inject God into his scientific works, such as *New Experiments Physico-Mechanicall, Touching the Spring of the Air, and its Effects (Made for the most part, in a New Pneumatical Engine* [1660]) and his numerous experiments on heat, cold, tastes, odors, magnetism, and electricity.[70]

Isaac Newton (1642–1727)

The other great figure of late-seventeenth-century natural philosophy in England was Sir Isaac Newton. Like Boyle, Newton was eminent in science but was equally, if not more, interested in religion; however, unlike Boyle, who published much on theology as well as science or natural philosophy, Newton said little about God and religion in published works, but had much to say about religion in his unpublished papers. Frank Manuel claims that "Newton's scrutiny of nature was directed almost exclusively to the knowledge of God and not to the increase of sensate pleasure or comfort. Science was pursued for what it could teach men about God, not for easement or commodiousness."[71] Does this signify that for Newton natural philosophy was about God? The evidence does not seem to support such an interpretation.

In 1687, Newton published the first edition of *The Mathematical Principles of Natural Philosophy (Philosophiae Naturalis Principia Mathematica)*, which established the laws of physics that remained dominant until the twentieth century.[72] In the first edition, which consists of 510 pages of text, Newton mentions God only once (in book 3), which makes a

---

[69] Ibid., 207–208.
[70] The treatises on all these experiments are in *The Works of Robert Boyle*, vol. 8 (2000).
[71] Manuel, *The Religion of Isaac Newton*, 48.
[72] The second edition of the *Principia* appeared in 1713; the third, and final, edition, in 1726. In what follows about Newton, I draw on my article (cited earlier), "God and Natural Philosophy: The Late Middle Ages and Sir Isaac Newton," 289–295.

recent claim that Newton's "natural philosophy was obsessed with God"[73] seem rather odd. It is a strange obsession indeed, where the obsessed displays virtually no symptoms of his obsession. Even this single mention was deleted in the later editions. In place of this isolated mention, Newton added his famous General Scholium to the end of the second edition (1713).

In the second edition of 1713, a work of 484 pages,[74] Newton saw fit to discourse on God in a separate section, which he titled the General Scholium, and which occupies the last four pages of the book. Here Newton praises the deity as the Universal Ruler and Supreme God and describes some of His attributes. Before turning to a discussion of gravity to conclude the General Scholium, Newton terminates his tribute to God with these words: "And thus much concerning God; to discourse of whom from the appearances of things, does certainly belong to Natural Philosophy."[75] During the Middle Ages, theology and natural philosophy were separate disciplines and discourses about God lay in the domain of theology and were excluded from natural philosophy. By the seventeenth century, however, the Protestant Reformation had obliterated the jurisdictional boundaries between theology and natural philosophy. When Newton wrote, it was thought wholly appropriate for natural philosophers to discourse about God "from the appearances of things." After all, that is what natural theology was all about: to discourse about God "from the appearances of things."

Despite the acknowledged equality of the two traditional disciplines in Newton's day, Newton found few occasions to discourse about God. Other than singing the praises of the deity, Newton found little to say about God in his scientific treatises. Indeed, even the General Scholium was added to the second edition of the *Principia*, some twenty-six years after the first edition, because Newton wished to fend off criticisms that his reliance on attractions and repulsions in his physics made his system appear too mechanical, and therefore much like the natural philosophy of Descartes.[76] In the first editions of his two greatest scientific works – *The Mathematical Principles of Natural Philosophy* and the *Opticks*, the first edition of which was published

73 Made by Andrew Cunningham, "How the *Principia* Got Its Name; Or, Taking Natural Philosophy Seriously," *History of Science* 29 (1991), 381.

74 The text of the third edition of 1726 is 530 pages.

75 The translation is from *Sir Isaac Newton's Mathematical Principles of Natural Philosophy and His System of the World*. Translated into English by Andrew Motte in 1729. The translations revised, and supplied with an historical and explanatory appendix, by Florian Cajori, 546. The *General Scholium* extends over pages 543–547.

76 Westfall explains that Newton wrote the *General Scholium* because "Newton and Newtonians were highly aware of the mounting tide of criticism of his natural philosophy and its concepts of attractions and repulsions. . . . " Richard S. Westfall, *Never at Rest: A Biography of Isaac Newton* (Cambridge: Cambridge University Press, 1980), 744.

in 1704 – Newton said virtually nothing about God. This left him vulnerable to criticisms that he had excluded God from the operations of the cosmos. It was probably to avoid charges of atheism and to make his natural philosophy seem less mechanical, that he added the General Scholium to the *Principia* in 1713, and added brief mentions of God in later editions of the *Opticks*. The desire to avoid possible accusations of atheism may have operated as a powerful incentive in the seventeenth and early eighteenth centuries for some natural philosophers to insert praiseworthy comments about God in their treatises. Nevertheless, most treatises on experimental and natural philosophy in this period were free of such remarks.

In light of Newton's assertion about God and natural philosophy in the General Scholium, that document raises a puzzling problem. Newton explains that only God could have produced the cosmos, because "Blind metaphysical necessity, which is certainly the same always and everywhere, could produce no variety of things."[77] Following his worshipful tribute to God, Newton, in the final two paragraph of his great work, admits that he has not yet discovered the cause of gravity. He rests content, therefore, to assert "that gravity does really exist, and act according to the laws which we have explained, and abundantly serves to account for all the motions of the celestial bodies, and of our sea."[78] This is a baffling statement. Why did Newton not attribute the cause of gravity to God? Such an attribution would almost seem to follow from his immediately preceding three-page discourse on God's power and attributes, as his brilliant biographer, Richard S. Westfall, recognized.[79] Newton's likely reason for not attributing the cause of gravity to God is that he was fully cognizant of the fact that such an attribution would have been to no avail. It would have explained nothing. Because Newton firmly believed that God had created our world and all its operations, he would have found it untenable to invoke God as an explanation for the cause of any particular effect. He would have assumed instead that God had provided a natural cause for every effect – including the effects of the force of gravity – and that it was the task of the natural philosopher to discover it. Medieval natural philosophers and theologian-natural philosophers had always assumed this as a basic truth about natural philosophy. During the late Middle Ages, scholastic natural philosophers believed that one learned about God from theology, not natural philosophy. From time to time, medieval popes ordered theologians to avoid introducing natural philosophy into theology, all to no avail. Indeed, it explains why, from the Middle Ages to the seventeenth-century, natural philosophy remained relatively free of theological encroachments.

---

[77] Florian Cajori's translation from the *General Scholium*, 546.
[78] Ibid., 547.   [79] Westfall, *Never at Rest*, 748.

THE RELATIONS BETWEEN NATURAL PHILOSOPHY AND SCIENCE IN THE SEVENTEENTH AND NINETEENTH CENTURIES

Andrew Cunningham and Perry Williams have claimed that natural philosophy and science were two wholly different enterprises that never coexisted: natural philosophy appeared first in the ancient world and continued to exist for many centuries until it was wholly replaced by science, or modern science, in the nineteenth century.[80] Similarly, Floris Cohen has argued that "the emancipation of science from an overarching entity called 'natural philosophy' is one defining characteristic of the Scientific Revolution."[81] I shall argue in this chapter, and in this book, that these are serious misunderstandings of the history of science. The virtual opposite of these claims is the more accurate description. The Scientific Revolution occurred because, after coexisting independently for many centuries, the exact sciences of optics, mechanics, and especially astronomy merged with natural philosophy in the seventeenth century. This momentous occurrence broadened the previously all-too narrow scope of the ancient and medieval exact sciences (see Chapter 2), which now, by virtue of natural philosophy, would seek physical causes for all sorts of natural phenomena, rather than being confined to mere calculation and quantification. Thus were the seeds planted for the flowering of the modern version of the exact physical sciences, and the many other sciences that emerged during the eighteenth and nineteenth centuries. As the "Great Mother of the Sciences," natural philosophy nourished within itself a multiplicity of specialized sciences, such as physics, chemistry, biology, and their numerous subdivisions. By the end of the nineteenth century, many of these sciences had reached sufficient maturity and development to depart natural philosophy and become independent entities. Thus it was that the brood of the Great Mother of the Sciences left the nest to flourish and thrive autonomously ever since. I shall now describe that momentous course of events.

The Medieval Background

During the late Middle Ages, astronomy, geometric optics, and mechanics (known as "the science of weights") were regarded by most natural philosophers as distinct from natural philosophy. Evidence of this can be seen at a number of levels in *The Book of Jordanus de Nemore on the Theory of Weight* (*Jordani de Nemore Liber de Ratione Ponderis*), composed in

---

[80] See Andrew Cunningham and Perry Williams, "De-centring the 'Big Picture': The Origins of Modern Science and the Modern Origins of Science," in *The British Journal for the History of Science* 26 (1993), 407–432.

[81] H. Floris Cohen, *The Scientific Revolution: A Historiographical Inquiry* (Chicago: University of Chicago Press, 1994), 167.

the thirteenth century. Jordanus's *Theory of Weights* reveals a spare, no-nonsense text in four parts that begins immediately with seven postulates (*suppositiones*) and then moves from theorem to theorem through twenty-seven pages of Latin text. Following the seventeenth theorem of part 4, the treatise concludes with the words: "Here ends the fourth part, and with it there ends the book of Jordanus on the theory of weight."[82] Modern physicists who inspect the treatise would undoubtedly conclude that it is a scientific work in mechanics. It is certainly not a treatise in natural philosophy. Blasius of Parma (ca. 1345–1416), whose treatise on weights follows immediately after that of Jordanus was almost, but not quite, as spare as Jordanus. In the opening words of his treatise, Blasius declares:

I. THE SCIENCE OF WEIGHTS IS SAID IN TRUTH TO BE SUBORDINATE TO NATURAL PHILOSOPHY.
Proof: The science of weights considers movements in a particular way, while natural philosophy takes movements under common scrutiny. Therefore, the supposition is true.[83]

By subordinating the science of weights to natural philosophy, Blasius indicates that he distinguishes the former from the latter. They are both concerned with motion, but treatises on the science of weights deal with it in a "particular way" that differentiates the two. It seems that this is just another way of saying that the *Science of Weights* is a "middle science," lying between natural philosophy and mathematics, but more natural philosophy than mathematics. In contrast with Blasius of Parma, and long before him, Domingo Gundisalvo (fl. 1140), classified the science of weights as subordinate to mathematics, rather than natural philosophy (Gundisalvo actually uses the expression "natural science"). He explains that:

Mathematics is also a universal science, since it contains seven arts under it: arithmetic, geometry, music, astrology, the science of aspects, the science of weights, and the science of devices (ingenia).[84]

During the Middle Ages, optics (the science of aspects), astronomy (cited by Gundisalvo as "astrology"), and the science of weights could be classified under mathematics or natural philosophy, but they belonged to neither. The

---

[82] For the Latin text and English translation by Ernest A. Moody, see *The Medieval Science of Weights ("Scientia de Ponderibus"): Treatises Ascribed to Euclid, Archimedes, Thabit ibn Qurra, Jordanus de Nemore and Blasius of Parma*, edited with Introductions, English translations, and Notes, by Ernest A. Moody and Marshall Clagett (Madison: University of Wisconsin Press, 1952), 174–227.

[83] Part I, Supposition 1, of *The Treatise on Weights of Master Blasius of Parma (Tractatus de Ponderibus Magistri Blasii de Parma)* in ibid., 239.

[84] From Grant, *A Source Book in Medieval Science*, 66. Gundisalvo's treatise on the classification of the sciences was titled *On the Division of Philosophy (De divisione philosophiae)* and was heavily dependent on a relatively recent translation from Arabic to Latin of al-Fārābī's (d. 950/951) *De scientiis (On the Sciences)*.

classification depended on how an author viewed the relationship of a particular middle science to mathematics or natural philosophy: if the motion of the bodies involved in the middle science seemed more significant than its mathematical aspects, it would be linked more closely to natural philosophy; if its mathematical features appeared more significant, it would be classified as subordinate to mathematics. But, strictly speaking, the middle science in question was not regarded as either mathematics or natural philosophy.

Francis Bacon: Natural Philosophy Is "The Great Mother of the Sciences"

In the final chapter of an earlier book published in 1996, I included a section titled "Natural Philosophy: The Mother of All Sciences."[85] I argued that following the middle sciences (mentioned in the preceding paragraph), "almost all other sciences – physics, chemistry, biology, geology, meteorology, and psychology, as well as their subdivisions and branches – emerged as independent disciplines from within the matrix of natural philosophy during the seventeenth to nineteenth centuries."[86] To my great surprise, I discovered that 376 years earlier, Francis Bacon had already used a virtually identical phrase to link natural philosophy with the sciences of his day. In 1620, he published the *New Organon (Novum Organum)*, as part of *"the Great Instauration,"* which was to consist of six parts, although not all were published. The second part was devoted to the *New Organon*, which was divided into two books, each consisting of numerous aphorisms, the first containing 130, the second 52. In the seventy-eighth aphorism of the first book, Bacon declares that of the approximately twenty-five centuries of recorded history, scarcely six centuries saw productive efforts in the sciences. During these six centuries, "we can properly count only three revolutions: the first with the Greeks; the second with the Romans; and the third with us, the Western European nations: and to each of these barely a couple of centuries can be properly allotted."[87] In the seventy-ninth aphorism, Bacon argues that in the ages when letters flourished, little intellectual energy was directed toward natural philosophy, even though the latter "should be regarded as the great mother of the sciences (*pro magna Scientiarum matre*). For all arts and sciences, if wrested from this root, can perhaps be refined and adapted to use, but they will not grow at all."[88] Because attention was not focused on natural philosophy even during the relatively brief ages of the three revolutions,

---

85 See Edward Grant, *The Foundations of Modern Science in the Middle Ages* (Cambridge: Cambridge University Press, 1996), 192–197.

86 Ibid., 193.

87 *The Oxford Francis Bacon* in *The Instauratio magna Part II: Novum organum and Associated Texts*, edited with introduction, notes, commentary, and facing-page translations by Graham Rees and Maria Wakely (Oxford: Clarendon Press, 2004), 123–125.

88 Ibid., 125.

"natural philosophy was for the most part neglected or obstructed during these three periods." Bacon uses the eightieth aphorism to present some startling characteristics about his vision of natural philosophy, once again referring to it as the "great mother of the sciences."

And this great mother of the sciences (*magna ista Scientiarum Mater*) has, with wonderful indignity, been forced into the role of a servant, dancing attendance on the business of medicine and mathematics.... Meanwhile let no one hope for great progress in the sciences (*scientijs*) (especially in the operative department) unless natural philosophy be extended to the particular sciences, and these in their turn reduced to natural philosophy. For hence it comes about that astronomy, optics, music, many of the mechanical arts, and medicine itself, and (which may surprise you) moral and political philosophy, and the science of logic have practically no depth but skate over the surface and variety of things; because once these are dispersed and set up as particular sciences, they are no longer nourished by natural philosophy; which could have given them new strength and growth at source and from a true knowledge of motions, rays, sounds, textures and schematisms of bodies, affections, and intellectual apprehensions. Since, therefore, the sciences have been cut off from their roots, it is no wonder that they do not grow.[89]

Bacon envisioned natural philosophy as the essential agent for the growth of the particular sciences within its domain and under its guidance. Indeed, he even denied that mathematics was a separate science, regarding it as subordinate to natural philosophy. Consequently, he rejected the traditional medieval idea of "middle sciences," thus denying that astronomy, perspective (optics), music, and other traditional middle sciences, were independent of natural philosophy, and located between that discipline and mathematics.[90] All of these disciplines, including mathematics, formed part of natural philosophy; but what kind of natural philosophy? What was Bacon seeking in his vision of natural philosophy? The puzzling nature of his quest can be seen in the brief, but important, ninety-sixth aphorism:

We have yet to find a pure natural philosophy; so far it has been infected and corrupted: in Aristotle's school by logic; in Plato's by natural theology; in Plato's second school – that of Proclus and others – by mathematics, which ought to round off natural philosophy and not generate or procreate it. But from a natural philosophy pure and quite without admixture we should hope for better things.[91]

Whatever Bacon may have had in mind by a "pure natural philosophy," it seems to have required that the particular sciences, such as astronomy, optics, mechanics, medicine, logic, and even moral and political philosophy, be subsumed under natural philosophy. When these sciences are separated

[89] Ibid., 127. The Latin terms are drawn from the Latin text on p. 126.
[90] See Sachiko Kusukawa, "Bacon's Classification of Knowledge," in Markku Peltonen, ed., *The Cambridge Companion to Bacon* (Cambridge: Cambridge University Press, 1996), 47–74; 59.
[91] *The Oxford Francis Bacon*, vol. xi, 153–155.

from natural philosophy, they do not grow or develop, because they are cut off from the broad issues and problems that are characteristic of natural philosophy. They "function as if cut off from their roots," because without natural philosophy they focus solely on specific problems that are only resolvable by mathematics. They become, and remain, too narrow in scope. To make the particular exact sciences more useful, it is therefore essential to apply natural philosophy to the particular sciences, which could only be achieved if they are integrated into the domain of natural philosophy.

Bacon rejected what had previously passed for natural philosophy. His apparent objective was to construct a new natural philosophy that would embrace all the particular sciences, whatever those might be. The latter could only grow and develop as part of natural philosophy. In a real sense, we might argue that Francis Bacon anticipated what was to come in the nineteenth century: he regarded natural philosophy as if it were equivalent to the generic term "science," which Bacon did not use, but that came into vogue in the nineteenth century as a general term to encompass the particular sciences (see the section on Dickinson College later in this chapter) and that also functioned as a synonym for the term natural philosophy. Thus for Bacon, the particular sciences existed in the seventeenth century as parts of natural philosophy, just as they did for many in the nineteenth century. Therefore, the concept of science existed in the seventeenth century – at least for Francis Bacon – when it was represented by the equivalent term natural philosophy. But long before Francis Bacon, there were those who suggested that the mathematical sciences merge in some fashion with natural philosophy.

THE REVOLUTION IN NATURAL PHILOSOPHY FROM THE MIDDLE AGES TO THE NINETEENTH CENTURY

Francis Bacon gave voice to the most significant problem that confronted natural philosophy in its lengthy history from Aristotle to the nineteenth century: what was its proper relationship with mathematics and the exact sciences? As we have seen, Bacon was convinced that as "The Great Mother of the sciences," natural philosophy had to embody within itself all of the exact sciences, as well as a number of other disciplines. These disciplines could only flourish as integral parts of natural philosophy. All medieval and early modern natural philosophers would undoubtedly have realized that by integrating the exact sciences into natural philosophy, traditional natural philosophy – that is, Aristotelian natural philosophy – would undergo a radical transformation. By the end of the seventeenth century, the transformation of natural philosophy was manifested in Isaac Newton's *Mathematical Principles of Natural Philosophy*, the very title of which reveals that a union of mathematics and natural philosophy had already occurred. Natural philosophy was transformed in such important ways that, as we shall see, it played an essential role in the emergence of a scientific revolution in the seventeenth-century. But long before Francis Bacon's call for the integration of the sciences and natural philosophy, there had been stirrings against Aristotle's version of a natural philosophy divorced from mathematics and the exact sciences. Bacon-like suggestions for the transformation of natural philosophy already had a long history. Indeed, we can do no better than to present the views of his famous medieval namesake, Roger Bacon.

In the first book of his *Communia Naturalium*, Bacon declares: "there ought to be one science of natural philosophy which treats everything common to natural things and this will be the first among the natural sciences. But there will be other great natural sciences many of which have subalternate sciences, just as is obvious in metaphysics."[92] Bacon then declares that "besides the science common to natural things there are seven special sciences, namely Perspective; Astronomy, [both] judicial [i.e., astrology] and operative [i.e. astronomy proper]; The Science of Weights of heavy and light bodies; Alchemy; Agriculture; Medicine; [and] Experimental Science."[93] Bacon goes on to explain why each of these seven special sciences is part of natural philosophy. Indeed, he speaks of eight sciences by the first of which he seems to understand natural philosophy in the most general sense, to which he would add the other seven special sciences. "The natural and common books cannot be known," Bacon argues, "without the other seven special sciences, nor, indeed, without mathematics."[94] Thus did Bacon's conception of natural philosophy embrace not only perspective, or the science of vision, the science of weights, and medicine, but also the two most important occult sciences in the Middle Ages: alchemy and astrology. Perspective, astronomy, and the science of weights were generally regarded as mathematical sciences and therefore characterized as middle sciences, belonging neither to natural philosophy nor to mathematics. Bacon, however, chose another path. If the mathematics was applied to natural physical entities capable of change, Bacon regarded that as natural philosophy, or a special science falling under natural philosophy. Few in the Middle Ages would have followed him and even fewer would have subsumed eight sciences under the rubric of natural philosophy, although, as we saw earlier (Chapter 7), Dominicus Gundissalinus included medicine as part of natural philosophy and Richard Kilwardby made Perspective a part of natural philosophy.

---

[92] "Quapropter manifestum est quod una debet esse sciencia naturalis philosophie, que omnia tractet communia naturalibus, et hec erit prima inter sciencias naturales. Set alie sciencie naturales erunt magne, quarum multe habent plures sub se sciencias, sicut in Methaphisica planum est" *Liber Primus Communium Naturalium Fratris Rogeri* in *Opera hactenus inedita Rogeri Baconi*, Fasc. II (Oxford: Clarendon Press, 1909), 5.

[93] "Declaravi...quod preter scienciam communem naturalibus sunt septem specialies: videlicet, Perspectiva: Astronomia, judiciaria et operativa: Sciencia ponderum de gravibus et levibus: Alkimia: Agricultura: Medicina: Sciencia Experimentalis." Ibid.

[94] "Deinde non possunt libri naturales et vulgati sciri sine aliis septem scienciis specialibus, nec eciam sine methamaticis." *Liber primus communium naturalium*, ch. 3, 11.

Roger Bacon was convinced that Aristotle's natural philosophy and the commentaries of Averroes on that natural philosophy were inadequate for the proper understanding of natural philosophy. This was partially the result of poor translations, but even those who heard lectures on Aristotle's works and actually read the works realized that "they could not know natural philosophy by the text of Aristotle and his Commentator [i.e., Averroes] and turned to the other seven natural sciences, and to mathematics, and to other authors of natural philosophy – for example, the books of Pliny, and Seneca, and many others. And thus did they arrive at knowledge of natural things, about which Aristotle, and his expositor (Averroes?) could not satisfy them."[95] Bacon's attitude toward Aristotle and Averroes was quite unusual and one may confidently assume that he had few followers.

The move toward a union of mathematical sciences and natural philosophy was not, however, the result of bold proposals by Roger Bacon and other possible medieval visionaries, but resulted more from various individuals who were actually engaged in one or more of the mathematical sciences. In optics, one of the greatest medieval perspectivists, Ibn al-Haytham (Alhazen) (ca. 956–1039), called for a union of mathematical optics and natural philosophy. In the preface to his highly influential *Optics*, al-Haytham declared that:

Our subject is obscure and the way leading to knowledge of its nature difficult; moreover, our inquiry requires a combination of the natural and the mathematical sciences. It is dependent on the natural sciences because vision is one of the senses and these belong to natural things. It is dependent on the mathematical sciences because sight perceives shape, position, magnitude, movement, and rest, in addition to its being characterized by straight lines; and since it is the mathematical sciences that investigate these things, the inquiry into our subject truly combines the natural and the mathematical sciences.[96]

For Ibn al-Haytham, the nature of vision was such that it required both "the natural and mathematical sciences." We have already seen that in the Latin West, Roger Bacon, who was a significant medieval writer on optics, advocated the union of Perspective – that is, optics – with natural philosophy. Others who wrote on optics and vision also may have shared this attitude.

In the discipline of mechanics, Jordanus de Nemore (fl. ca. 1220) is the most notable and significant mechanician of the Middle Ages. He joined two statical traditions that had been independent in antiquity, namely, the Archimedean and Aristotelian.[97] "Archimedean statics depended, above all,

---

[95] Ibid., ch. 3, 12–13.

[96] *The Optics of Ibn al-Haytham Books I–III*. Translated with Introduction and Commentary by A. I. Sabra Vol. I: Translations; Vol. 2: Introduction, Commentary, Glossaries, Concordance, Indices (London: Warburg Institute, University of London, 1989), Vol. 1, 4, para. 2. Also see Sabra's enlightening comments on this passage in Vol. 2, p. 4.

[97] See Moody and Clagett, *The Medieval Science of Weights*, 4.

on the determinations of centers of gravity. Demonstrations were wholly geometrical in character, theorems being inferred from postulates or axioms in the Euclidean manner."[98] The Aristotelian approach was more dynamical in character and lacked the rigorous Archimedean mathematical approach. This tradition was associated with a work titled *Mechanical Problems*, which was falsely attributed to Aristotle. In his numerous treatises, Jordanus, who also was an excellent mathematician, brought together the two traditions and thus in a real sense joined important aspects of natural philosophy with mathematics, even though as we saw, his treatises were rigorously mathematical. Opposition to Jordanus's methodology appears in some later Italian mechanicians. That Jordanus's path was not always preferred is evident in a commentary in 1581 by Filippo Pigafetta on Guido Ubaldo's *Mechanicorum liber*, composed in 1577. In his commentary, Pigafetta refers to Jordanus and his followers as the "moderns." He explains that "these predecessors of ours are to be understood as being the modern writers on this subject" and cites Jordanus as a famous master who has been "much followed in his teachings." He goes on to say that "our author," Guido Ubaldo, "has tried in every way to travel the road of the good ancient Greeks, masters of the sciences, and in particular that of Archimedes of Syracuse (the most famous prince of mathematicians) and Pappus of Alexandria. . . ."[99] Thus, it is clear that Ubaldo sought to follow the pure, ancient Greek mathematical tradition rather than mixing the ancient Greek tradition with that of the moderns. Other Renaissance mechanicians followed a similar path. Nevertheless, although "Benedetti, Galileo, and other founders of modern mechanics professed to follow Archimedes and to reject Aristotle, the mechanics which they created was conceived in dynamical terms, after the pattern of the Aristotelian tradition, and not in the rigorous but limited form which characterized the statics of Archimedes."[100]

A significant and widespread union between mathematics and medieval natural philosophy also occurred with respect to problems of motion in both dynamics and kinematics. Although some of this appears in traditional commentaries and questions on natural philosophy, many of the most important relationships occurred in separate tractates on themes in natural philosophy. In Chapter 8, in the section titled "Beyond Aristotle," I described how mathematics was used to describe intensive and remissive variations in all sorts of qualities. By treating velocity as a quality, Nicole Oresme formulated a geometric proof of the mean speed theorem, while natural philosophers at Merton College, Oxford, arrived at the same proof by the use of arithmetic. Problems about qualitative variations and how speeds varied in accordance

---

[98] Ibid.
[99] Translated by Stillman Drake in *Mechanics in Sixteenth-Century Italy*. Translated and Annotated by Stillman Drake and I. E. Drabkin (Madison: University of Wisconsin Press, 1969), 295.
[100] Moody and Clagett, *The Medieval Science of Weights*, 20.

with a ratio of a motive force to a resistance were derived from Aristotle's *Physics*. Aristotle's resolutions of such problems were done with virtually no mathematics. It was medieval natural philosophers who applied mathematics to these problems and formed another significant union between mathematics and natural philosophy.

Thus, in all the ways described here, natural philosophy was united with mathematics during the late Middle Ages, and therefore long before the seventeenth century, the century in which natural philosophy and mathematics were joined in their most dramatic and permanent union. Thus far, the sciences in which mathematics was applied to problems of natural philosophy are all aspects of physics. Astronomy during the Middle Ages was not joined with natural philosophy.[101] It remained a mathematical discipline concerned with calculating planetary positions. This is true of Ptolemy's *Almagest*, the fundamental treatise of ancient and medieval astronomy, as well as of Campanus of Novara's (ca. 1220–1296) *Theory of the Planets* (*Theorica planetarum*), a widely used text in the Middle Ages. Indeed, this is also true for Nicholas Copernicus, whose monumental treatise *On the Revolutions of the Celestial Orbs* (*De revolutionibus orbium caelestium*) had very little that was natural philosophy and almost everything else that was mathematical astronomy. Indeed, in his famous preface to Pope Paul III, Copernicus declares "mathemata mathematicis scribuntur," which can be, and has been, translated as "mathematics is written for mathematicians."[102] However, Edward Rosen also has translated the phrase as "astronomy is written for astronomers."[103] Either translation is acceptable, because mathematics and astronomy often were used synonymously. It seems that Copernicus was writing for those who had knowledge and understanding of mathematics and astronomy.[104] He was not writing for natural philosophers, even though he had to speak about the earth's location and motions, topics that would normally belong to natural philosophy. Copernicus's *On the Revolutions* was intended as a technical work in mathematical astronomy. Thomas Kuhn was convinced that "it was the reform of mathematical astronomy that alone compelled him to move the earth."[105] In any event, Copernicus's *On the*

[101] See my discussion of the relations between natural philosophy and astronomy in Grant, *Planets, Stars, and Orbs: The Medieval Cosmos, 1200–1687*, 36–39.

[102] A. M. Duncan has given this translation in his *Copernicus: On the Revolutions of the Heavenly Spheres*, a new translation from the Latin, with an introduction and notes (Newton Abbott, Devon: David & Charles, 1976), 27. A slight variation on this version appears in Thomas S. Kuhn, *The Copernican Revolution: Planetary Astronomy in the Development of Western Thought* (New York: Vintage Books, 1959), 143.

[103] See *Nicholas Copernicus "On the Revolutions,"* ed. Jerzy Dobrzycki; Translation and Commentary by Edward Rosen (London: Macmillan Press Ltd., 1978), 5.

[104] See Robert S. Westman, "The Copernicans and the Churches," in David C. Lindberg and Ronald L. Numbers, eds., *God and Nature: Historical Essays on the Encounter between Christianity and Science* (Berkeley: University of California Press, 1986), 80.

[105] Kuhn, *The Copernican Revolution*, 143.

Revolutions had little to do with natural philosophy, as that was understood in the sixteenth century, and everything to do with mathematical astronomy.

Copernicus's dramatic claim that the earth really rotated daily on its axis as it moved annually around the sun, caused enough discussion, however, so that "towards the end of the sixteenth century the demarcation between astronomical and natural philosophical study of the heavens is steadily eroded."[106] Thus was the way prepared for Johannes Kepler, who, more than anyone else, saw the need to unify natural philosophy and mathematical astronomy. In his *Epitome of Copernican Astronomy* (*Epitome astronomiae Copernicanae*), published in 1618, Kepler declared, in a section titled *"What is astronomy?"* that

It is a part of physics [i.e., natural philosophy], because it seeks the causes of things and natural occurrences, because the motion of the heavenly bodies is amongst its subjects, and because one of its purposes is to inquire into the form of the structure of the universe and its parts. . . . *Concerning the causes of hypotheses.* What, then, is the third part of the task of an astronomer? The third part, physics [i.e., natural philosophy] is popularly deemed unnecessary for the astronomer, but truly it is in the highest degree relevant to the purpose of this branch of philosophy, and cannot, indeed, be dispensed with by the astronomer. For astronomers should not have absolute freedom to think up anything they please without reason; on the contrary you should be able to give *causas probabiles* [i.e. probable causes] for your hypotheses which you propose as the true causes of the appearances, and thus establish in advance the principles of your astronomy in a higher science, namely physics or metaphysics.[107]

With Kepler, the justification for uniting mathematical astronomy with natural philosophy was given its most authoritative rationale. He fused the two in his famous astronomical treatises.[108] The science of astronomy was not only concerned with mathematically determining the positions of planets but was now equally engaged in seeking the physical causes of their motions. Thus was the foundation laid for the emergence of the science of modern astronomy, which is primarily concerned with the application of mathematics to determine and predict planetary positions, and with cosmology, which had always been a part of natural philosophy and was concerned with the material and physical aspects of celestial bodies.

[106] N. Jardine, *The Birth of History and Philosophy of Science: Kepler's "A Defence of Tycho Against Ursus with Essays on its Provenance and Significance"* (Cambridge: Cambridge University Press, 1984), 244. On pages 244–245, Jardine identifies some of those who were involved in bringing mathematical astronomy and natural philosophy together. Indeed he devotes most of chapter 7, "The Status of Astronomy," to the gradual fusion of the two disciplines in the latter half of the sixteenth century.

[107] Translated by N. Jardine in ibid., 250. I have added the bracketed phrases. See also Grant, *Planets, Stars, and Orbs*, 39.

[108] For a fine, brief account of Kepler's impact on astronomy, see John Henry, *The Scientific Revolution and the Origins of Modern Science* (Great Britain: Macmillan Press Ltd., 1997), 13–14.

The disciplinary unions just described with respect to optics, mechanics, and astronomy, permit the inference that natural philosophy had become much more mathematical than ever before. Indeed, in principle, it was not supposed to include mathematics at all. In the Aristotelian system, the application of mathematics was restricted to the middle sciences. By the seventeenth century, it was applied to natural philosophy on a considerable scale. In his study of Kepler, James Voelkel argues that it was Kepler who "substituted a physical approach to astronomy – 'celestial physics,' as he named it – in which theories of planetary motion were derived from the physical consideration of the cause of their motion." In Voelkel's judgment, "the unification of physics and astronomy in which Kepler played a leading role represents the most important conceptual change in science during the period."[109] What Kepler began in the first half of the seventeenth century, Isaac Newton brought to a fruitful climax in the latter half of the century.

Isaac Newton Again

As mentioned earlier in this chapter, the title of Newton's great treatise – *The Mathematical Principles of Natural Philosophy* (*Philosophiae Naturalis Principia Mathematica*) – seemingly proclaims a union between mathematics and natural philosophy. Because of the title of his book, it has been assumed that Newton himself was convinced that he was doing natural philosophy, not science or modern physics. As we shall see, however, this is open to serious doubt. If Newton really believed he was doing natural philosophy, then he also must have been aware that his kind of natural philosophy marked a radical departure from the long history of that discipline. He would have known this from the very title he chose. To devote a treatise to ascertaining the mathematical principles of natural philosophy qualified as a virtual contradiction in terms. Why? Because natural philosophy in the medieval Aristotelian tradition did not – and could not – have mathematical principles. As we saw in the second chapter, Aristotle divided theoretical knowledge into three parts: metaphysics; mathematics; and natural philosophy. Mathematics is concerned solely with unchangeable entities that are abstractions from physical bodies, while natural philosophy is applied only to changeable bodies. During the fourteenth century, scholastic natural philosophers frequently applied mathematics to problems of motion and variations in qualities (known as "the intension and remission of forms"), but these applications were never regarded as illustrations of "mathematical principles of natural philosophy."

It appears that Newton regarded his treatise as if it were revealing the mathematical structure of physical nature, rather than as the mere application of mathematics to nature, as virtually all previous natural philosophers

[109] James R. Voelkel, *The Composition of Kepler's "Astronomia Nova"* (Princeton, NJ: Princeton University Press, 2001), 1.

would have perceived it. In effect, Newton's *Principia* was nothing less than an early version of modern physics. There were numerous other treatises in antiquity and the Middle Ages that, as we shall see, would rightly qualify as examples of mathematical physics and which were not regarded as instances of natural philosophy. But Newton was the first to regard natural philosophy as inherently mathematical, so much so that he could speak of its "mathematical principles."

In judging what Newton was really doing in his *Mathematical Principles of Natural Philosophy*, we should not be misled by the title.[110] Newton might have used either of two medieval and early modern synonyms for natural philosophy, namely, "natural science" (*scientia naturalis*) or "physics" (*physica*), to produce, respectively, the titles "Mathematical Principles of Natural Science" and "Mathematical Principles of Physics." Whatever he might have named his treatise, Newton was mathematizing natural philosophy and, depending on the subject matter, also was producing particular sciences. Whatever terms he may have used in his title, a glance at the approximately 530 pages of the *Principia* can leave no doubt that Newton was doing mathematical physics. We need not engage in mental gymnastics about the real meaning of the expression "natural philosophy." And we should not be mislead by the fact that because the term "scientist" was only introduced in the nineteenth century, therefore no activity we would want to call science could have occurred before that century simply because "scientists" may have been called by another name: natural philosophers.

On June 20, 1686, Newton sent a letter to Edmond Halley (ca. 1656–1743), the famous astronomer and mathematician. In his letter, Newton informs Halley that he will suppress the third book of the *Principia*, which involved the application of mathematics and dynamics to celestial bodies, including comets. He was fearful that book 3 would involve him in controversy. "Philosophy," Newton explains, "is such an impertinently litigious Lady that a man had as good be engaged in Law suits as have to do with her. I found it so formerly & now I no sooner come near to her again but she gives me warning." And then, in an important and revealing passage, Newton explains that:

The first two books without the third will not so well bear the title of *Philosophiae naturalis Principia Mathematica* & therefore I had altered it to this *De motu corporum libri duo*: but upon second thoughts I retain the former title. Twill help the sale of the book which I ought not to diminish now tis yours.[111]

---

[110] In what follows on Newton, I draw heavily on my article "God and Natural Philosophy: The Late Middle Ages and Sir Isaac Newton," *Early Science and Medicine*, 5, no. 3 (2000), 291–298.

[111] Cited from I. Bernard Cohen, *Introduction to Newton's 'Principia'*, 133. Cohen cites the passage from H. W. Turnbull, ed., *The Correspondence of Isaac Newton*, vol. 2: 1676–1687 (Cambridge, 1960), 437.

In this remarkable letter, Newton acknowledges that the title *Mathematical Principles of Natural Philosophy* was not appropriate for the first two books without the third, and therefore seriously considered abandoning the third book and naming his treatise *Two Books on the Motion of Bodies (De motu corporum libri duo)*. Newton fully realized that the first two books taken separately were too mathematical to qualify as natural philosophy and could not be called *Two Books on Natural Philosophy*. Newton was, however, willing to retain the title *Mathematical Principles of Natural Philosophy* because the book would sell better! We may rightly conclude, therefore, that Newton did not retain the famous title of his book because he thought he was doing natural philosophy but did so because book sales would undoubtedly be better than if the book were titled *Two Books on the Motion of Bodies*.

Halley was displeased with Newton's decision and ultimately convinced him to retain the third book along with the original title for the entire book. He believed that without book 3, the first two books would only be understood by those knowledgeable about mathematics, but "those that will call themselves philosophers without Mathematicks, which are by far the greater number,"[112] would not understand them. Newton's third book is also heavily mathematical and hardly qualifies as natural philosophy, although it more nearly does than the first two books, as Newton rightly understood. With its major emphasis on celestial mechanics and cosmology, the third book, as Halley foresaw, would have had more appeal to interested readers than the first two forbiddingly mathematical books. But even the third book was far too mathematical and dynamical to qualify as traditional natural philosophy. Thus, the argument that Newton did not believe he was doing natural philosophy in the *Principia* gains credibility from Newton himself.

And yet none of this affects the great changes Newton made in natural philosophy. His private thoughts about the matter expressed in letters that only a few could have known about cannot alter the fact that the title of his monumental treatise led all who knew about it to believe that mathematical principles formed an inherent part of natural philosophy. The title reflects what had been going on since the late sixteenth century: the serious incorporation of mathematics and mathematical sciences into natural philosophy. What Newton presented to the world was a natural philosophy that sought to describe physical causes and forces in mathematical terms. Galileo had done so earlier in the century. Natural philosophy as Newton did it in the *Principia* is equivalent to what would be called mathematical physics two centuries later. Moreover, Newton was also doing what we would today call "science" – the science of physics.

In his biography of Newton, Richard Westfall provides further evidence that Newton was doing science. "The changes" Newton made to later editions of the *Principia* "only repeated Newton's deepest convictions about

---

[112] Cohen, ibid.

the nature of the scientific enterprise, that if its business began with observation, only the derivation of exact quantitative relations concealed in observations deserved the name of science." Westfall adds that "James Jurin once reminded Martin Folkes of a saying frequently in Newton's mouth: 'That Natural History might indeed furnish materials for Natural Philosophy; but, however, Natural History was not Natural Philosophy.'"[113] What Newton might have intended by such a sentiment is unclear. But he may have divorced natural history from natural philosophy because knowledge and observations about life forms – humans, animals, and plants, the domain of natural history – was largely nonquantitative and inexact. Because Newton sought to make natural philosophy as mathematically exact as possible, he divorced it from natural history. This would be another indication that Newton's natural philosophy was more akin to mathematical physics than anything else. This gains further support from the fact that "throughout the eighteenth century in France, *physique* had consisted of two separate disciplines: *physique générale* and *physique particulière*. After mid-century the former meant Newtonian mechanics, while the latter connoted experimental science in general, but sometimes meant specific studies in heat, light, sound, electricity, and magnetism."[114]

Natural Philosophy as a Synonym for Physics and Science in the Nineteenth Century

By extending the application of the term natural philosophy to the mathematical sciences, most notably mathematical physics, Newton may have begun a tradition that reached fruition in the nineteenth century when natural philosophy came to be called physics, and, often enough, science in general. There is little doubt that physics in particular, and perhaps science in general, evolved from natural philosophy. "Once created out of natural philosophy, though, wherever that may have occurred, physics rapidly became a professional discipline, something that natural philosophy had never been."[115] Thus, the physics, or science, that evolved from natural philosophy differed in various ways from the latter. And yet, there was no rapid displacement of natural philosophy by the terms physics or science. Indeed, the two terms coexisted throughout the nineteenth century and often were used synonymously. Let me now cite examples and instances of the usage of

---

[113] Westfall, *Never at Rest*, 731. James Jurin and Martin Folkes were members of the Royal Society.

[114] Jed Z. Buchwald and Sungook Hong, "Physics," in David Cahan, ed., *From Natural Philosophy to the Sciences: Writing the History of Nineteenth-Century Science* (Chicago: University of Chicago Press, 2003), 163–195; 168.

[115] Ibid.

the expression natural philosophy in the nineteenth century and how that term related to the terms physics or science.

Natural philosophy was widely taught in virtually all colleges in the United States during the nineteenth century. In a catalogue issued by Dickinson College (Carlisle, Pennsylvania) for the period 1845–1846, we read that:

> *Natural philosophy* may be considered as the science which examines the general and permanent properties of bodies; the laws which govern them, and the reciprocal action which these bodies are capable of exerting upon each other, at greater or less distances, without changing their matter.[116]

We see that natural philosophy is here equated with science. Among the sciences included under natural philosophy at Dickinson College, and many other colleges, are mechanics, hydrostatics, hydraulics, pneumatics, acoustics, optics, astronomy, electricity, galvanism, magnetism, and chromatics.[117] To this list we may add crystallography and chemistry, which appear in John William Frederick Herschel's *Preliminary Discourse on the Study of Natural Philosophy*, first published in 1830.[118] At Dartmouth College, there were Professors of Natural Philosophy almost to the end of the nineteenth century. Indeed, professorial titles at Dartmouth, and elsewhere, sometimes encompassed more than just natural philosophy. They might embrace natural philosophy and astronomy (Professor of Natural Philosophy and Astronomy) or natural philosophy and mathematics (Professor of Natural Philosophy and Mathematics).[119] At Dartmouth,

> it was evident toward the latter part of the 19th century that the natural sciences were becoming a competitive research field and that excellence in teaching must somehow be geared to the rapid change in subject matter; thus it became incumbent on any first-rate teacher to keep up, at least in some measure, with the new knowledge by carrying out research of his own. It was in this era that for the first time in the history of Dartmouth the College had not a single professor of natural philosophy, but a professor of physics gathering around him a staff of young men who could share with him the teaching load, and also help him in his research projects.[120]

From this passage, we may plausibly infer that physics was regarded as a more professional version of natural philosophy and its name eventually supplanted it. But for a long while, the two terms were used synonymously. That

---

[116] Quoted from Stanley Guralnick, *Science and the Ante-Bellum American College* (Philadelphia: American Philosophical Society, 1975), Memoirs, vol. 109, 60. Also cited in Grant, *Foundations*, 193, and Grant, "God and Natural Philosophy: The Late Middle Ages and Sir Isaac Newton," 296.

[117] Guralnick, ibid., 61.

[118] For the reference, see Grant, "God and Natural Philosophy," 296, n. 31.

[119] See Sanborn C. Brown and Leonard M. Rieser, *Natural Philosophy at Dartmouth: From Surveyors' Chains to the Pressure of Light* (Hanover, NH: University Press of New England, 1974), 87.

[120] Ibid., 101.

natural philosophy could stand for physics, just as well as the term physics itself is evident from a famous two-volume book on physics by William Thomson (Lord Kelvin) (1824–1907) and Peter Guthrie Tait (1831–1901) written in the early 1860s under the title *Treatise on Natural Philosophy*, which considered the themes of kinematics and dynamics. Jed Buchwald explains that in their *Treatise on Natural Philosophy* "Thomson and Tait presented in full the kinematics of point particles and the dynamics of motion under force; they placed heavy emphasis upon the dynamics of material media; and they made detailed use both of Lagrangean mechanics and the conservation of energy."[121] Buchwald concludes: "*The Treatise on Natural Philosophy* introduced a new generation of British and American physical scientists to the details and concepts of mechanics." Along with James Clerk Maxwell's (1831–1879) later *Treatise on Electricity and Magnetism* (two vols.; Oxford, 1873), Buchwald regards The *Treatise on Natural Philosophy* as the most influential physical text of the second half of the nineteenth century. How would anyone who regards the disciplines of natural philosophy and science as mutually exclusive, explain that Thomson and Tait were doing modern physics, and therefore modern science, under the rubric of natural philosophy?

Indeed, chemistry also was used synonymously with natural philosophy. In a Web site at the College of William and Mary ("Chemistry at William and Mary – Three Centuries"), Thomas Jefferson is quoted as saying: "What are the objects of an useful American Education? Classical knowledge, modern languages, chiefly French, Spanish, and Italian; mathematics, natural philosophy, natural history, civil history, and ethics. In natural philosophy I mean to include chemistry and agriculture, and in natural history to include botany, as well as the other branches of those departments."[122] We are further informed that "chemistry continued as part of the curriculum through the nineteenth century under the professorship of natural philosophy." In 1905, however, "chemistry became an autonomous department with its own professor. Physics and biology also became departments at this time. The traditional rubric of natural philosophy was abandoned." Furthermore, from the Dickinson College catalogue, it seems obvious that natural philosophy also was a synonym for science in general, as the terms natural philosophy and science were employed to subsume much the same list of independent sciences.

There can be no doubt that, throughout the nineteenth century, the terms natural philosophy, physics, and science were used synonymously, as was chemistry, although perhaps less frequently. Our knowledge of the synonymous use of these terms is derived from the scientists, or natural philosophers,

---

[121] See Jed Buchwald, "Thomson, Sir William," in *Dictionary of Scientific Biography*, 13 (1976), 386. Also cited in Grant, "God and Natural Philosophy," 297.
[122] In this section, I draw on the following Web site: http://www.chem.wm.edu/history.html.

themselves. It is they who identified themselves as natural philosophers doing physics, or chemistry, or science in general. The nineteenth century scholars we have mentioned here – especially William Thomson (Lord Kelvin) and his colleague Peter Guthrie Tait – leave little doubt that they were doing science.

THE CONTINUITY OF HISTORY AND THE PROBLEM OF NAMES AND TERMINOLOGY

In retrospect, the major thesis of this study is that although the exact sciences and natural philosophy were largely distinct from one another in the late Middle Ages, the horizons of each began to expand as they gradually came together. The fusion manifested itself bilaterally: natural philosophy influenced the exact sciences to seek the physical causes of relevant phenomena, and thereby broaden the scope of their activities; as this occurred, natural philosophy was inevitably mathematized and its scope expanded. By this process, natural philosophy was the basic instrument in the development of our many modern sciences. But before the fusion of the exact sciences and natural philosophy, when Aristotelian natural philosophy was dominant, there were numerous discussions in natural philosophy that would qualify as relevant to science. This is especially true for the nonmathematical sciences, as was shown earlier (see the end of Chapter 8: "Was Aristotelian Natural Philosophy Science?"). The same may be said for the exact sciences before their union with natural philosophy. Astronomical, optical, and statics treatises by Greek and medieval authors may have been narrowly quantitative before their union with natural philosophy, but they certainly should be regarded as treatises in the domain of "science."

If we distinguish between, or separate, science and natural philosophy, or physics and natural philosophy we mistakenly allow mere names, or labels, to determine whether an author or experimenter was engaged in one or the other activity. Shouldn't the substantive content of a treatise be regarded as far more important than the name we attach to it? Whatever Isaac Newton may have titled his great work, we must get behind the terms in the title and examine what he was actually doing.[123] Whether Newton was doing things that are truly representable by the terms science or physics, or by the expression natural philosophy, is analogous to the manner in which Aristotle dealt with the term logic. No one would contest the claim that Aristotle is the founder, and even inventor, of formal syllogistic logic. Aristotle, however, did not use the term "logic" but called what he was doing by the name of "analytics." It was not until five centuries later that Alexander of Aphrodisias applied the term logic to what Aristotle did. Does this signify that by using the term "analytics" instead of logic, Aristotle was not really doing logic, but something called "analytics"? Of course not, for just as Aristotle's

---

[123] I am here following my account in Grant, "God and Natural Philosophy," 295.

analytics and logic are identical, so also is the content of Newton's *Principia* the same as physics, perhaps even modern physics – despite being called natural philosophy. Similarly, Aristotle did not use the term biology, which was not invented until the nineteenth century, but does anyone doubt that Aristotle was doing biology when he wrote *On the Parts of Animals*, *The Generation of Animals*, and the *History of Animals*? Those who have doubts that Aristotle should be regarded as a biologist should know that in a letter to William Ogle, who had just translated Aristotle's *Parts of Animals*, Charles Darwin declared, after reading less than a quarter of the book: "From quotations I had seen, I had a high notion of Aristotle's merits, but I had not the most remote notion what a wonderful man he was. Linnaeus and Cuvier have been my two gods, though in very different ways, but they were mere schoolboys to old Aristotle."[124]

It makes little sense to distinguish between natural philosophy and science as if they were incommensurable activities. The different terms that are used to describe activities must not be allowed to mask and obscure the activities themselves. We should here be guided by the wise words of Shakespeare's Juliet, who asks:

> What's in a name? that which we call a rose
> By any other name would smell as sweet.

Much that most modern scientists would call science – albeit early, or premodern science – was embedded in natural philosophy, as that discipline was studied and written about by Aristotle and his medieval followers, who took it in new directions and to new heights. When we see discussions about dynamics (impetus theory), kinematics, earthquakes, mountain formation, comets, descriptions of plants and animals, and numerous other subjects, embedded in treatises on natural philosophy, we may rightly regard those specific discussions as part of the history of particular sciences, and of the history of science in general. What label we attach to these discussions – science or natural philosophy, or something else – is of little consequence (as we saw, the terms were used synonymously in the nineteenth century). Natural philosophy forms a proper part of the history of science. Indeed, were we to omit it, our histories of science would be seriously incomplete and defective. A balanced approach is essential, as we find it in David C. Lindberg's judiciously titled survey of the history of science: *The Beginnings of Western Science: The European Scientific Tradition in Philosophical, Religious, and Institutional Context, 600 B.C. to A.D. 1450*. Within the compass of that title, Lindberg describes all that he regarded as relevant to *The Beginnings of Western Science*, using the term science as synonymous with natural philosophy. "I will make regular use of the expression 'natural philosophy',"

---

[124] Cited by George Sarton, *A History of Science: Ancient Science through the Golden Age of Greece*, 545.

he declares, "either to denote the scientific enterprise as a whole or to signify its more philosophical side. The term 'science' will also be employed, most often as a synonym for 'natural philosophy,' sometimes to designate the more technical aspects of natural philosophy, and occasionally simply because conventional usage calls for that term in a certain context."[125] As I have declared here and elsewhere, natural philosophy is the "Mother of all Sciences," or as Francis Bacon expressed it more dramatically centuries before, "The Great Mother of the Sciences." Without natural philosophy, there would probably be no modern science, as I argue in the Conclusion.

Those who would deny that anything we would want to call science ever existed before the late eighteenth and nineteenth centuries, effectively deny the continuity of the history of science by insisting that Newton was only doing natural philosophy in his *Principia* and, in no way, doing mathematical physics. Andrew Cunningham's insistence that no one did science before the late eighteenth and nineteenth centuries is a fundamental mistake. The error stems from a lack of awareness of the role of the "middle sciences," namely, astronomy, optics (or perspective), and mechanics (or statics; or "the science of weights"). During the Middle Ages, natural philosophy and mathematics were regarded as distinct subjects. When mathematics was applied to certain aspects of natural philosophy, a new kind of theoretical knowledge emerged that was called a "middle science" (*media scientia*). The middle sciences were so called because they were regarded as lying between mathematics and natural philosophy and were assumed independent of both, although they were often regarded as lying closer to the one than the other. They were judged to be exact sciences, and included astronomy, optics, and statics, or mechanics. Cunningham, however, mistakenly declares that these sciences were regarded as belonging to mathematics, and, therefore, were not sciences.[126]

But even if the middle sciences had not been distinguished from mathematics and natural philosophy, there is every reason to believe that science had been practiced from the ancient world to the seventeenth and eighteenth centuries. We must not be misled by the titles of works, or by the fact that early science was different, usually radically different, than what has evolved into modern science. Indeed, many modern sciences were not identified or characterized until the modern era. But this does not mean that earlier historical versions of those sciences were not practiced in premodern history, often going back to the ancient Greeks, or even as far back as the ancient Egyptians and Babylonians. If science is strictly a modern affair, then there can be no history of science that stretches back through the ages, and,

---

[125] Lindberg, *The Beginnings of Western Science*, 4.

[126] In "Getting the Game Right," 379, n. 9, Cunningham declares that "some of the disciplines we now count under science would have been placed under Mathematics (for instance optics, astronomy, mechanics)."

consequently, science can have had no continuous history. This is an absurd notion, because it presupposes that science began in the nineteenth century and did so by springing fully developed from the minds of its nineteenth-century practitioners. It excludes Isaac Newton as a scientist simply because that term was not invented until the nineteenth century. But all modern physicists look on Newton as the one who laid the foundations of physics, the discipline they practice. By his insistence that there was no science before the late eighteenth and nineteenth centuries, Andrew Cunningham raises the continuity/discontinuity issue, an issue he believes is "an old battle which may not be worth fighting any longer"; in his judgment, it is "a misplaced concept."[127]

But when the concept of historical continuity is rejected, it becomes essential to fight the battle. For if history, including the history of science, is not regarded as continuous, history will become little more than a sequence of unrelated, incommensurable, quantized temporal packets. History is akin to the relationship between the human embryo and the full-blown adult. The latter comes from the former by a complex, lengthy process, although one might not reach that conclusion by a superficial inspection of appearances. We can always find reasons to distinguish things, but historians know that appearances are often deceptive. New things and innovations occur all the time, but they are connected to what went before. They did not spring into being like Athena from the head of Zeus. We ignore these innumerable connections and become fashionably iconoclastic at our peril.

The physical and biological sciences, along with mathematics, have continuous histories from the ancient world to the present. Geometry, algebra, and trigonometry have deep roots in ancient Mesopotamia and Greece. The calculus that Newton invented would not have been possible without the long tradition of mathematical accomplishment that had been attained by his day. Indeed, his achievements in mathematical physics would have been impossible without the innumerable contributions that made mathematics what it was when he turned his attention to problems of motion; nor would it have been possible without the significant work that had been done in mechanics by the likes of Archimedes, Hero of Alexandria, Jordanus de Nemore, and Galileo. If Newton and his fellow scientists, or natural philosophers (however you may wish to call them) had not made their great contributions to physical science, the physicists and astronomers of the nineteenth and twentieth centuries – even to the present day – might still be struggling to reach the level of seventeenth-century science.

---

[127] Cunningham, "The Identity of Natural Philosophy. A Response to Edward Grant," *Early Science and Medicine*, vol. 5, no. 3 (2000), 277.

Conclusion

As we look back on the history of science, it seems that a strong case can be made for the assumption that the emergence of modern science was in some significant and meaningful sense dependent on the existence of a well-developed natural philosophy. To establish this thesis, we must first go back to the late Middle Ages, when natural philosophy reached its mature development after it became a required subject in the medieval universities. Did this mature and institutionalized natural philosophy possess the requisite characteristics that would enable science in general, and the particular sciences that comprise it, to emerge in the centuries to follow?

Is this perhaps a pseudo-issue? Was it not true that during the late Middle Ages exact, or "middle," sciences, such as astronomy, optics, and mechanics, already existed independently of, but concurrently with, natural philosophy? In truth, these exact sciences had themselves once been a part of natural philosophy as far back as the period just before Aristotle. In his classification of theoretical knowledge, Aristotle classified the exact sciences as independent of natural philosophy, because they were seemingly as much mathematical as they were natural philosophy; but they were neither mathematics nor natural philosophy. Nor were they sciences in the later modern sense. Why is this so? Precisely because they were mathematical disciplines that were only supposed to focus on limited problems that could be resolved only by mathematics but were not to be resolved by natural philosophy. For example, astronomers concerned themselves largely with the positions and paths of celestial bodies – data that could be represented numerically and by geometric figures – but left to natural philosophy all discussions of the composition and behavior of these bodies. In general, cosmic problems were the domain of natural philosophy, whereas planetary positions were the responsibility of mathematical astronomy. To evolve into some form of modern science, the exact mathematical sciences had to be integrated with the relevant subject matter in natural philosophy. Only then would astronomy, optics, and the numerous particular sciences to come be transformed into something closely resembling modern science. From this, we may infer that important features of natural philosophy had to be joined with the exact sciences. The consequences of such a union would be a natural philosophy made more mathematical; and exact sciences that considered problems ranging beyond mere quantification and that embraced more cosmic concerns, such as the composition of the planets. Although, as we saw, mathematics was applied to problems of motion in natural philosophy, the union of natural philosophy and the exact sciences did not begin seriously until the seventeenth century. Nevertheless, natural philosophy contributed vital attributes that prepared the way for modern science to emerge. Whatever these attributes might be, they had to be broadly and deeply disseminated within medieval society; and they had to be of a kind without which modern science could not have come into being.

With the establishment of the universities of Paris, Oxford, and Bologna by 1200, the institutional foundation was laid within which the essential attributes for the development of modern science could take root. The medieval university was the vehicle for the development of natural philosophy, which became the basic course of study for all students: for those who sought only a Master's degree in arts, as well as for those who wished to matriculate for doctorates in law, medicine, or theology. By 1500, there were approximately sixty-four active universities spread across Europe from Cracow in the East to Lisbon in the West; from Uppsala in the North to Catania in the South.[1] In any given year from approximately 1250 to 1550, we may rightly assume that thousands of students matriculated at these universities. By the time they completed their coursework, they should have been quite familiar with natural philosophy. They had a worldview shaped by Aristotelian natural philosophy, a worldview they carried with them wherever their careers took them. For the first time in history, a large number of scholars with similar training in natural philosophy, and therefore with a reasonable level of contemporary scientific knowledge, were absorbed into the broader reaches of European society. Motivated by a love of learning, the aim of a university education was to acquire knowledge for its own sake, much as Aristotle, and numerous other Greek philosophers, had envisioned. As the core disciplines of learning in the medieval universities, natural philosophy and logic were studied for their own sakes to gain knowledge of the natural world, knowledge that was always understood to be rational. Medieval academic life was driven by, among other things, a "belief in a world order, created by God, rational, accessible to human reason, to be explained by human reason and to be mastered by it; this belief underlies scientific and scholarly research as the attempt to understand this rational order of God's creation."[2] The emphasis on human reason is underscored by the fundamental role assigned to the study of logic in the medieval university.

The development of natural philosophy with its emphasis on reason and its inquiring spirit was the major activity of universities, which, in turn, were

---

[1] For a map showing the distributions of universities around 1500, see Jacques Verger, "Patterns" in H. de Ridder-Symoens, *Universities in the Middle Ages*, vol. 1, 74.
[2] Walter Rüegg, "Themes," in ibid., 32.

an exclusive invention, or creation, of the Late Latin Middle Ages in Western Europe. As Jacques Verger explains:

> It is no doubt true that other civilizations, prior to, or wholly alien to, the medieval West, such as the Roman Empire, Byzantium, Islam, or China, were familiar with forms of higher education which a number of historians, for the sake of convenience, have sometimes described as universities. Yet a closer look makes it plain that the institutional reality was altogether different and, no matter what has been said on the subject, there is no real link such as would justify us in associating them with medieval universities in the West. Until there is definite proof to the contrary, these latter must be regarded as the sole source of the model which gradually spread through the whole of Europe and then to the whole world.[3]

Over some four centuries, medieval natural philosophers transmitted a legacy to their non-Aristotelian, and largely anti-Aristotelian, successors in the early modern period, a legacy that was unacknowledged. That legacy was a pervasive and deep-seated spirit of inquiry that was a natural consequence of the widespread and intensive emphasis on reason that began in the Middle Ages. With the exception of revealed truths, reason was the ultimate arbiter for most intellectual arguments and controversies in medieval universities. It was quite natural for scholars immersed in a university environment to employ reason to probe into subject areas that had not been explored before, as well as to discuss possibilities that had not previously been seriously entertained. Reason and the spirit of inquiry appear to be natural companions. The spirit of inquiry that took hold in the Middle Ages may be aptly described as the spirit of "probing and poking around," a spirit that manifests itself through an urge to apply reason to almost every kind of question and problem that confronts scholars of any particular period. Indeed, "probing and poking around" inevitably triggers an irresistible urge to raise new questions, which eventually give rise to even more questions. The spirit of "probing and poking around" may be appropriately characterized as nothing less than the spirit of scientific inquiry.[4]

In the Middle Ages, reason was joined to an analytic questioning technique that was ubiquitous in university education and therefore widespread among the literate class. Questions were posed in natural philosophy that asked about the structure and operation of the physical world that Aristotle had described. Questions also were posed in theology about every aspect of faith and revelation. But the probing character of medieval questions went far beyond the straightforward and routine. Scholastic natural philosophers and theologians asked questions not only about what is but also about what could be but probably wasn't. Theologians exercised their logical talents by inquiring about what God could and could not do, or what He could

---

[3] Jacques Verger, "Patterns," in ibid., 35.
[4] What follows is largely taken from my article "What Was Natural Philosophy in the Late Middle Ages?" in *History of Universities*, vol. 20, part 2 (2005), 12–46.

and could not know. The criterion for judging God's infinite power was simple: if the claim or action led to a contradiction, God could not do it; if no contradiction was involved, God could do it. Every question in the scholastic arsenal produced pro and contra arguments that were intended to include all plausible and feasible positions.

What makes the "probing and poking around" approach so important is the fact that it was institutionalized in the medieval universities where it was the modus operandi for more than four centuries. Thus, a spirit of inquiry took deep and extensive root in Western Europe. The myriad questions that were raised reflected the desires of an intellectual class that sought to know as much as they could by reason alone. The structural form of the question as it was used in the medieval universities was meant to provide a definitive answer to each question raised, although scholars might arrive at different, and conflicting, answers. Even if modern critics judge the questions and their responses to be trivial, or of little utility, those who posed the questions and answered them regarded their efforts as of great importance. They were, after all, solving questions that ostensibly informed their contemporaries about the inner and outer workings of the world, as these were understood at the time. Not only did they provide their audience with answers to such questions, but they also included refutations of the arguments they found wanting.

And yet, despite the "probing and poking around" that produced numerous departures from Aristotle's natural philosophy, the intense questioning and probing did not transform medieval Aristotelian natural philosophy into a new way of doing science. The thought experiments, hypothetical questions, and the questions about what God could or could not do, or what He knows or does not know, which were so characteristic of the Middle Ages, were largely abandoned by the natural philosophers who produced the Scientific Revolution. The numerous departures from Aristotle's physics and cosmology by medieval natural philosophers were never incorporated into Aristotle's natural philosophy. No serious effort was ever made to transform and update the Aristotelian worldview. The numerous medieval departures and innovations were left as part of an unwieldy mass of unintegrated and conflicting ideas. The hundreds of medieval questions on the works of Aristotle were left as a mass of independent, but unrelated conclusions. If progress was to be made, the Aristotelian worldview had to be abandoned, as it was in the seventeenth century.

But if they abandoned Aristotle's explanations of cosmic operations, nonscholastic natural philosophers also proceeded by way of questions. But the questions were now often only in their minds to guide them in their research and inquiries. The literary tradition of explicating a text by questions came to an end. The results nonscholastic researchers published might not explicitly include the questions that guided the researcher and led to those results. Moreover, the questions they posed to themselves and to others were rarely about hypothetical, or imaginary, conditions, or about God's power to do or not to do some particular act, but were about the real world. Also noteworthy is the fact that natural philosophers in the seventeenth century answered the questions they posed to nature by appeals to observation, or by means of experiments, or by the application of mathematics. This became the way scientists would proceed to the present day. Nonscholastic natural philosophers and scientists of the sixteenth and seventeenth centuries devised superior methods and techniques for resolving the problems that their scholastic predecessors and contemporaries had grappled with.

Although scientists in the various sciences have evolved different techniques and procedures for answering the never-ending parade of questions they generate, and without which modern science could not exist, the spirit of inquiry remains essentially what it was in the Middle Ages: an effort to advance a subject by probing and poking around with one or more questions to which answers are sought, after which more questions are posed, in a never-ending process. We are a questioning society that constantly seeks answers to queries about virtually everything, especially about nature, religion, government, and society.

The questioning method is the driving force in science, social science, and technology. Ironically, it is absent from modern theology, which no longer raises the kinds of questions that theologians in the Middle Ages characteristically posed. It would be difficult to imagine modern theologians asking about the limits of God's power and determining those limits by application of the law of noncontradiction. Not only did the scholars in the Middle Ages lay the basis for our probing society by means of an unending stream of questions, but they used reason as the fundamental criterion for arriving at their answers. By the seventeenth century, natural philosophers saw that "pure" reason alone was often inadequate, and they devised the experimental method to furnish evidence that reason alone could not provide. It was in this spirit that Isaac Newton began his work on the *Opticks* by proclaiming to his readers: "My Design in this Book is not to explain the Properties of Light by Hypotheses, but to propose and prove them by Reason and Experiments."[5]

Natural philosophy is the "Great Mother of the Sciences" because she provided the exact sciences with the cosmic issues they required in order to grapple with wide-ranging problems. The exact sciences had to be applied to "why" problems, not just to "how" problems. Even if the "why" answers were inadequate, it was nevertheless essential to ask them in order to determine whether some kind of answers could be provided. Thus, although astronomy, optics, and mechanics had been independent sciences after the time of Hellenistic Greek science, they needed, once again, to rejoin natural

[5] Quoted by Westfall, *Never at Rest: A Biography of Isaac Newton* (Cambridge: Cambridge University Press, 1980), from p. 1 of Newton's *Opticks*.

philosophy to enlarge their horizons and grapple with more cosmic problems. In this regard, Francis Bacon was right to declare that the particular sciences, among which he mentions astronomy and optics, do not grow when "they are no longer nourished by natural philosophy" (see Chapter 10). What Bacon failed to envision was the time (in the nineteenth century) when each of the particular sciences would mature and develop to the point where it could, and would, emerge as an independent science able to exist by itself, no longer requiring an association with natural philosophy.

Natural philosophy played an essential role in the development of science. It ranged over the whole of nature. Questions were posed about the operations of the physical world for which precise answers could not be given. Before the seventeenth century, the problems that were considered by the exact sciences were limited in scope and largely confined to quantification. It was by means of Aristotelian natural philosophy, as developed in the late Middle Ages, that scholars sought to comprehend and explain the operations of the physical world. In light of what has been said in the last few paragraphs, one may reasonably conclude that natural philosophy can flourish while the level of the exact sciences is rather low; but that the exact sciences will not advance significantly without a vibrant natural philosophy. A number of ancient civilizations – for example, Babylonia and India – attained reasonably high levels of proficiency in mathematics but had no well-developed natural philosophy. They did not sustain their mathematical achievements.

The civilization of Islam is undoubtedly the best example of this phenomenon. By 1300, scholars living in lands dominated by the Islamic religion had arrived at the highest level of achievement in the exact sciences of mathematics, astronomy, optics, and mechanics. In addition, as we saw (Chapter 4), Islamic natural philosophers, using Aristotle's natural philosophy as their point of departure, produced a high level of analysis of the physical world, bringing natural philosophy to new heights. And yet, because of the many controversial ideas that permeated natural philosophy and that brought it into conflict with the Islamic religion, natural philosophy was usually regarded with suspicion and hostility. Moreover, it was obvious that natural philosophers relied heavily on reason rather than revelation, with the attendant danger that they might apply reason to the interpretation of the Qur'an. In brief, natural philosophy was an alien discipline, a "foreign science," which was potentially dangerous for Islamic beliefs and traditions. Natural philosophy was never absorbed into Islamic culture and remained a peripheral intellectual activity, largely viewed with suspicion. With the marginalization of natural philosophy, it was not long before the exact sciences – which were not under suspicion – lost the momentum they had acquired in the earlier Islamic centuries, and eventually faltered and stagnated. Without a viable natural philosophy to expand the horizons of the exact sciences, they gradually withered and virtually disappeared.

If modern science has progressed unrecognizably beyond anything known or contemplated in the natural philosophy and science of the Middle Ages, modern scientists are, nonetheless, heirs to the remarkable achievements of their medieval predecessors. The idea, and the habit, of applying reason to resolve the innumerable questions about our world, and of always raising new questions, did not come to modern science from out of the void. Nor did it originate with the great scientific minds of the sixteenth and seventeenth centuries, from the likes of Copernicus, Galileo, Kepler, Descartes, and Newton. It came out of the Middle Ages from many faceless scholastic logicians, natural philosophers, and theologians, in the manner I have described. If you are skeptical about the medieval role in the advent of early modern science, I ask you to consider this question: Could a Scientific Revolution have occurred in the seventeenth century if the level of science and natural philosophy in Western Europe had remained what it was in the first half of the twelfth century? That is, could the dramatic changes in science and natural philosophy have occurred in the seventeenth century if medieval natural philosophers had not absorbed and developed the new Greco-Arabic science and natural philosophy that had been translated into Latin in the twelfth and thirteenth centuries? The response is obvious: no, it could not have occurred. We ought, therefore, to conclude that something important occurred between approximately 1200 and 1600 that proved conducive for the emergence of the Scientific Revolution. Without the level that medieval natural philosophy attained, with its overwhelming emphasis on reason and analysis, and without the important questions that were first raised in the Middle Ages about other worlds, space, motion, the infinite, and without the kinds of answers they gave, we might, today, still be waiting for Galileo and Newton.

Bibliography

Abelard, Peter. *A Dialogue of a Philosopher with a Jew, and a Christian*. Translated by Pierre J. Payer. Toronto: Pontifical Institute of Mediaeval Studies, 1979.
————. *Peter Abailard Sic Et Non, A Critical Edition*. Edited by Blanche Boyer and Richard McKeon. Chicago: University of Chicago Press, 1976, 1977.
Adelard of Bath. *Adelard of Bath, Conversations with His Nephew On the Same and the Different, Questions on Natural Science, and On Birds*. Edited and translated by Charles Burnett, with the collaboration of Italo Ronca, Pedro Mantas Espana and Baudouin van den Abeele. Cambridge: Cambridge University Press, 1998.
Albert of Saxony. *Questiones et decisiones physicales insignium virorum: Alberti de Saxonia in octo libros Physicorum; tres Libros De celo et mundo; duos libros De generatione et corruptione; Thimonis in quatuor libros Meteororum; Buridani in tres libros De anima; librum De sensu et sensato; librum De memoria et reminiscentia; librum De somno et vigilia; librum De longitudine et brevitate vite; librum De iuventute et senectute Aristotelis. Recognitae rursus et emendatae summa accuratione et iudicio Magistri Georgii Lokert Scotia quo sunt tractatus proportionum additi*. Paris: vaenundantur in aedibus Iodici Badii Ascensii et Conradi Resch, 1518.
Albertus Magnus (Albert the Great). *Alberti Magni Ordinis Fratrum Praedicatorum Episcopi Opera Omnia, Tomus IV: Pars I: Physica, Pars I, libri 1–4*. Ed. Paul Hossfeld. Aschendorff: Monasterii Westfalorum, 1987.
Alexander of Hales. *Summa Theologica*, Tomus II: Prima Pars secundi libri. Florence: Collegium S. Bonaventurae, 1928.
Anawati, G. C. "Philosophy, Theology, and Mysticism." In Joseph Schacht with C. E. Bosworth, eds., *The Legacy of Islam*. Second edition. Oxford: Clarendon Press, 1974, 350–391.
Anawati, G. C., and Albert Z. Iskandar. "Hunayn ibn Ishaq al-Ibadi, abu Zayd, known in the Latin West as Johannitius." In Charles C. Gillispie, ed., *Dictionary of Scientific Biography*, vol. 15, Supplement 1 (1978), 230–249.
Anonymous Latin Treatise on Natural Philosophy. Bibliothèque Nationale 6752, fols. 1r–236r.
Anselm, Saint. *Proslogium; Monologium; An Appendix in Behalf of the Fool Guanilon; and Cur Deus Homo*. Translated from the Latin by Sidney Norton Deane, with an Introduction, Bibliography, and Reprints of the Opinions of the Leading Philosophers and Writers on the Ontological Argument. La Salle, IL: Open Court Publishing Co., 1944.
Arberry, Arthur J., trans. *Avicenna on Theology*. London: John Murray, 1951.
————. *Revelation and Reason in Islam*. London: George Allen & Unwin Ltd., 1957.
Aristotle. *The Complete Works of Aristotle*. The Revised Oxford Translation. 2 vols. Edited by Jonathan Barnes. Princeton: Princeton University Press, 1984.
On Generation and Corruption, trans. H. H. Joachim.
On the Heavens, trans. J. L. Stocks.
On the Soul, trans. J. A. Smith.
Metaphysics, trans. W. D. Ross.

Physics, trans. R. P. Hardie and R. K. Gaye.
History of Animals, d'A. W. Thompson.
Parts of Animals, trans. W. A. Ogle.
Prior Analytics, trans. A. J. Jenkinson.
Posterior Analytics, trans. Jonathan Barnes.
Sophistical Refutations, trans. W. A. Pickard-Cambridge.
De Interpretatione, trans. J. L. Ackrill.
Meteorology, trans. E. W. Webster.
Sense and Sensibilia, trans. J. I. Beare.
On Memory, trans. J. I. Beare.
On Sleep, trans. J. I. Beare.
On Dreams, trans. J. I. Beare.
On Divination in Sleep, trans. J. I. Beare.
On Length and Shortness of Life, trans. G. R. T. Ross.
On Youth, Old Age, Life and Death, and Respiration, trans. G. R. T. Ross.
Armstrong, A. H. "Plotinus." In A. H. Armstrong, ed., *The Cambridge History of Later Greek and Early Medieval Philosophy*. (Cambridge: Cambridge University Press, 1970, ch. 12 ("Life: Plotinus and the Religion and Superstition of His Time"), 195–210.
Arnaldez, R. "Ibn Rushd." In *The Encyclopaedia of Islam*, New edition, edited by B. Lewis, V. L. Ménage, Ch. Pellat, and J. Schacht. Vol. III (H–IRAM). Leiden: E. J. Brill, 1986, 909–920.
Asztalos, Monika. "The Faculty of Theology." In H. de Ridder-Symoens, ed., *A History of the University in Europe*, Vol. 1: *Universities in the Middle Ages*. Cambridge: Cambridge University Press, 1992, ch. 13, 409–441.
Augustine, Saint. *The Literal Meaning of Genesis: De Genesi ad litteram*. Edited and translated by John Hammond Taylor. In Johannes Quasten, Walter J. Burghardt, and Thomas Comerford Lawler, eds., *Ancient Christian Writers: The Works of the Fathers in Translation*, 2 vols. (vols. 41, 42). New York: Newman, 1982.
———. *The Works of Saint Augustine, A Translation for the 21st Century*. Edited by John E. Rotelle, O.S.A.: Part 1, Vol. 11: *Teaching Christianity (De doctrina Christiana)*, introduction, translation, and notes by Edmund Hill, O. P. Hyde Park, NY: New City Press, 1996.
Aureoli, Peter. *Petri Aureoli Verberii Ordinis Minorum Archiepiscopi Aquensis S. R. E. Cardinalis Commentariorum in primum [-quartum] librum Sententiarum, pars prima [-quarta]*. 2 vols. Rome: Aloysius Zannetti, 1596–1605.
Averroes (Ibn Rushd). *Averrois Cordubensis Commentarium Medium in Aristotelis De anima libros*, recensuit F. Stuart Crawford. Cambridge, MA: The Mediaeval Academy of America, 1953.
———. *Averroes on Aristotle's "De Generatione et Corruptione" Middle Commentary and Epitome*. Translated from the original Arabic and the Hebrew and Latin versions, with notes and introduction by Samuel Kurland. Cambridge, MA: The Mediaeval Academy of America, 1958.
———. *Averroes' Tahafut al-Tahafut (The Incoherence of the Incoherence)*. Translated from the Arabic with introduction and notes by Simon van den Bergh. 2 vols. Oxford: University Press, 1954.
Bacon, Francis. *Francis Bacon: The Advancement of Learning and New Atlantis*. Edited by Arthur Johnston. Oxford: Clarendon Press, 1974.
———. *The Oxford Francis Bacon*. In *The Instauratio magna Part II: Novum organum and Associated Texts*. Edited with introduction, notes, commentaries, and facing-page translations by Graham Rees and Maria Wakely. Oxford: Clarendon Press, 2004. *See also* Sargent, Rose-Mary.
Bacon, Roger. *Liber Primus Communium Naturalium Fratris Rogeri*. Edited by Robert Steele in *Opera hactenus inedita Rogeri Baconi*, Fasc. II. Oxford: Clarendon Press, 1909.

Barnes, Jonathan. *Aristotle*. Oxford: Oxford University Press, 1982.

———. "Life and Work." In Jonathan Barnes, ed., *The Cambridge Companion to Aristotle*. Cambridge: Cambridge University Press, 1995, 1–26.

Basil, Saint. *Saint Basil Exegetic Homilies*. Translated by Sister Agnes Clare Way. Vol. 46 of *The Fathers of the Church: A New Translation*. Washington, DC: Catholic University of America Press, 1963.

Belloni, Luigi. "Redi, Francesco." In *Dictionary of Scientific Biography*, vol. 11 (1975), 341–343.

Berman, Harold J. *Law and Revolution: The Formation of the Western Legal Tradition*. Cambridge, MA: Harvard University Press, 1983.

Bernard of Clairvaux. *The Life and Letters of St. Bernard of Clairvaux*. Newly translated by Bruno Scott James. London: Burns Oates, 1953.

Bernard Silvestris. *The "Cosmographia" of Bernardus Silvestris*. A Translation with introduction and notes by Winthrop Wetherbee. New York: Columbia University Press, 1973.

Bonaventure, Saint. *Opera Omnia*. Ad Claras Aquas [Quaracchi]: Collegium S. Bonaventurae, 1882–1901, vol. 2: *Commentaria in quattuor libros Sententiarum Magistri Petri Lombardi: In Secundum librum Sententiarum* (1885).

Boyle, Robert. *The Works of Robert Boyle*. Edited by Michael Hunter and Edward B. Davis. 14 vol. London: Pickering & Chatto, 1999–2000.

Brooke, John Hedley. *Science and Religion: Some Historical Perspectives*. Cambridge: Cambridge University Press, 1991.

Brown, Sanborn C., and Leonard M. Rieser. *Natural Philosophy at Dartmouth: From Surveyors' Chains to the Pressure of Light*. Hanover, NH: University Press of New England, 1974.

Buchwald, Jed. "Thomson, Sir William," in *Dictionary of Scientific Biography*, 13 (1976), 374–388.

Buchwald, Jed Z., and Sungook Hong. "Physics." In David Cahan, ed. *From Natural Philosophy to the Sciences: Writing the History of Nineteenth-Century Science*. Chicago: University of Chicago Press, 2003, 163–195.

Buridan, John. *Acutissimi philosophi reverendi Magistri Johannis Buridani subtilissime questiones super octo Phisicorum libros Aristotelis diligenter recognite et revise a Magistro Johanne Dullaert de Gandavo antea nusquam impresse*. Paris, 1509. Facsimile, entitled *Johannes Buridanus, Kommentar zur Aristotelischen Physik*. Frankfurt: Minerva, 1964.

———. *Johannes Buridanus, Kommentar zur Aristotelischen Metaphysik*. Paris, 1588; reprinted Frankfurt: Minerva, 1964.

———. *Ioannis Buridani Quaestiones super libris quattuor De caelo et mundo*. Edited by Ernest A. Moody. Cambridge, MA: The Mediaeval Academy of America, 1942.

Burnett, Charles. *See* Pseudo-Bede.

Burton, D. E., ed. and trans. *Nicole Oresme's "On Seeing the Stars" (De visione stellarum): A critical edition of Oresme's treatise on optics and atmospheric refraction, with an introduction, commentary, and English translation*. Ph.D. diss., Indiana University, Bloomington, 2000.

Callus, D. A., O. P. "Introduction of Aristotelian Learning to Oxford." In *Proceedings of the British Academy* 29 (1943), 229–281.

Campanus of Novara. *Campanus of Novara and Medieval Planetary Theory: "Theorica planetarum."* Edited with an introduction, English translation, and commentary by Francis Benjamin, Jr., and G. J. Toomer. Madison: University of Wisconsin Press, 1971.

Caroti, Stefano. *See* Oresme, Nicole.

Case, Thomas. "Aristotle." In *The Encyclopaedia Britannica, A Dictionary of Arts, Sciences, Literature and General Information*, Eleventh edition. Cambridge: University Press, 1910, vol. 2: Andros to Austria, 501–522.

Clagett, Marshall. "Some General Aspects of Physics in the Middle Ages." *Isis* 39 (May 1948), 29–44.

———. *Greek Science in Antiquity*. London: Abelard-Schuman, 1957.

_____. *The Science of Mechanics in the Middle Ages*. Madison: University of Wisconsin Press, 1959.

_____. "Adelard of Bath." In Charles C. Gillispie, ed., *Dictionary of Scientific Biography*, 18 vols. New York: Charles Scribner's Sons, 1970–1990, vol. 1 (1970), 61–64.

_____. *Archimedes in the Middle Ages*. Vol. 2: *The Translations from the Greek by William of Moerbeke*. Philadelphia: American Philosophical Society, 1976.

_____. *Ancient Egyptian Science: A Source Book*, 3 vols. Philadelphia: American Philosophical Society, 1989–1999.

_____. *See also* Moody, E. A.; Oresme, Nicole.

Clement of Alexandria. *Miscellanies*. Translated in *The Ante-Nicene Fathers: Translations of the Writings of the Fathers Down to A.D. 325*. Vol. 2: *Fathers of the Second Century: Hermas, Tatian, Athenagoras, Theophilus, and Clement of Alexandria (Entire)*. American edition, chronologically arranged, with notes, prefaces, and elucidations by A. Cleveland Coxe, D. D. Grand Rapids, MI: Wm. B. Eerdmans Publishing Co., 1983.

Cobban, Alan B. *The Medieval English Universities: Oxford and Cambridge to c. 1500*. Berkeley: University of California Press, 1988.

Cochrane, Louise. *Adelard of Bath: The First English Scientist*. London: British Museum Press, 1994.

Cohen, H. Floris. *The Scientific Revolution: A Historiographical Inquiry*. Chicago: University of Chicago Press, 1994.

Cohen, I. Bernard. *Introduction to Newton's "Principia."* Cambridge: University Press, 1971.

Cohen, Morris R., and I. E. Drabkin, eds., *A Source Book in Greek Science*. New York: McGraw-Hill Book Company, Inc., 1948.

Cooper, L. *The Poetics of Aristotle*. Revised edition. Ithaca, NY: Cornell University Press, 1956.

Copenhaver, Brian P. "Natural Magic, Hermetism, and Occultism in Early Modern Science." In David C. Lindberg and Robert S. Westman, eds., *Reappraisals of the Scientific Revolution*. Cambridge: Cambridge University Press, 1990, 261–301.

Copernicus, Nicholas. *Copernicus: On the Revolutions of the Heavenly Spheres*. A New Translation from the Latin, with an introduction and notes by A. M. Duncan. Newton Abbott, Devon: David & Charles, 1976.

_____. *Nicholas Copernicus "On the Revolutions."* Edited by Jerzy Dobrzycki. Translation and commentary by Edward Rosen. London: The Macmillan Press Ltd., 1978.

Copleston, Frederick, S. J. *A History of Philosophy*. Vol. 1: *Ancient Philosophy. The Bellarmine Series*. No place or date.

Courtenay, W. A. "Theology and Theologians from Ockham to Wyclif." In J. I. Catto and Ralph Evans, eds., *The History of the University of Oxford*. Vol. 2: *Late Medieval Oxford*. Oxford: Clarendon Press, 1992, 1–34.

Crombie, A. C. *Medieval and Early Modern Science*. 2 vols. Garden City, NY: Doubleday, 1959.

_____. "Grosseteste, Robert," in Charles C. Gillispie ed., *Dictionary of Scientific Biography*, 18 vols. New York: Charles Scribner's Sons, 1970–1990, vol. 5 (1972), 548–554.

_____. *Styles of Scientific Thinking in the European Tradition: The history of argument and explanation especially in the mathematical and biomedical sciences and arts*. 3 vols. London: Duckworth, 1994.

Crosby, H. Lamar, Jr., ed. and trans. *Thomas of Bradwardine His "Tractatus de Proportionibus: Its Significance for the Development of Mathematical Physics*. Madison: University of Wisconsin Press, 1955.

Cunningham, Andrew. "Getting the Game Right: Some Plain Words on the Identity and Invention of Science." *Studies in History and Philosophy of Science* 19 (1988), 365–389.

_____. "How the Principia Got Its Name; Or, Taking Natural Philosophy Seriously." *History of Science* 29 (1991), 377–392.

_____. "The Identity of Natural Philosophy. A Response to Edward Grant." In *Early Science and Medicine: A Journal for the Study of Science, Technology and Medicine in the Pre-Modern Period*, vol. 5, no. 3 (2000), 259–278.

————. "A Last Word." *Early Science and Medicine: A Journal for the Study of Science, Technology and Medicine in the Pre-Modern Period*, vol. 5, no. 3 (2000), 299–300.

Cunningham, Andrew, and Perry Williams. "De-Centring the 'Big Picture': The Origins of Modern Science and the Modern Origins of Science." *The British Journal for the History of Science* 26 (1993), 407–432.

Darwin, Francis. *The Life and Letters of Charles Darwin, Including an Autobiographical Chapter.* 3 vols. Second edition. London: John Murray, 1887.

De Wulf, Maurice. "The Teaching of Philosophy and the Classification of the Sciences in the Thirteenth Century." *The Philosophical Review* 27 (1918), 356–373.

Dick, Steven J. *Plurality of Worlds: The Extraterrestrial Debate from Democritus to Kant.* Cambridge: Cambridge University Press, 1982.

Dictionary of Scientific Biography. See Gillispie, Charles C.

Dod, Bernard G. "Aristoteles latinus." In Norman Kretzmann, Anthony Kenny, and Jan Pinborg, eds., *The Cambridge History of Later Medieval Philosophy from the Rediscovery of Aristotle to the Disintegration of Scholasticism 1100–1600.* Cambridge: Cambridge University Press, 1982, 45–79.

Dodge, Bayard, ed. and trans. *The Fihrist of al-Nadim: A Tenth-Century Survey of Muslim Culture.* 2 vols. New York: Columbia University Press, 1970.

Drake, Stillman, and Drabkin, I. E. *Mechanics in Sixteenth-Century Italy.* Translated and annotated by Stillman Drake and I. E. Drabkin. Madison: University of Wisconsin Press, 1969.

Drake, Stillman. "Galilei, Galileo." In *Dictionary of Scientific Biography*, vol. 5 (1972), 237–249.

Eamon, William. "Magic and the Occult." In Gary B. Ferngren, ed., *The History of Science and Religion in the Western Tradition: An Encyclopedia.* New York: Garland Publishing, Inc., 2000, 533–540.

Eichholz, David E. "Pliny (Gaius Plinius Secundus)." In Charles C. Gillispie, ed., *Dictionary of Scientific Biography*, vol. 11 (1975), 38–40.

Elford, Dorothy. "William of Conches." In Peter Dronke, ed., *A History of Twelfth-Century Western Philosophy.* Cambridge: Cambridge University Press, 1988, 308–327.

Eriugena, John Scotus. *On the Division of Nature* in *Periphyseon (De divisione naturae) liber primus.* Ed. and trans. I. P. Sheldon-Williams, with the collaboration of Ludwig Bieler. In *Scriptores Latini Hiberniae*, 7. Dublin: Dublin Institute for Advanced Studies, 1968.

Fakhry, Majid. *A History of Islamic Philosophy.* New York: Columbia University Press, 1970.

Ferruolo, Stephen C. *The Origins of the University: The Schools of Paris and Their Critics, 1100–1215.* Stanford, CA: Stanford University Press, 1985.

Fontaine, Jacques. *Isidore de Seville: Traité de la Nature.* Bordeaux: Féret et fils, 1960.

Fortin, Ernest, and Peter D. O'Neill, trans. "Condemnation of 219 Propositions." In Ralph Lerner and Muhsin Mahdi, eds. *Medieval Political Philosophy: A Sourcebook.* Ithaca, NY: Cornell University Press, 1963, 335–354.

Freeman, Kathleen. *Ancilla to the Pre-Socratic Philosophers: A complete translation of the Fragments in Diels, "Fragmente der Vorsokratiker."* Oxford: Basil Blackwell, 1948.

Furley, David, trans. *Place, Void, and Eternity; Philoponus: Corollaries on Place and Void.* Translated by David Furley; *with Simplicius: Against Philoponus on the Eternity of the World*, translated by Christian Wildberg. Ithaca, NY: Cornell University Press, 1991.

Galileo Galilei. *Discoveries and Opinions of Galileo, Including: The Starry Messenger (1610); Letters on Sunspots (1613); Letter to the Grand Duchess Christina (1615); And Excerpts from The Assayer (1623).* Translated with an introduction and notes by Stillman Drake. New York: Doubleday Anchor Books, 1957.

————. *Sidereus Nuncius or the Sidereal Messenger Galileo Galilei.* Translated with introduction, conclusion, and notes by Albert Van Helden. Chicago: University of Chicago Press, 1989.

Ghazali, al-. *Al-Ghazali's Tahafut al-Falasifah [Incoherence of the Philosophers]*. Translated into English by Sabih Ahmad Kamali. Pakistan Philosophical Congress Publication, no. 3, 1963.

Gieysztor, Aleksander. "Management and Resources." In Hilde Ridder-Symoens, ed., *A History of the University in Europe*: Vol. I: *Universities in the Middle Ages*, ch. 4, 108–143.

Gillispie, Charles C., ed. *Dictionary of Scientific Biography*. 18 vols. New York: Charles Scribner's Sons, 1970–1990.

Gilson, Etienne. *A History of Christian Philosophy in the Middle Ages*. London: Sheed and Ward, 1955.

Gingerich, Owen. *The Book Nobody Read: Chasing the Revolutions of Nicolaus Copernicus*. New York: Walker and Co., 2004.

Gliozzi, Mario. "Torricelli, Evangelista." In *Dictionary of Scientific Biography*, vol. 13 (1976), 433–440.

Goichon, A. M. "Ibn Sina." In B. Lewis, V. L. Ménage, Ch. Pellat, and J. Schacht, eds., *The Encyclopaedia of Islam*, New Edition, vol. III, H-IRAM. Leiden: E. J. Brill, 1986, 941–947.

Goldziher, Ignaz. "The Attitude of Orthodox Islam toward the Ancient Sciences." In Merlin Swartz, ed., *Studies on Islam*. New York: Oxford University Press, 1981, 185–215.

Goode, Erica. "How Culture Molds Habits of Thought." In the *Science Times* section of the *New York Times* for August 8, 2000.

Goodman, L. E. "The Translation of Greek Materials into Arabic." In M. J. L. Young, J. D. Latham and R. B. Serjeant, eds., *Religion, Learning and Science in the 'Abbasid Period*. Cambridge: Cambridge University Press, 1990, 477–497.

———. "Al-Razi, Abu Bakr Muhammad B. Zakariyya, known to the Latins as Rhazes." In C. E. Bosorth, E. van Donzel, W. P. Heinrichs and G. Lecomte, eds., *The Encyclopaedia of Islam*, vol. 8 (NED-SAM). Leiden: E. J. Brill, 1995, 474–477.

Gottschalk, Hans B. "The Earliest Aristotelian Commentators." In Richard Sorabji, ed., *Aristotle Transformed: The Ancient Commentators and Their Influence*. Ithaca, NY: Cornell University Press, 1990, 55–81.

Grant, Edward. *A Source Book in Medieval Science*. Cambridge, MA: Harvard University Press, 1974.

———. *Physical Science in the Middle Ages*. Cambridge: Cambridge University Press, 1977.

———. "Scientific Thought in Fourteenth-Century Paris: Jean Buridan and Nicole Oresme." In Madeleine Pelner Cosman and Bruce Chandler, eds., *Machaut's World: Science and Art in the Fourteenth Century*. In *Annals of the New York Academy of Sciences*, Vol. 314. New York: The New York Academy of Sciences, 1978, 105–124.

———. "The Condemnation of 1277, God's Absolute Power, and Physical Thought in the Late Middle Ages." *Viator*, vol. 10 (1979), 211–244.

———. "Ways to Interpret the Terms 'Aristotelian' and 'Aristotelianism' in Medieval and Renaissance Natural Philosophy." In *History of Science* 25 (1987), 335–358.

———. "Celestial Orbs in the Latin Middle Ages." In *Isis* 78 (June 1987), 153–173. Reprinted in Michael H. Shank, ed., *The Scientific Enterprise in Antiquity and the Middle Ages*. Chicago: University of Chicago Press, 2000, 183–203.

———. "The Unusual Structure and Organization of Albert of Saxony's *Questions on De caelo*." In Joël Biard, ed., *Itinéraires d'Albert de Saxe: Paris-Vienne au xive siècle* (Paris: Librairie philosophique J. Vrin, 1991), 205–217.

———. *Much Ado about Nothing: Theories of Space and Vacuum from the Middle Ages to the Scientific Revolution*. Cambridge: Cambridge University Press, 1981.

———. *Planets, Stars, and Orbs: The Medieval Cosmos, 1200–1687*. Cambridge: Cambridge University Press, 1994.

———. *The Foundations of Modern Science in the Middle Ages, Their Religious, Institutional, and Intellectual Contexts*. Cambridge: Cambridge University Press, 1996.

———. "God, Science, and Natural Philosophy in the Late Middle Ages." In Nauta and Vanderjagt, *Between Demonstration and Imagination Essays in the History of Science and Philosophy Presented to John D. North*. Leiden: Brill, 1999, 243–267.

———. "God and Natural Philosophy: The Late Middle Ages and Sir Isaac Newton." In *Early Science and Medicine: A Journal for the Study of Science, Technology and Medicine in the Pre-Modern Period*, vol. 5, no. 3 (2000), 279–298.

———. *God and Reason in the Middle Ages*. Cambridge: Cambridge University Press, 2001.

———. "Medieval Natural Philosophy: Empiricism without Observation." In Cees Leijenhorst, Christoph Lüthy, and Johannes M. M. H. Thijssen, eds., *The Dynamics of Aristotelian Natural Philosophy from Antiquity to the Seventeenth Century*. Leiden: Brill, 2002, 141–168.

———. "Scientific Imagination in the Middle Ages." In *Perspectives on Science*, vol. 12, no. 4 (2004), 394–423.

———. "What Was Natural Philosophy in the Late Middle Ages?" In *History of Universities*, vol. 20, part 2 (2005), 12–46. *See also* Oresme, Nicole.

Gregory of Rimini. *Lectura super primum et secundum Sententiarum*. 7 vols. Berlin: Walter de Gruyter, 1979–1987.

Guralnick, Stanley. *Science and the Ante-Bellum American College*, Philadelphia: American Philosophical Society, 1975. [Memoirs, vol. 109].

Hadot, Ilsetraut. *Le problème du néoplatonisme alexandrin: Hiéroclès et Simplicius*. Paris: Études augustiniennes, 1978.

Hall, A. R. *The Scientific Revolution 1500–1800: The Formation of the Modern Scientific Attitude*. London: Longmans, Green and Co., 1954.

Hall, Marie Boas, ed. *Robert Boyle on Natural Philosophy: An Essay with Selections from His Writings*. Bloomington: Indiana University Press, 1965.

Hankinson, R. J. "Science." In Jonathan Barnes, ed., *The Cambridge Companion to Aristotle*, Cambridge: Cambridge University Press, 1995, 140–167.

Hansen, Bert. "Magic, Bookish (Western European)." In Joseph R. Strayer, ed., *Dictionary of the Middle Ages*, vol. 8. New York: Charles Scribner's Sons, 1987, 31–40.

Häring. Nikolaus M. "Thierry of Chartres." In Charles C. Gillispie, ed., *Dictionary of Scientific Biography*, 18 vols. New York: Charles Scribner's Sons, 1970–1990, vol. 13 (1976), 339–341.

Haskins, Charles Homer. *Studies in the History of Mediaeval Science*. Cambridge, MA: Harvard University Press, 1924.

Ibn al-Haytham. *The Optics of Ibn al-Haytham. Books I–III: On Direct Vision*. Translated with introduction and commentary by A. I. Sabra. 2 vols. London: The Warburg Institute, University of London, 1989.

Heidel, Alexander. *The Babylonian Genesis: The Story of Creation*. Second edition. Chicago: University of Chicago Press, 1951.

Henry, John. *The Scientific Revolution and the Origins of Modern Science*. London: Macmillan Press Ltd. 1997.

Hesse, Mary. "Bacon, Francis." In Charles C. Gillispie, ed., *Dictionary of Scientific Biography*, 18 vols. New York: Charles Scribner's Sons, 1970–1990, vol. 1 (1970), 372–377.

Hobbes, Thomas. *Leviathan. Or the Matter, Form, and Power of a Commonwealth Ecclesiastical and Civil.* In *The English Works of Thomas Hobbes of Malmesbury*, Now First Collected by Sir William Molesworth. 11 vols. London: John Bohn, 1839–1845; reprint, Darmstadt: Scientia Verlag, Aalen, 1966, vol. 3.

Hoffmann, Philippe. "Simplicius' Polemics." In Richard Sorabji, ed., *Philoponus and the Rejection of Aristotelian Science*. Ithaca, NY: Cornell University Press, 1987, 57–83.

Hollister, C. Warren. *Medieval Europe: A Short History*. Seventh edition. New York: McGraw-Hill, 1994.

Holopainen, Toivo J. *Dialectic and Theology in the Eleventh Century*. Leiden: E. J. Brill, 1996.

Hooykaas, R. "Beeckman, Isaac." In *Dictionary of Scientific Biography*, vol. 1 (1970), 566–568.

Hourani, George F. *On the Harmony of Religion and Philosophy. A Translation with introduction and notes, of Ibn Rushd's Kitab fasl al-maqal, with its appendix (Damima) and an extract from Kitab al-kashf' an manahij al-adilla.* London: Luzac, 1976.

House, Humphry. *Aristotle's Poetics: A Course of Eight Lectures.* London: R. Hart-Davis, 1956.

Howell, Kenneth. *God's Two Books: Copernican Cosmology and Biblical Interpretation in Early Modern Science.* Notre Dame, IN: University of Notre Dame Press, 2002.

Huff, Toby. *The Rise of Early Modern Science: Islam, China, and the West.* Second edition. Cambridge: Cambridge University Press, 2003.

Hugh of Saint Victor. *The "Didascalicon" of Hugh of St. Victor: A Medieval Guide to the Arts.* Translated from the Latin with an introduction and notes by Jerome Taylor. New York: Columbia University Press, 1961.

Hyman, Arthur, and James J. Walsh, eds. *Philosophy in the Middle Ages: The Christian, Islamic, and Jewish Traditions.* Indianapolis: Hackett Publishing Company, 1973.

Iskandar, Albert Z. *See* Anawati, G. C.

Jardine, N. *The Birth of History and Philosophy of Science: Kepler's "A Defence of Tycho Against Ursus" with Essays on its Provenance and Significance.* Cambridge: Cambridge University Press, 1984.

John of Damascus, Saint. *The Fount of Knowledge.* Translated by Frederic H. Chase, Jr. In *Saint John of Damascus, Writings.* In *The Fathers of the Church, A New Translation,* vol. 37. New York: Fathers of the Church, Inc., 1958.

John of Salisbury. *The "Metalogicon" of John of Salisbury: A Twelfth-Century Defense of the Verbal and Logical Arts of the Trivium.* Translated with an introduction and notes by Daniel D. McGarry. Gloucester, MA: Peter Smith, 1971 [ca. 1955].

Joly, Robert. "Hippocrates of Cos." In Charles C. Gillispie, ed., *Dictionary of Scientific Biography.* 18 vols. New York: Charles Scribner's Sons, 1970–1990, vol. 6 (1972), 418–431.

Khaldun, Ibn. *The Muqaddimah: An Introduction to History.* Translated from the Arabic by Franz Rosenthal. 3 vols. Princeton, NJ: Princeton University Press, 1958; corrected, 1967.

Kilwardby, Robert. *Robert Kilwardby O. P. De ortu scientiarum.* Edited by Albert P. Judy O. P. London: Published jointly by the British Academy and The Pontifical Institute of Mediaeval Studies, 1976.

Kirk, G. S., and J. E. Raven. *The Presocratic Philosophers: A Critical History with a Selection of Texts.* Cambridge: Cambridge University Press, 1957.

Knowles, David. *The Evolution of Medieval Thought.* Baltimore: Helicon Press, 1962.

Krafft, Fritz. "Guericke (Gericke), Otto von." In *Dictionary of Scientific Biography,* vol. 5 (1972), 574–576.

Kren, Claudia. "Gundissalinus, Dominicus." In *Dictionary of Scientific Biography,* vol. 5 (1972), 591–593. *See also* Oresme, Nicole.

Kuhn, Thomas S. *The Copernican Revolution: Planetary Astronomy in the Development of Western Thought.* New York: Vintage Books, 1959.

Kusukawa, Sachiko. "Bacon's Classification of Knowledge." In Markku Peltonen, ed., *The Cambridge Companion to Bacon.* Cambridge: Cambridge University Press, 1996, 47–74.

Leff, Gordon. *Paris and Oxford Universities in the Thirteenth and Fourteenth Centuries.* New York: John Wiley & Sons, 1968.

————. "The *Trivium* and the Three Philosophies." In H. de Ridder-Symoens, ed., *A History of the University in Europe,* Vol. 1: *Universities in the Middle Ages.* Cambridge: Cambridge University Press, 1992, 307–336.

Lemay, Richard. *Abu Mashar and Latin Aristotelianism in the Twelfth Century: The Recovery of Aristotle's Natural Philosophy through Arabic Astrology.* Publication of the Faculty of Arts and Sciences, Oriental Series No. 28. Beirut: American University of Beirut, 1962.

————. "Gerard of Cremona." In *Dictionary of Scientific Biography,* vol. 15, Supplement I (1978), 173–192.

Lerner, Ralph, and Muhsin Mahdi, eds. *Medieval Political Philosophy: A Sourcebook*. Ithaca, NY: Cornell University Press, 1963.

Lewes, George Henry. *Aristotle: A Chapter from the History of Science, Including Analyses of Aristotle's Scientific Writings*. London: Smith, Elder and Co., 1864.

Lindberg, David C. "The Transmission of Greek and Arabic Learning to the West." In David C. Lindberg, ed., *Science in the Middle Ages*. Chicago: University of Chicago Press, 1978, 52–90.

————. ed. and trans. *Roger Bacon's Philosophy of Nature: A Critical Edition, with English Translation, Introduction, and Notes, of "De multiplicatione specierum" and "De speculis comburentibus."* Oxford: Clarendon Press, 1983.

————. *The Beginnings of Western Science: The European Scientific Tradition in Philosophical, Religious, and Institutional Context, 600 B.C. to A.D. 1450*. Chicago: University of Chicago Press, 1992.

Lloyd, G. E. R. *Aristotle: The Growth and Structure of His Thought*. Cambridge: Cambridge University Press, 1968.

————. *Early Greek Science: Thales to Aristotle*. New York: W. W. Norton, 1970.

————. *Greek Science After Aristotle*. New York: W. W. Norton, 1973.

Locke, John. *An Essay Concerning Human Understanding*. Edited with an introduction by Peter H. Nidditch. Oxford: Clarendon Press, 1975.

Lohr, Charles H., S. J., "Medieval Latin Aristotle Commentaries: Authors A-F." In *Traditio* 23 (1967), 313–413.

Lones, Thomas E. *Aristotle's Researches in Natural Science*. London: West, Newman & Co., 1912.

Lucretius. *Lucretius "On the Nature of the Universe."* Translated and introduced by R. E. Latham. Harmondsworth, Middlesex, England: Penguin Books, 1968; first published 1951.

Maccagnolo, Enzo. "David of Dinant and the Beginnings of Aristotelianism in Paris." Translated by Jonathan Hunt in Peter Dronke, ed., *A History of Twelfth-Century Western Philosophy*. Cambridge: Cambridge University Press, 1988, 429–442.

Macrobius. *Macrobius: Commentary on the Dream of Scipio*. Translated with an introduction and notes by William Harris Stahl. New York: Columbia University Press, 1952.

Mahdi, Muhsin. "Al-Farabi, Abu Nasr Muhammad ibn Muhammad ibn Tarkhan ibn Awzalagh." In *Dictionary of Scientific Biography*, vol. 4 (1971), 523–526.

Makdisi, George. *The Rise of Colleges: Institutions of Learning in Islam and the West*. Edinburgh: Edinburgh University Press, 1981.

Manuel, Frank E. *The Religion of Isaac Newton: The Fremantle Lectures 1973*. Oxford: Clarendon Press, 1974.

Marenbon, John. *From the Circle of Alcuin to the School of Auxerre: Logic, Theology and Philosophy in the Early Middle Ages*. Cambridge: Cambridge University Press, 1981.

Marrone, Steven P. *William of Auvergne and Robert Grosseteste: New Ideas of Truth in the Early Thirteenth Century*. Princeton, NJ: Princeton University Press, 1983.

Marshall, Peter, ed. *Nicholas Oresme's 'Questiones super libros Aristotelis De anima: A Critical Edition with Introduction and Commentary* by Peter Marshall. Ph.D. diss., Cornell University, 1980.

McCluskey, Stephen C. *Astronomies and Cultures in Early Medieval Europe*. Cambridge: Cambridge University Press, 1998.

McVaugh, Michael. "A List of Translations Made from Arabic into Latin in the Twelfth Century: Gerard of Cremona (ca. 1114–1187)." In Edward Grant, ed., *A Source Book in Medieval Science*. Cambridge, MA: Harvard University Press, 1974, 35–38.

Mercer, Christia. "The Vitality and Importance of Early Modern Aristotelianism." In Tom Sorrell, ed., *The Rise of Modern Philosophy: The Tension between the New and Traditional Philosophies from Machiavelli to Leibniz*. Oxford: Clarendon Press, 1993, 33–67.

Merlan, Philip. "Alexander of Aphrodisias." In Charles C. Gillispie, ed., *Dictionary of Scientific Biography*, 18 vols. New York: Charles Scribner's Sons, 1970–1990, vol. 1 (1970), 117–120.

Millen, Ron. "The Manifestation of Occult Qualities in the Scientific Revolution." In Margaret J. Osler and Paul Lawrence Farber, eds., *Religion, Science, and Worldview: Essays in Honor of Richard S. Westfall*. Cambridge: Cambridge University Press, 1985, 185–216.

Minio-Paluello, Lorenzo. "Abailard, Pierre." In Charles C. Gillispie, ed. *Dictionary of Scientific Biography*, 18 vols. New York: Charles Scribner's Sons, 1970–1990, vol. 1 (1970), 1–4.

––––––. "James of Venice." In Charles C. Gillispie, ed., *Dictionary of Scientific Biography*, 18 vols. New York: Charles Scribner's Sons, 1970–1990, vol. 7 (1973), 65–67.

––––––. "Michael Scot." In Charles C. Gillispie, ed., *Dictionary of Scientific Biography*, 18 vols. New York: Charles Scribner's Sons, 1970–1990, vol. 9 (1974), 361–365.

––––––. "Moerbeke, William of." In Charles C. Gillispie, ed., *Dictionary of Scientific Biography*, 18 vols. New York: Charles Scribner's Sons, 1970–1990, vol. 9 (1974), 434–440.

Montgomery, Scott. *Science in Translation: Movements of Knowledge through Culture and Time*. Chicago: University of Chicago Press, 2000.

Moody, Ernest A., and Marshall Clagett, eds. *The Medieval Science of Weights ("Scientia de Ponderibus"): Treatises Ascribed to Euclid, Archimedes, Thabit ibn Qurra, Jordanus de Nemore and Blasius of Parma*. Edited with introductions, English translations, and notes, by Ernest A. Moody and Marshall Clagett. Madison: University of Wisconsin Press, 1952.

Mottahedeh, Roy. *The Mantle of the Prophet*. New York: Pantheon Books, 1985.

Murdoch, John E. "From Social into Intellectual Factors: An Aspect of the Unitary Character of Late Medieval Learning." In John E. Murdoch and Edith Dudley Sylla, eds., *The Cultural Context of Medieval Learning*. Dordrecht, Holland: D. Reidel, 1975, 271–339.

––––––. "The Analytic Character of Late Medieval Learning: Natural Philosophy Without Nature." In Lawrence D. Roberts, ed., *Approaches to Nature in the Middle Ages*. Binghamton, NY: Center for Medieval & Early Renaissance Studies, 1982, 171–213.

Neugebauer, O. *The Exact Sciences in Antiquity*. Princeton, NJ: Princeton University Press, 1952.

Newton, Isaac. *Sir Isaac Newton's Mathematical Principles of Natural Philosophy and His System of the World*. Trans. into English by Andrew Motte in 1729. The translations revised and supplied with an historical and explanatory appendix by Florian Cajori. Berkeley: University of California Press, 1947.

––––––. *Opticks or a Treatise of the Reflections, Refractions, Inflections & Colours of Light, Sir Isaac Newton*. Based on the fourth edition, London 1730; with a foreword by Albert Einstein; and introduction by Sir Edmund Whittaker; a preface by I. Bernard Cohen; and an analytical table of contents prepared by Duane H. D. Roller. New York: Dover Publications, 1952.

––––––. *The Correspondence of Isaac Newton*. Vol. 2: 1676–1687. Edited by H. W. Turnbull. Cambridge: Cambridge University Press (published for the Royal Society), 1960.

Oesterle, J. A. "Poetics (Aristotelian)." In the *New Catholic Encyclopedia*. Second edition. Vol. XI. Washington, DC: Catholic University of America, 2003, 433–435.

O'Leary, De Lacy. *How Greek Science Passed to the Arabs*. London: Routledge & Kegan Paul, 1949.

Olivi, Peter John. *Fr. Petrus Iohannis Olivi, O. F. M. Quaestiones in Secundum Librum Sententiarum*. Edited by Bernard Jansen, S.I. 3 vols. Ad Claras Aquas [Quaracchi]: Ex Typographia Collegii S. Bonaventurae, 1922–1926.

O'Malley, C. D. "Vesalius, Andreas." In Charles C. Gillispie ed., *Dictionary of Scientific Biography*, 18 vols. New York: Charles Scribner's Sons, 1970–1990, vol. 14 (1976), 3–12.

Ong, Walter. *Ramus, Method, and the Decay of Dialogue: From the Art of Discourse to the Art of Reason*. Cambridge, MA: Harvard University Press, 1958.

Oresme, Nicole. *The "Questiones super De celo" of Nicole Oresme*. Edited and translated by Claudia Kren. Ph.D diss., University of Wisconsin, 1965.

————. *Nicole Oresme "De proportionibus proportionum" and "Ad pauca respicientes."* Edited with introductions, English translations, and critical notes by Edward Grant. Madison/Milwaukee: University of Wisconsin Press, 1966.

————. *Nicole Oresme and the Medieval Geometry of Qualities, A Treatise on the Uniformity and Difformity of Intensities Known as "Tractatus de configurationibus qualitatum et motuum."* Edited with an introduction, English translations, and commentary by Marshall Clagett. Madison: University of Wisconsin Press, 1968.

————. *Nicole Oresme: Le Livre du ciel et du monde.* Edited by Albert D. Menut and Alexander J. Denomy; translated with an introduction by Albert D. Menut. Madison: University of Wisconsin Press, 1968.

————. *Nicole Oresme and the Kinematics of Circular Motion. "Tractatus de commensurabilitate vel incommensurabilitate motuum celi."* Edited with an introduction, English translation, and commentary by Edward Grant. Madison: University of Wisconsin Press, 1971.

————. *Nicole Oresme Quaestiones super De generatione et corruptione.* Edited by Stefano Caroti. München: Verlag der Bayerischen Akademie der Wissenschaften, 1996. *See also* Burton, D.E., and Marshall, Peter.

Organ, Troy W. *An Index to Aristotle in English Translation.* New York: Gordian Press, 1966.

Owen, G. E. L. "Zeno and the Mathematicians." *Proceedings of the Aristotelian Society,* 1957–58, 199–222; reprinted in Wesley C. Salmon, *Zeno's Paradoxes* (Indianapolis: Library of Liberal Arts, Bobbs-Merrill, 1970), 139–163.

Palencia, Angel González. *Al-Farabi, Catalogo de las ciencias.* Second edition. Madrid: Instituto Miguel Asin, 1953.

Pedersen, J., and G. Makdisi. "Madrasa." In *The Encyclopaedia of Islam.* New edition. Edited by C. E. Bosworth, E. van Donzel, B. Lewis, and Ch. Pellat. Vol. 5 (KHE–MAHI). Leiden: E. J. Brill, 1986, 1123–1134.

Pedersen, Olaf. *A Survey of the Almagest.* Odense: Odense University Press, 1974.

Peters, F. E. *Aristotle and the Arabs: The Aristotelian Tradition in Islam.* New York: New York University Press, 1968.

————. *Aristoteles Arabus: The Oriental Translations and Commentaries on the Aristotelian "Corpus."* Leiden: E. J. Brill, 1968.

Piltz, Anders. *The World of Medieval Learning.* Translated into English by David Jones. Totowa, NJ: Barnes & Noble Books, 1981.

Pines, Shlomo. "Al-Razi, Abu Bakr Muhammad ibn Zakariyya (Rhazes)." In Charles C. Gillispie, ed., *Dictionary of Scientific Biography,* 18 vols. New York: Charles Scribner's Sons, 1970–1990, vol. 11 (1975), 323–326.

————. *Studies in Islamic Atomism.* Translated from German by Michael Schwarz, edited by Tzvi Langermann. Jerusalem: The Magnes Press, The Hebrew University, 1997.

Pingree, David. "Abu Ma'shar al-Balkhi, Jafar ibn Muhammad." In Charles C. Gillispie, ed., *Dictionary of Scientific Biography,* 18 vols. New York: Charles Scribner's Sons, 1970–1990, vol. 1 (1970), 32–39.

Plato. *Plato's Cosmology: The "Timaeus" of Plato translated with a running commentary* by Francis MacDonald Cornford. New York: The Liberal Arts Press, 1957.

————. *Plato: The Laws.* Translated with an Introduction and Notes by Trevor J. Saunders. Preface by R. F. Stalley. London: Penguin Books, 1970; Preface 2004.

Pseudo-Bede. *Pseudo-Bede: "De mundi celestis terrestrisque constitutione."* A Treatise on the Universe and the Soul. Edited and translated by Charles Burnett, with the collaboration of members of a seminar group at the Warburg Institute, London. London: The Warburg Institute, University of London, 1985.

Ptolemy, Claudius. *Tetrabiblos.* Edited and translated F. E. Robbins. Loeb Classical Library. Cambridge, MA: Harvard University Press, 1948.

Richard of Middleton. *Clarissimi theologie magistri Ricardi de Media Villa...Super quatuor libros Sententiarum Petri Lombardi questiones subtilissimae.* 4 vols. Brixia [Brescia], 1591. Facsimile. Frankfurt: Minerva, 1963.

Ridder-Symoens, Hilde de, ed. *A History of the University in Europe.* Vol. I: *Universities in the Middle Ages.* Cambridge: Cambridge University Press, 1992.

Rienstra, M. Howard. "Porta, Giambattista Della." In Charles C. Gillispie, ed., *Dictionary of Scientific Biography,* 18 vols. New York: Charles Scribner's Sons, 1970–1990, vol. 11 (1975), 95–98.

Rosenthal, Franz. "Ibn Khaldun." In Charles C. Gillispie, ed., *Dictionary of Scientific Biography,* 18 vols. New York: Charles Scribner's Sons, 1970–1990, vol. 7 (1973), 320–323.

Rüegg, Walter. "Themes." In Hilde de Ridder-Symoens, ed., *A History of the University in Europe: Vol. I: Universities in the Middle Ages,* ch. 1, 3–34.

Russell, Bertrand. *A History of Western Philosophy.* New York: Simon and Schuster, 1945.

———. *Our Knowledge of the External World.* New York: Mentor Books, New American Library, 1960.

Sabra, A. I. "Science and Philosophy in Medieval Islamic Theology." In *Zeitschrift für Geschichte der Arabisch-Islamischen Wissenschaften,* vol. 9 (1994), 1–42. *See also* Ibn al-Haytham.

Sambursky, S. "John Philoponus." In Charles C. Gillispie, ed., *Dictionary of Scientific Biography,* 18 vols. New York: Charles Scribner's Sons, 1970–1980, vol. 7 (1973), 134–139.

Sargent, Rose-Mary. *Francis Bacon: Selected Philosophical Works.* Edited, with introduction, by Rose-Mary Sargent. Indianapolis: Hackett Publishing Co., 1999.

Sarton, George. *Introduction to the History of Science.* 3 vols. Baltimore: Williams & Wilkins, 1927–1948.

———. *A History of Science: Ancient Science through the Golden Age of Greece.* Cambridge, MA: Harvard University Press, 1952.

Schmitt, Charles B. "Pseudo-Aristotle in the Latin Middle Ages." In Jill Kraye, W. F. Ryan and C. B. Schmitt, eds., *Pseudo-Aristotle in the Middle Ages.* London: The Warburg Institute, University of London, 1986, 3–14.

Scriba, Christoph J. "Wallis, John," in Charles C. Gillispie, ed., *Dictionary of Scientific Biography,* 18 vols. New York: Charles Scribner's Sons, 1970–1990, vol. 14 (1976), 146–155.

Sextus Empiricus. *Sextus Empiricus with an English Translation* by the Rev. R. G. Bury: *Against the Physicists; Against the Ethicists.* London: William Heinemann Ltd.; Cambridge, MA: Harvard University Press, 1936.

———. *Sextus Empiricus: Outlines of Pyrrhonism* with an English translation by R.G. Bury. Cambridge, MA: Harvard University Press, 1933.

Shakir, M. H. *The Qur'an.* Translated by M. H. Shakir. Eleventh U.S. edition. Elmhurst, NY: Tahrike Tarsile Qur'an, Inc., 1999.

Shapin, Steven. *The Scientific Revolution.* Chicago: University of Chicago Press, 1996.

Sharples, Robert W. "The School of Alexander?" In Richard Sorabji, ed., *Aristotle Transformed: The Ancient Commentators and Their Influence.* Ithaca, NY: Cornell University Press, 1990, 83–111.

Shumaker, Wayne. *The Occult Sciences in the Renaissance: A Study in Intellectual Patterns.* Berkeley: University of California Press, 1972.

Smoller, Laura Ackerman. *History, Prophecy, and the Stars: The Christian Astrology of Pierre d'Ailly, 1350–1420.* Princeton, NJ: Princeton University Press, 1994.

Solmsen, Friedrich. *Plato's Theology.* Ithaca, NY: Cornell University Press, 1942.

Sorabji, Richard, ed. *Philoponus and the Rejection of Aristotelian Science.* Ithaca, NY: Cornell University Press, 1987.

———, ed. *Aristotle Transformed: The Ancient Commentators and Their Influence.* Ithaca, NY: Cornell University Press, 1990.

_____. "The Ancient Commentators on Aristotle." In Richard Sorabji, ed., *Aristotle Transformed: The Ancient Commentators and Their Influence*, 1990, 1–30.

Spade, Paul Vincent. *Thoughts, Words, and Things: An Introduction to Late Mediaeval Logic and Semantic Theory.* Version 1.1. August 9, 2002 (Copyright 2002 by Paul Vincent Spade), ch. 2 ("Thumbnail Sketch of the History of Logic to the End of the Middle Ages"), Section B: "Aristotelian Logic." Web Address: http://pvspade.com/Logic/docs/thoughts1_1a.pdf.

Stahl, William H. *Roman Science: Origins, Development and Influence to the Later Middle Ages.* Madison: University of Wisconsin Press, 1962.

Staley, Kevin. "Al-Kindi on Creation: Aristotle's Challenge to Islam." In *The Journal of the History of Ideas*, vol. 50, no. 3 (July–Sept. 1989), 355–370.

Steneck, Nicholas H. *Science and Creation in the Middle Ages: Henry of Langenstein (d. 1397) on Genesis.* Notre Dame, IN: University of Notre Dame Press, 1976.

Stock, Brian. *Myth and Science in the Twelfth Century: A Study of Bernard Silvester.* Princeton, NJ: Princeton University Press, 1972.

Sylla, Edith D. "Autonomous and Handmaiden Science: St. Thomas Aquinas and William of Ockham on the Physics of the Eucharist." In *The Cultural Context of Medieval Learning*, edited with an Introduction by John E. Murdoch and Edith Dudley Sylla. Dordrecht, Holland: D. Reidel, 1975, 349–396.

Synan, E. "Introduction: Albertus Magnus and the Sciences." In J. A. Weisheipl O. P., ed., *Albertus Magnus and the Sciences: Commemorative Essays 1980.* Toronto: The Pontifical Institute of Mediaeval Studies, 1980, 1–12.

Taylor, A. E. *Aristotle.* New York: Dover Publications, 1955; reprint of revised edition of 1919.

Taylor, Jerome. *See* Hugh of Saint Victor.

Taylor, Richard C. "Averroes." In Jorge J. E. Gracia and Timothy B. Noone, eds. *A Companion to Medieval Philosophy in the Middle Ages.* Oxford: Blackwell Publishing Ltd. 2003, 182–195.

Tertullian. *On Prescription against Heretics.* Translated by Peter Holmes in *The Ante-Nicene Fathers.* Edited by Alexander Roberts and James Donaldson, 10 vols. New York: Charles Scribner's Sons, 1896–1903, vol. 3.

Teske, Roland J., S. J., *William of Auvergne: The Universe of Creatures. Selections translated from the Latin with an Introduction and Notes.* Milwaukee: Marquette University Press, 1998.

Themon Judaeus. *Questiones et decisiones physicales insignium virorum: Alberti de Saxonia in octo libros Physicorum; tres libros De celo et mundo; duos libros De generatione et corruptione; Thimonis in quatuor libros Meteororum; Buridani in tres libros De anima; librum De sensu et sensato; librum De memoria et reminiscentia; librum De somno et vigilia; librum De longitudine et brevitate vite; librum De iuventute et senectute Aristotelis. Recognitae rursus et emendatae summa accuratione et iudicio Magistri Georgii Lokert Scotia quo sunt tractatus proportionum additi.* Paris: vaenundantur in aedibus Iodici Badii Ascensii et Conradi Resch, 1518.

Thijssen, J. M. M. H. *Censure and Heresy at the University of Paris, 1200–1400.* Philadelphia: University of Pennsylvania Press, 1998.

Thomas Aquinas, O. P. *Scriptum super libros Sententiarum Magistri Petri Lombardi Episcopi Parisiensis.* New edition, 4 vols. Paris: P. Lethielleux, 1929–1947.

_____. *The Letter of Saint Thomas Aquinas "De Occultis Operibus Naturae Ad Quemdam Militem Ultramontanum."* Translated by Joseph Bernard McAllister, M.A., S.T.B. PhD diss. Washington, DC: Catholic University of America Press, 1939.

_____. *Introduction to Saint Thomas Aquinas.* Edited, with an introduction, by Anton C. Pegis. New York: Modern Library, 1948.

_____. *S. Thomae Aquinatis in Aristotelis Libros De caelo et mundo, De generatione et corruptione, Meteorologicorum Expositio.* Edited by Raymundi M. Spiazzi. Turin: Marietti, 1952.

_____. *S. Thomae Aquinatis In octo libros De physico auditu sive Physicorum Aristotelis Commentaria*. Edited P. Fr. Angel and M. Pirotta O. P. Neapoli (Italia): M. d'Auria Pontificius Editor, 1953.

_____. *Commentary on Aristotle's "Physics" by St. Thomas Aquinas*. Translated by Richard J. Blackwell, Richard J. Spath and W. Edmund Thirlkel. New Haven, CT: Yale University Press, 1963.

Thorndike, Lynn. *A History of Magic and Experimental Science During the First Thirteen Centuries of Our Era*. 8 vols. New York: Columbia University Press, 1923–1958.

_____. *University Records and Life in the Middle Ages*. New York: Columbia University Press, 1944.

_____, ed. and trans. *The "Sphere" of Sacrobosco and Its Commentators*. Chicago: University of Chicago Press, 1949.

_____. "Newness and Novelty in Seventeenth-Century Science and Medicine." In Philip P. Wiener and Aaron Noland, eds., *Roots of Scientific Thought: A Cultural Perspective*. New York: Basic Books Publishers, 1957, 443–457.

Toynbee, Arnold. *A Study of History*, 12 vols. Oxford: Oxford University Press, 1934–1960.

Van Den Bergh, Simon. *See* Averroes.

Vasiliev, A. A. *History of the Byzantine Empire 324–1453*. Madison: University of Wisconsin Press, 1952.

Verbeke, G. "Simplicius." In Charles C. Gillispie, ed., *Dictionary of Scientific Biography*, 18 vols. New York: Charles Scribner's Sons, 1970–1990, vol. 12 (1975), 440–443.

_____. "Themistius." In Charles C. Gillispie, ed., *Dictionary of Scientific Biography*, 18 vols. New York: Charles Scribner's Sons, 1970–1990, vol. 13 (1976), 307–309.

Vlastos, Gregory. "Zeno's Race Course." *Journal of the History of Philosophy*, vol. 4, no. 2 (April 1966), 95–108.

Voelkel, James R. *The Composition of Kepler's "Astronomia Nova."* Princeton, NJ: Princeton University Press, 2001.

Walbridge, John. "Logic in the Islamic Intellectual Tradition: The Recent Centuries." In *Islamic Studies*, 39, no. 1 (Spring 2000), 55–75.

Wallace, William A., O. P. *Causality and Scientific Explanation*, 2 vols. Ann Arbor: University of Michigan Press, vol. 1, 1972; vol. 2, 1974.

_____. "William of Auvergne (Guilielmus Arvernus or Alvernus)," in Charles C. Gillispie, ed., *Dictionary of Scientific Biography*, 18 vols. New York: Charles Scribner's Sons, 1970–1990, vol. 14 (1976), 388–389.

Walzer, R. "New Studies on Al-Kindi." In Richard Walzer, *Greek into Arabic: Essays on Islamic Philosophy*. Cambridge, MA: Harvard University Press, 1962, 175–205.

_____. "Early Islamic Philosophy." In A. H. Armstrong, ed., *The Cambridge History of Later Greek and Early Medieval Philosophy*. Cambridge: Cambridge University Press, 1970, 643–669.

_____. "Aristutalis." In *The Encyclopaedia of Islam*. New edition, Prepared by a Number of Leading Orientalists. Edited by H. A. R. Gibb, J. H. Kramers, E. Lévi-Provençal, and J. Schacht. Vol. I (A-B). Leiden: E. J. Brill, 1986, 630–633.

_____. "Al-Farabi Abu Nasr Muhammad B. Muhammad B. Tarkhan B. Awzalagh (Uzlugh?)." In *The Encyclopaedia of Islam*, New edition. Prepared by a Number of Leading Orientalists. Edited by B. Lewis, Ch. Pellat, and J. Schacht. Vol. II (C-G). Leiden: E. J. Brill, 1991, 778–781.

Waterlow, Sarah. *Nature, Change, and Agency in Aristotle's "Physics."* Oxford: Clarendon Press, 1982.

Watt, M. Montgomery. *The Faith and Practice of al-Ghazali*. London: George Allen and Unwin Ltd., 1953.

Weisheipl, James A., O. P. *Friar Thomas d'Aquino: His Life, Thought, and Work*. Garden City, NY: Doubleday & Co., 1974.

Westerink, L. G. "The Alexandrian Commentators and the Introductions to Their Commentaries." In Richard Sorabji, ed., *Aristotle Transformed: The Ancient Commentators and Their Influence*. Ithaca, NY: Cornell University Press, 1990, 325–348.

Westfall, Richard S. *The Construction of Modern Science: Mechanisms and Mechanics*. New York: John Wiley & Sons, 1971.

———. "Hooke, Robert." In Charles C. Gillispie, ed., *Dictionary of Scientific Biography*, 18 vols. New York: Charles Scribner's Sons, 1970–1990, vol. 6 (1972), 481–488.

———. *Never at Rest: A Biography of Isaac Newton*. Cambridge: Cambridge University Press, 1980.

Westman, Robert S. "The Copernicans and the Churches," in David C. Lindberg and Ronald L. Numbers, eds., *God and Nature: Historical Essays on the Encounter between Christianity and Science*. Berkeley: University of California Press, 1986, 76–113.

White, Lynn, Jr. "Technology and Invention in the Middle Ages." In *Medieval Religion and Technology: Collected Essays*. Berkeley: University of California Press, 1978, 1–22. Reprinted from *Speculum*, 15 (1940), 141–159.

Wilbur, James B., and Harold J. Allen. *The Worlds of the Early Greek Philosophers*. Buffalo: Prometheus Books, 1979.

Wildberg, Christian. *See* Furley, David.

William of Auvergne. *Guilielmi Alverni Episcopi Parisiensis, mathematici perfectissimi, eximii philosophi, ac theology praesantissimi, Opera Omnia* Tomus Primus. Parisiis: Apud Ludovicum Billaine, 1674: *De Universo ... In duas partes principales divisum*. *See* Teske, Roland J.

William of Conches. *William of Conches: A Dialogue on Natural Philosophy ("Dragmaticon Philosophiae")*. Translation of the New Latin Critical Text with a Short Introduction and Explanatory Notes by Italo Ronca and Matthew Curr. Notre Dame, IN: University of Notre Dame Press, 1997.

Wilshire, Leland Edward. "The Condemnations of 1277 and the Intellectual Climate of the Medieval University." In Nancy Van Deusen, ed., *The Intellectual Climate of the Early University: Essays in Honor of Otto Grundler*, Studies in Medieval Culture, XXXIX. Kalamazoo, MI: Medieval Institute Publications, Western Michigan University, 1997, 151–193.

Wolfson, Harry A. *Crescas' Critique of Aristotle: Problems of Aristotle's "Physics" in Jewish and Arabic Philosophy*. Cambridge, MA: Harvard University Press, 1929.

———. "Revised Plan for the Publication of a *Corpus Commentariorum Averrois in Aristotelem*." *Speculum* 38 (Jan. 1963), 88–104.

Zeller, Eduard. *Outlines of the History of Greek Philosophy*. Thirteenth edition. Revised by Wilhelm Nestle and translated by L. R. Palmer, Trinity College, Cambridge. New York: Meridian Books, 1955; thirteenth edition revised and published 1931.

Index

Abelard, Peter: logician and theologian, 113; showed disagreements of Church Fathers, 114; taught students to regard all texts as open to criticism, 114; *Yes and No (Sic et Non)*, 113, 272; 107, 242

Abū Ma'shar. *See* Albumasar

Abū Ya'qūb Yūsuf, caliph, 78

Abu'l Barakat: described impressed force, 194

Académie Royale des Sciences, 290

Academy, 22

Academy of the Lynxes (Accademia dei Lincei), 289

Accademia del Cimento, 290

acceleration, of falling bodies, 196

Achilles argument, 11

Adelard of Bath: chapter topics in *Natural Questions*, 118; detailed description of *Natural Questions*, 118–122; did not include account of creation of world, 122; drew on Plato's *Timaeus*, 122; emphasized reason and rationality, 121; goes from "lowest to highest" beginning with plants, 118; and his nephew, 120–122; his version of natural philosophy titled *Natural Questions*, 117; list of questions, 118–120; translations by, 117; used dialogue format for natural philosophy, 116–117, 126; 105, 129, 133, 242

Adrastus, 100

air, as basic substance, 9

al-Rāzī (Rhazes), 286

Albert of Saxony: 107 questions in his *Questions on the Physics* and *Questions on Generation and Corruption*, 187; and God's absolute power, 260; and the upward motion of a projectile, 217n; assumed if vacuum existed, bodies could move in it, 231; differentiated vacuum into places corresponding to each

of the four elements, 230; example of mixed bodies falling in vacuum, 208–209; experience of water in straw following air upward, 223; nature abhors a vacuum, 230; on the empyrean heaven, 264; on the fall of unequal, homogeneous mixed bodies, 209–210; on the possibility of other worlds, 203; rejects moment of rest, 196; works available in early printed editions, 274; *Questions on Aristotle's Physics*, 192; *Questions on the Heavens*, 203; *Questions on the Physics*, 184n, 229; 175, 185n, 258

Albertus Magnus (Albert the Great): accepted threefold division of philosophy, 163; assumed with Aristotle that other worlds are impossible, 252; believed that theology should not be intruded into natural philosophy, 252; his conception of natural philosophy, 164; judgment on Aristotle, 193; on the nature of science (natural philosophy), 163–164; proclaimed superiority of senses, 216; rarely mentions God in his *Commentary on the Physics*, 258; on vacuum, 181; on whether natural philosophy is about God and faith, 251–252; *Commentary on Aristotle's Physics*, 180, 258; 153, 210, 233, 252, 257

Albumasar (Abū Ma'shar): sought to provide a scientific foundation for astrology, 132; *Introduction to Astronomy (Astrology)*, 124, 129, 132; 72

Alexander Neckham (or Nequam), 147

Alexander of Aphrodisias: commentator on Aristotle, 57; 319

Alexander of Hales, 247, 262

Alexander the Great, 28

Alfred of Sareshel, 147

algebra, 6

Alhazen. *See* Ibn al-Haytham

College catalogue, 317; difficulties of injecting theology and faith into it, 250; forbidden in Paris in 1210, 143; formats used to convey, 116–117; forms proper part of history of science, 320; four categories in which God and faith are mentioned, 259–260; had great influence on theology, 263; how it differed from astronomy, 323; is the Mother of all Sciences, 321; Macrobius joins Pythagorean number lore to, 99; Macrobius links soul and number to, 99; manifested itself before it was named, 42; new, based on tractates, 288; no term for it prior to the twelfth century, 97; peripheral activity viewed with suspicion in Islam, 93–94; played significant role in the emergence of modern science, 323; primarily concerned with mobile bodies, 234; regarded as if equivalent to term "science," 307; study of, in Islam done privately, 92; subject of, is mobile bodies, 159; two kinds of disagreements with Aristotle's, 214; united with mathematics as applied to physics, 311; views of Thomas Aquinas on, 164; was a "foreign science" in Islam, 92, 328; was it science? 234–235; was the basic tool in the development of modern sciences, 319; why it was not about God and the faith, 249–250; without nature, 233

natural science: as synonym for natural philosophy, 164; divided into eight parts, 156–158, 169; its subject matter is mobile body, 163, 164; synonym for natural philosophy, 162, 163; 234

Natural Theology, 297, 301

nature: abhors a vacuum, 223, 230; common course of, 257; definition of, 126; discovery of, 7–8; domain of, 1; as opposed to art, 22; as understood by Aristotle, 41

Neleus of Scepsis, 29

Neoplatonism, 53, 274

Neoplatonists: among Latin Encyclopedists, 97; viewed Aristotle's writings favorably, 53

Nesbitt, Richard: explanation of cultural differences between East Asians and European Americans, 94

Nestorians: translated Aristotle's logical works from Greek into Syriac, 62; 61

Nestorius, 61

new: in book titles, 278

Newton, Sir Isaac: and the objective of his *Opticks*, 327; and universal theory of gravitation, 282, 291; banned religious subjects from *Royal Society* meetings, 297; concept of momentum, 196; disapproved of occult causes in his *Opticks*, 293; found few occasions to discourse about God in his scientific treatises, 300, 301; held that discoursing about God belongs to natural philosophy, 301; identified God with infinite space, 282; in the first edition of *The Mathematical Principles of Natural Philosophy*, God is mentioned only once, 300; in the second edition of his *Principia*, he discussed God in a four-page General Scholium, 301; kept Book of Nature separate from Book of Scripture, 295; physicists regard Newton as laying the foundations of physics, 322; problems he faced in arriving at the title of the *Principia*, 314–315; regarded his *Principia* as revealing the mathematical structure of physical nature, 313; retained title of *Principia* because he believed it would sell better, 315; the title of his *Principia* seems to proclaim a union between mathematics and natural philosophy, 307, 313; was doing mathematical physics in the *Principia*; was doing the science of physics in his *Principia*, 315; what he was actually doing was more important than the way he titled his work, 319; why he did not attribute the cause of gravity to God, 302; *The Mathematical Principles of Natural Philosophy*, 282, 287, 298, 300, 301, 307; *Opticks*, 287, 293, 301; 283, 285, 322, 329

Nicholas of Autrecourt, 223

Nicomachus, 27

nonmathematical sciences, 235

Noys: Divine Providence, 123

Nun, 2

oath of 1272, 254, 255

observation: little direct, in medieval natural philosophy, 231; 283

occult causes, were discovered by experiment and use of scientific instruments, 291

occult sciences, 170–171

Ogdoad, 2